Mathematical Engineering

Today, the development of high-tech systems is unthinkable without mathematical modeling and analysis of system behavior. As such, many fields in the modern engineering sciences (e.g. control engineering, communications engineering, mechanical engineering, and robotics) call for sophisticated mathematical methods in order to solve the tasks at hand.

The series Mathematical Engineering presents new or heretofore little-known methods to support engineers in finding suitable answers to their questions, presenting those methods in such manner as to make them ideally comprehensible and applicable in practice.

Therefore, the primary focus is—without neglecting mathematical accuracy—on comprehensibility and real-world applicability.

To submit a proposal or request further information, please use the PDF Proposal Form or contact directly: *Dr. Jan-Philip Schmidt, Publishing Editor (jan-philip.schmidt@springer.com).*

More information about this series at http://www.springer.com/series/8445

Livija Cveticanin

Strong Nonlinear Oscillators

Analytical Solutions

Second Edition

Springer

Livija Cveticanin
Faculty of Technical Sciences
University of Novi Sad
Novi Sad
Serbia

ISSN 2192-4732 ISSN 2192-4740 (electronic)
Mathematical Engineering
ISBN 978-3-319-58825-4 ISBN 978-3-319-58826-1 (eBook)
DOI 10.1007/978-3-319-58826-1

Library of Congress Control Number: 2017939540

Printed on acid-free paper

This Springer imprint is published by Springer Nature
The registered company is Springer International Publishing AG
The registered company address is: Gewerbestrasse 11, 6330 Cham, Switzerland

Preface to Second Edition

The intention of the Second Edition of this book is to extend the previously published one with the new results of investigation in the matter. Besides, the application of the suggested methodologies on some practical problems is presented.

In this book, a new chapter is added in which vibrations of the axially purely nonlinear rod are considered. A new method, based on Hamiltonian approach, for the determination of free vibrations of the oscillator is considered. Now, there is a variety of procedures for solving free strong nonlinear oscillators in this book. Which of the method would be applied depends on the user.

In this book, the comparison between two oscillators with symmetric and asymmetric nonlinearity is given. The type of the model depends on the real physical problem which has to be described. Vibrations in an optomechanical system are discussed.

Forced vibration of the oscillator excited with the excitation force in the form of the Ateb periodic function is also discussed. A procedure for excitation design and derivation of amplitude–frequency equation is considered.

For the oscillator with two degrees of freedom, the generalization of the solving procedure is done. Based on the obtained results, vibrations of the vocal cord are analysed. To make the text more understandable, two new appendices are added: one, the Fourier series of the ca Ateb function, and the second, inverse incomplete Beta function.

I thank the publisher for the offer to publish the recent version of this book.

Livija Cveticanin
Obuda University
Budapest, Hungary

Preface to First Edition

This book is the result of my long-time investigations and interest in the field of nonlinear vibration. The intention of this text is to give the approximate analytical solution procedures for strong nonlinear oscillators and to explain some of the phenomena which occur in such systems. This book considers the free and forced vibrations, takes the positive and negative damping of Van der Pol type, analyses the criteria for deterministic chaos and investigates the parametrically excited vibration in one-degree-of-freedom oscillators. Special attention is given to vibration properties of the two-mass system with two-degrees-of-freedom. The oscillation of the rotor, modelled as an one mass system with two-degrees-of-freedom, is also discussed. The ideal and non-ideal nonlinear mechanical systems are also treated where the jump phenomena, the Sommerfeld effect and the control of the system are included. The basic part for all considerations is a pure nonlinear oscillator whose order of nonlinearity is any positive rational number (integer or non-integer). This type of nonlinearity is the generalization for the previously discussed linear or pure cubic oscillators and oscillators with small nonlinearity. All the suggested solution procedures are based on the exact or approximate solution of the strong nonlinear differential equation which is the mathematical model of the corresponding oscillator.

I hope that this book will be suitable to be a textbook for the students in nonlinear vibrations, but also for those who are researching the nonlinear phenomena in oscillatory systems in mechanics, mechanical devices, electromechanical systems, electric circuits, physics, chemistry, etc.

This book has an intention to give some practical information for engineers and technicians dealing with the problem of vibration and its elimination. The results of investigation show that independently on the amplitude and frequency of excitation force by proper treatment of the strong nonlinear system, the vibration level may be kept at a small level. Namely, not only in mechanical systems like cutting machines with periodical motion of the cutting tools, presses, supports for machines, seats in vehicles, etc., but also in electronics (electromechanical devices such as micro-actuators and micro oscillators), the requirement of small oscillations but without introducing dampers, which cause energy dissipation and decreasing of the efficiency of machines, can be achieved by proper use of the nonlinear properties of the system.

The results published in this book are applicable for the improvement in designing, for example, of music instruments and their parts such as the hammers in a piano. At the other side, the investigation is also of potential interest for modelling human voice production in cases where the vocal cords and voice producing fold are damaged.

Finally, I have to thank Prof. Anatoly Andreevich Martynyuk, member of the Ukrainian Academy of Sciences, for his invitation to write and publish this book.

Novi Sad, Serbia Livija Cveticanin

Contents

Chapter 1
Introduction

The problem of nonlinear oscillators occupies many researchers. Namely, the nonlinear oscillations occur in many real systems from macro to nano in size, and are the basic or auxiliary motions which follow the main motion. Thus, nonlinear oscillations are evident in many fields of science, not only in physics, mechanics and mathematics, but also in electronics, chemistry, biology, astronomy. To explain the phenomena in nonlinear oscillators, like bifurcation and deterministic chaos, etc., a proper mathematical model of the problem has to be formed. Usually, it is a strong nonlinear second order differential equation which has to be solved. Recently, for certain parameter values, the numerical computation of the problem is done. The numerical results are very accurate, but valid only for certain numerical parameters. Very often the obtained results are applicable for solving some technical problems, but are very pure for deep qualitative analysis of the problem. It is the reason, that we need the analytical solution of the mathematical problem which has to be transparent and suitable for discussion. The most of solution procedures which are developed require the linearization of the oscillator, or to take the nonlinearity to be small. Only the models which represent the small perturbed version of the linear one are tested. Recently, a few mathematical procedures for solving of the strong nonlinear differential equations of the oscillators are developed. The most of them are applicable only for some certain type of nonlinearity and are not valid in general. Unfortunately, the solution for an oscillator where the nonlinearity is of any type is not developed, yet.

The intention of this book is to give the approximate analytical solution procedures for the strong nonlinear oscillator where the basic solution corresponds to the pure nonlinear oscillator with any positive order of nonlinearity (in the differential equation the order of nonlinearity is a rational number: integer or non-integer). The free and forced vibrations of the strong nonlinear oscillator are obtained. The parametrically excited one-degree-of-freedom strong nonlinear oscillator is considered. The regular motion and the deterministic chaos of the pure nonlinear oscillator with a small linear damping and small periodical excitation is also analyzed. The control

L. Cveticanin, *Strong Nonlinear Oscillators*, Mathematical Engineering,
DOI 10.1007/978-3-319-58826-1_1

of the chaotic motion is discussed. The solution procedure for the two-degree-of-freedom oscillators is developed and the results are applied for analyzing of the rotor motion. Besides, ideal and non-ideal mechanical nonlinear systems are investigated. The jump phenomena, Sommerfeld effect and their elimination is widely discussed.

The book is organized in 9 chapters.

After the Introduction in Chap. 2 the physical and mathematical models of nonlinear oscillatory systems are considered. There is a significant number of systems whose behavior can be described as a nonlinear oscillator. Some of them can be described as pure nonlinear oscillators. The oscillators where the elastic force is given as a strong nonlinear function of any rational order (integer or non-integer) is named 'pure nonlinear oscillator'. This type of oscillator is the basic to be investigated in this book. The nonlinearity is strong. In the oscillator some additional forces may act: positive viscous or negative Van der Pol damping force, friction force, periodical or constant excitation force, etc.

In Chap. 3 the solution procedures for the pure nonlinear oscillators are shown. First the qualitative analysis of the mathematical model of the pure nonlinear oscillator is done. It is proved that the motion is periodical and the exact period of vibration for this type of oscillators is determined. The exact solution of the conservative pure nonlinear oscillator is obtained in the form of the Ateb periodic function. Based on the exact period of vibration and on the known amplitude, frequency and maximal velocity of vibration some approximate solution procedures are developed. The methods are compared. The advantages and disadvantages for all of methods are emphasized. In this chapter the approximate solution for the pure nonlinear oscillator with linear viscous damping is presented, too.

In Chap. 4 the pure nonlinear oscillator with additional small linear or nonlinear terms is considered. The mathematical model is a nonlinear second order differential equation with strong and weak nonlinear terms. The approximate solution methods use the exact or approximate solution of the pure nonlinear oscillator. The solution is assumed in the form of an Ateb function, Jacobi elliptic function or trigonometric function. For all of the solution is common that the parameters are time variable and their variation is obtained according to the additional terms in the differential equation. Besides, the Lindstedt–Poincaré method developed for the linear oscillator with small nonlinearity is adopted and modified for the strong nonlinear differential equation. The homotopy perturbation procedure for the strong nonlinear differential equation is also given. The averaging solution procedure is applied for the pure nonlinear oscillator with linear damping and small nonlinearity of Van der Pol type.

In Chap. 5, the nonstationary oscillators are considered. The parameters of the oscillator are time variable functions: periodically varied in time and nonperiodical but continually changeable with the slow time. Analytical solution procedures are based on the exact or almost exact solutions of the constant parameter valued differential equation. The solution is the perturbed version of that where the Ateb, Jacobi elliptic or trigonometric function are the basic. The special attention is given to mass variable systems, where due to mass variation a reactive force acts. The influence of the reactive force on the motion of the system is analyzed. The Levi–Civita and Van der Pol oscillators with time variable parameters are also considered.

In Chap. 6, the forced vibrations of the pure nonlinear oscillator are presented. The influence of the constant and also of the periodical excitation force is considered. Varying the parameters of the excitation force different nonlinear phenomena in the oscillator are discovered and discussed. Forced vibrations have been considered as well where the external excitation is designed to yield the closed-form solution in terms of the Ateb or Jacobi cn function. Frequency-response curves have been presented to illustrate the response in the frequency domain and the forced vibrations in the time domain.

Chapter 7 investigates the two-degree-of-freedom systems. Two types of oscillators are considered: one, described with two coupled second order differential equations and the other, with only one expressed in the form of a complex function. Namely, for the latter oscillator the two coupled second order differential equations are simplified to only one second order differential equation by introducing of a complex function. The first mathematical model corresponds to the two-mass system, while the second, to the one-mass system with two-degrees-of- freedom.

In Chap. 8 the phenomena of deterministic chaos is analyzed. The pure nonlinear oscillator with a small viscous damping and small periodic excitation force is considered. The Melnikov's criteria for chaos is applied for obtaining of the critical parameters for chaos. Numerical simulation is done. The Lyapunov exponent is calculated and is used to prove the existence of chaos. In this Chapter the vibration of the non-ideal mechanical system is given with a system of two coupled strong nonlinear second order differential equations. The resonance in the system is determined. The approximate solving procedure presented in the previous Chapter is adopted for the resonant condition of the system. For the non-ideal system, beside the chaos and its control, the Sommerfeld effect and its elimination is shown.

In Chap. 9 the axial vibration of a rod with strong nonlinearity is considered. The model is a clamped-free rod with a strongly nonlinear elastic property. Axial vibration is described by a strong nonlinear partial differential equation. A solution of the equation is constructed for special initial conditions by using the method of separation of variables. Using boundary and initial conditions, a frequency of vibration is obtained which is described in the exact analytic form for the axially vibrating purely nonlinear clamped-free rod. Solution illustration and simulation is given. The procedure suggested in this Chapter is applied for calculation of the frequency of the longissimus dorsi muscle of a cow. The influence of elasticity order and elasticity coefficient on the frequency property of vibration is tested.

At the end of every chapter a Reference list is given.

The book ends with six Appendixes: A the periodical Ateb function, B the Fourier series of the ca Ateb function, C the averaging of the Ateb functions, D the Jacobi elliptic function, E the Euler's integrals and F the inverse incomplete Beta function.

Chapter 2
Nonlinear Oscillators

In this book the pure nonlinear oscillator is considered. The pure nonlinear oscillator has a pure nonlinearity. The nonlinear function $f(x)$, which depends on the variable $x \in (-\infty, +\infty)$ and is a continual one, is defined as a pure nonlinear if it satisfies the condition that it has no linear approximation in any neighborhood of $x = 0$ (Mickens 2010). In general the pure nonlinearity is expressed as

$$f(x) = c_\alpha^2 x \, |x|^{\alpha-1}, \tag{2.1}$$

where c_α^2 is a positive constant which need not to be small, and the order of nonlinearity $\alpha \in \mathbb{R}_+$ is the positive rational number written as a termination decimal or as an exact fraction, $\alpha \in \mathbb{Q}_+ = \left\{\frac{m}{n} > 0 : m \in \mathbb{Z}, n \in \mathbb{Z}, n \neq 0\right\}$ and $\mathbb{Z}$ is integer. The absolute value of the variable x is used in order to ensure the function (2.1) to be an odd one: for $x > 0$ the nonlinearity $f(x)$ is positive, and for $x < 0$ the nonlinearity is negative. Namely, $|\pm x| > 0$, independently on the sign of x. Analyzing the first derivative $C \equiv df/dx = \alpha c_\alpha^2 \, |x|^{\alpha-1}$, two types of pure nonlinearities are evident: hard, for $\alpha > 1$ when C continually increases with x, and soft, for $\alpha < 1$ when C continually decreases. For $\alpha = 1$, we have $C = c_\alpha^2 = const.$ Thus, for various types of nonlinearity (2.1), the pure nonlinear oscillators with hard, soft and linear properties are obtained.

In this chapter the examples for various nonlinearities (2.1) and corresponding pure nonlinear oscillators are given. The mathematical modeling of the pure nonlinear oscillator is expressed. Solution procedures for solving pure nonlinear differential equation for some special cases of nonlinearity are presented.

L. Cveticanin, *Strong Nonlinear Oscillators*, Mathematical Engineering,
DOI 10.1007/978-3-319-58826-1_2

2.1 Physical Models

Experimental investigation on a significant number of materials, for example: aluminum, titanium and other aircraft materials (Prathap and Varadan 1976), copper and copper alloys (Lo and Gupta 1978), aluminum alloys and annealed copper (Lewis and Monasa 1982), wood (Haslach 1985), ceramic materials (Colm and Clark 1988), hydrophilic polymers (Haslach 1992), composites (Chen and Gibson 1998), polyurethane foam (Patten et al. 1998), felt (Russell and Rossing 1998), etc., show that force decreases or increases more rapidly than the deflection, i.e. the stress–strain properties of the material are strong nonlinear. Then, the nonlinear dependence of the restoring force on the deflection is usually modelled as a polynomial whose exponent is any positive integer. However, there are examples of the systems, for which this exponent can be of non-integer order. Very often the experimentally obtained stress–strain diagrams are mathematically approximated with an one term of a polynome whose coefficient and order correspond to experimentally obtained data. The convenient function has the form (2.1). The order of nonlinearity is an integer or a non-integer, i.e., any rational number. In praxis, due to simplicity, it is assumed that the elastic properties of the materials are linear, or weak nonlinear. It has been emphasized that the assumption of the linear elasticity is correct only for small deformation and is not correct for large deformation force.

The nonlinearity (2.1) must not be the result of physical properties of the material (as it is previously mentioned), but also of the geometry of the system: shape and dimensions of the body, type of laying, loading etc. The examples with geometrical nonlinearity are the helicoidal and the conical springs made of the material with linear property, but due to geometry, the function (2.1) is nonlinear (Lou et al. 2009). The nonlinearities may be caused by physical effects such as the contacting of coils in a compressed coil spring, or by excessively straining the spring material (Beards 1995). The concept of the passive vibration isolator is based on the geometric nonlinear property of the system. Thus, three linear elastic springs fixed at one end and connected to each other at the other end represent a passive vibration isolator with quasi-zero stiffness characteristic whose property is nonlinear (Alabudzev et al. 1989; Carela et al. 2007; Kovacic et al. 2008a; Gatti et al. 2010). Such a nonlinear effect can be realized by application of the permanent magnets, too (Nijse 2001, Xing et al. 2005). The order of nonlinearity is usually assumed to be quadratic. Hence, Ravindra and Mallik (1994) applied the pure cubic nonlinearity for vibroisolating system. Dymnikov (1972) supposed the deformation characteristic of a radially loaded rubber cylinder to be cubic, too. Rivin (2003) presented the vibration isolators made of wire-mesh and felt materials, while Ibrahim (2008) introduces cable isolators for vibrations. The passive vibration isolators with their nonlinear properties are important for protecting of the buildings during the earthquake (Araky et al. 2010). Besides, these isolators can protect the ships from the sea waves excited vibrations (Xiong et al. 2005). Finally, it is important to mention that in contrary to the physical nonlinearity of the material, the geometric nonlinearity can be eliminated by proper design (Jutte 2008).

The practical application of the integer or non-integer order nonlinearity is evident in engineering (in micro-electro-mechanical systems (MEMS), nano-electro-mechanical devices, nanometer switches, vibration-, acoustic- and impact isolators (Bondar 1978; Pilipchuk 2007, 2010; Afsharfard and Farshidianfar 2012), snap-through mechanisms, etc.), but also for explaining phenomena in structural mechanics, nanotechnology, chemistry and physics.

Nowadays, the relevance of nanotechnology is well recognized, so new developments and applications based on nonlinear dynamics are reached in an interdisciplinary framework. The most common structure which is applied in nanotechnology is the system of nano-oscillators which represents the micro-electro-mechanical system (MEMS). The term MEMS refers to mechanical microstructures (on the order to 10–1000 μm), such as sensors, valves, gears, gyroscopes microbridges, electric microactuators etc. The MEMS are suitable to be modeled by one or more nano-masses connected by one or more nonlinear springs (see for example, Polo et al. 2009, 2010; Jones and Nenadic 2013). Usually, the nonlinearity in these systems is assumed to be cubic (Mojahedi et al. 2001) or quadratic, for example for micromirrors (Burns and Bright 1997). In microactuators the order of nonlinearity is in the interval [2,7] (Cortopassi and Englander 2010). The simulation values obtained for the system are compared with results measured on experimental devices (see de Sudipto and Aluru 2006a, b). The difference between the results is obvious due to the fact that the nonlinear property is not modelled in the correct manner, i.e., the order of nonlinearity is neither quadratic nor cubic, as is usually considered. Such an assumption does not correspond to the order of nonlinearity of the real system. The proposal is to design an adequate model of the system and to obtain the position and velocity time distribution with nanometer accuracy, which would be the starting parameters for control of actual MEMS devices. Only the correct input parameters with excellent precision would give the correct control laws and accurate motion of the MEMS. The mechanical model has to be improved and the accurate order of the nonlinearity (which may be a positive integer and/or non-integer) to be considered. Such modification in the mechanical model of MEMS would give not only the correct qualitative behavior of the system but also the most accurate quantitative results.

The nonlinearity described with (2.1) is registered in vehicle hanging, seats, and vehicle tires. In vehicle hanging the nonlinearity is of order 3/2 (Zhu and Ishitoby 2004), and for tires it is in the interval [2.5,3] (Dixon 1996). Supports for machines, cutting machines with periodical motion of the cutting tools, presses, etc., have also nonlinear properties. Nonlinearity is detected in music instruments (hammers in piano, for example). The human voice producing folds (voice cords) exhibit nonlinear property, too.

The vibrating system with nonlinearity (2.1) represents the pure nonlinear oscillator whose investigation is of prior interest in science and engineering.

2.2 Mathematical Models

Since 1918, when Duffing published his results in oscillator with cubic nonlinearity, a significant number of investigation in the free vibrations of the one-degree-of-freedom strong nonlinear undamped and damped oscillators is done. The most often investigated oscillator is with pure cubic nonlinearity ($\alpha = 3$), whose mathematical model is

$$\ddot{x} + c_3^2 x^3 = \varepsilon f(x, \dot{x}), \tag{2.2}$$

with the initial conditions

$$x(0) = A, \qquad \dot{x}(0) = 0. \tag{2.3}$$

where c_3^2 is a positive constant and εf is a small additional function in comparison to the strong nonlinearity. Usually, it is impossible to find the exact analytical solution of (2.2) with initial conditions (2.3), and various approximate analytical solving methods are developed. Let us mention some of them: elliptic perturbation method (Mickens and Oyedeji 1985), averaging procedure (Coppola and Rand 1990), variable amplitude and phase method (Yuste and Bejarano 1986, 1990; Cveticanin 2009a), Galerkin's method (Chen and Gibson 1998), harmonic balance method (Chen 2003; Leung et al. 2012), modified Lindstedt–Poincar é method (Cheng et al. (1991), series expansion method (Kovacic and Brennan 2008), etc. The pure Duffing oscillator with additional damping term is also widely investigated. Trueba et al. (2000) and Sharma et al. (2012) considered the Duffing oscillator with positive linear and cubic term and small linear and cubic damping. The influence of the linear damping is investigated by Waluya and van Horssen (2003) and also by Cveticanin (2011). Cveticanin (2004, 2008, 2009b) and Akinpelu (2011) considered the influence of the quadratic damping on the vibrations of the Duffing oscillator. Siewe et al. (2009) and also Kanai and Yabuno (2012) extended the investigation to the so called Rayleigh–Duffing oscillator with cubic and linear damping terms. The role of nonlinear damping in soft Duffing oscillator with a simultaneous presence of viscous damping has been discussed in Ravindra and Mallik (1994) and Sanjuan (1999). Baltanas et al. (2001) have studied the effect of a nonlinear damping term, proportional to the power of velocity, on the dynamics of the double-well Duffing oscillator.

Finally, it has to be mentioned that a wide range of approximate solutions for (2.2) is given in Chap. 4 "Analysis Technique for the Various Forms of the Duffing Equation" in the book entitled: "The Duffing Equation: Nonlinear Oscillators and their Behaviour" edited by Brennan and Kovacic.

Cveticanin in her papers (1998, 2001, 2005a, b) considered the Duffing oscillator with complex-valued function. She adopted the previously mentioned methods for the differential equation with complex variable.

The oscillator with quadratic nonlinearity is also intensively studied. In Chen et al. (1998) and Chen and Cheung (1996) the elliptic perturbation method is applied for solving of the second order differential equation with quadratic nonlinearity

$$\ddot{x} + c_1^2 x + c_2^2 x^2 = \varepsilon f(x, \dot{x}), \tag{2.4}$$

where c_1^2 and c_2^2 are the constants of the linear and quadratic term. Cveticanin (2003) considered the oscillator with quadratic term and an additional constant F

$$\ddot{x} + c_1^2 x + c_2^2 x\,|x| = F.$$

For the small nonlinearity the approximate solution is assumed in the form of a trigonometric function. A similar model was considered by Mickens (1981)

$$\ddot{x} + c_1^2 x - c_2^2 x^2 = F, \tag{2.5}$$

who gives the solution by applying of the power series solution procedure. The expression (2.5) describes some phenomena in general reliability and also in solid-state physics. The solution for the pure nonlinear oscillator with quadratic term

$$\ddot{x} + c_1^2 x + c_2^2 x\,|x| = \varepsilon f(x, \dot{x}), \tag{2.6}$$

is determined assuming the same methods as for the Duffing equation (Cveticanin 2004).

Combining the both nonlinearities, the quadratic and cubic one, the mixed parity oscillator is formed (Hu 2007). The harmonic balance method is seen to be the most simple solution procedure.

Recently, the more general type of pure nonlinear oscillators is investigated: the nonlinearity is an integer order and $\alpha > 1$. Thus, Andrianov (2002), Andrianov and Awrejcewicz (2003a), Cveticanin (2011), Kovacic (2011) and Cveticanin et al. (2012) consider the oscillators with pure nonlinearity of any integer, specially of higher order i.e., $\alpha \gg 1$. Opposite, Awrejcewicz and Andrianov (2002) considered the oscillations of the nonlinear system where the order of nonlinearity is extremely small, i.e., $\alpha \approx 0$. Applying the harmonic balance method the approximate analytic solution is obtained. Andrianov and Awrejcewicz (2003b) analyzed the asymptotic behavior of the oscillator with damping and high power form nonlinearity.

Mickens (2001) was the first to investigate the pure nonlinear oscillators with non-integer order. Cooper and Mickens (2002) considered the oscillator with $x^{4/3}$ potential. The generalized harmonic balance/numerical method for determining analytical approximations is applied. Ozis and Yildirm (2007) applied the modified Lindsedt–Poincaré method for solving the differential equation with nonlinearity of order $1/3$. van Horssen (2003) generalized the problem and investigated the oscillator with the order of nonlinearity $\alpha < 1$. The mathematical model of the oscillator is

$$\ddot{x} + x^{1/(2n+1)} = 0, \tag{2.7}$$

where $n = 1, 2, 3, \ldots$ is a positive integer. van Hoorsen assumed the approximate analytical solution in the form of a trigonometric function. Applying the harmonic balance method, he obtained the approximate value of the frequency of vibration as

$$\Omega(x_0) = \left[\frac{2^{2n}}{\binom{2n+1}{n} x_0^{2n}} \right]^{1/(4n+2)}. \tag{2.8}$$

Belendez et al. (2007) obtained higher-order approximations applying a modified He's homotopy perturbation method. Hu and Xiong (2003) extended the consideration to the system

$$\ddot{x} + x^{(2m+2)/(2n+1)} = 0, \tag{2.9}$$

where $n = 1, 2, 3, ...,$ and $m = 1, 2, 3, ...$ are positive integers. Using the same solution procedure the following approximate frequency of vibration is obtained

$$\Omega(x_0) = \left[\frac{2^{2(2n-m)} \binom{2(m-n)+1}{m-n}}{\binom{2n+1}{n} x_0^{2(2n-m)}} \right]^{1/(4n+2)}. \tag{2.10}$$

Gottlieb (2003) compared this value with the exact one. Andrianov and Awrejcewicz (2003a) and also Andrianov and van Horssen (2006) considered the nonlinear oscillator (2.9) extended with a negative damping of Van der Pol type.

Nowadays, the oscillator model (2.9) is generalized and the nonlinearity of any rational number α is studied. Various methods for obtaining the frequency of oscillation are developed: the improved Lindstedt–Poincaré method (Amore and Aranda 2005), the series expansion method (Kovacic and Brennan 2008), the adopted Lindstedt–Poincaré method (Belendez et al. 2007; Cveticanin et al. 2010), the non-simultaneous variational approach (Kovacic et al. 2010), Hamiltonian approach (Cveticanin et al. 2010), the modified Lindstedt–Poincaré method (Cheng et al. 1991; He 2002a, b; Ozis and Yildirm 2007), the decomposition method (Kermani and Dehestani 2013), etc. Very often the solution procedures applied for pure nonlinear oscillators require addition of a linear term into the differential equation. The equation is transformed to the model with strong linearity and a weak nonlinearity, which is already widely investigated and a numerous solution procedures are developed. The most often applied solution methods are: the Krylov and Bogolubov (1943), Bogolubov and Mitropolski (1963) methods, the multiple scale method (Nayfeh and Mook 1979), perturbation method for certain nonlinear oscillators (Burton 1984), the method of straightforward expansion, the Lindstedt–Poinceré method, the homotopy perturbation technique (He 1998a, b; Cveticanin 2006, 2009a), the homotopy analysis method (Liao and Tan 2007), combined equivalent linearization and averaging perturbation method (Mickens and Oyedeji 1985; Mickens 2003), the iteration procedure for calculating approximations to periodic solutions (Mickens 2005, 2006), the method of slowly varying amplitude and phase (Cveticanin 2009b; Mickens 2010), etc. The mentioned methods are appropriate for solving strong linear and additional weak nonlinear complex-valued differential equations of vibration (Cveticanin 1992, 1993), too.

Unfortunately, there are numerous oscillators where the nonlinearity is much stronger than the linearity and even the oscillator is purely nonlinear. For such systems the application of the aforementioned methods is not possible. Namely, the differential equation is without a linear term and also the linearization of the equation is not possible due to the property of the system. These oscillators are not the perturbed versions of the linear ones and their behavior is far of those obtained for linear ones. To exceed this problem, in this book the solution procedures for pure strong nonlinear differential equation, which describe the oscillatory motion of pure nonlinear oscillator, are presented.

References

Afsharfard, A., & Farshidianfar, A. (2012). Design of nonlinear impact dampers based on acoustic and damping behavior. *International Journal of Mechanical Sciences*, *65*, 125–133.
Akinpelu, F. O. (2011). The nonlineary fractionally oscillator with strong quadratic damping force. *Research Journal of Applied Sciences*, *6*, 398–404.
Alabudzev, P., Grichin, A., Kim, L., Migirenko, G., Chon, V., & Stepanov, P. (1989). *Vibration protecting and measuring systems with quasi-zero stiffness*. New York: Hemisphere Publishing.
Amore, P., & Aranda, A. (2005). Improved Lindstedt–Poincaré method for the solution of nonlinear problems. *Journal of Sound and Vibration*, *283*, 1115–1136.
Andrianov, I. V. (2002). Asymptotics of nonlinear dynamical systems with high degree of nonlinearity. *Doklady RAN*, *386*, 165–168.
Andrianov, I. V., & Awrejcewicz, J. (2003a). Asymptotic approaches to strongly nonlinear dynamical systems. *Journal on Systems Analysis Modelling Simulation*, *43*, 255–268.
Andrianov, I. V., & Awrejcewicz, J. (2003b). Asymptotical behaviour of a system with damping and high power-form nonl-linearity. *Journal of Sound and Vibration*, *267*, 1169–1174.
Andrianov, I. A., & van Horssen, D. T. (2006). Analytical approximations of the period of a generalized nonlinear van der Pol oscillator. *Journal of Sound and Vibration*, *295*, 1099–1104.
Araky, Y., Asai, T., & Masui, T. (2010). Vertical vibration isolator having piecewise-constant restoring force. *Earthquake Engineering Structural Dynamics*, *38*, 1505–1523.
Awrejcewicz, L., & Andrianov, I. V. (2002). Oscillations of non-linear system with restoring force close to sign(x). *Journal of Sound and Vibration*, *252*, 962–966.
Baltanas, J. P., Trueba, J. L., & Sanjuan, M. A. F. (2001). Energy dissipation in nonlinearly damped Duffing oscillator. *Physica D*, *159*, 22–34.
Beards, C. (1995). *Engineering vibration analysis with application to control systems*. London: Edward Arnold.
Belendez, A., Pascual, C., Gallego, S., Ortuño, M., & Neipp, V. (2007). Application of a modified He's homotopy perturbation method to obtain higher-order approximations of an $x^{1/3}$ force nonlinear oscillator. *Physics Letters A*, *371*, 421–426.
Bondar, N. G. (1978). *Nonlinear vibrations excited by impulses*. Kiev-Doneck: Visca skola. (in Russian).
Bogolubov, N., & Mitropolski, J. A. (1963). *Asymptotical methods in the theory of nonlinear oscillations*. Delhi: Hindustan Publishing Co.
Burns, D. M., & Bright, V. M. (1997). Nonlinear flexures for stable deflection of an electrostatically actuated micromirror. In: *Proceedings of SPIE: Microelectronics Structures and MEMS for Optical Processing III*, *3226*, 125–135.

Burton, T. D. (1984). A perturbation method for certain non-linear oscillators. *International Journal of Non-Linear Mechanics, 19*, 397–407.

Carela, A., Brennan, M. J., & Waters, T. P. (2007). Static analysis of a passive vibration isolator with quasi-zero stiffness characteristic. *Journal of Sound and Vibration, 301*, 678–689.

Chen, S. H., Yang, X. M., & Cheung, Y. K. (1998). Periodic solutions of strongly quadratic nonlinear oscillators by the elliptic perturbation method. *Journal of Sound and Vibration, 212*, 771–780.

Chen, S. H., & Cheung, Y. K. (1996). An elliptic perturbation method for certain strongly non-linear oscillators. *Journal of Sound and Vibration, 192*, 453–464.

Chen, W. H., & Gibson, R. F. (1998). Property distribution determination for nonuniform composite beams from vibration response measurements and Galerkin's method. *Journal of Applied Mechanics, ASME, 65*, 127–133.

Chen, Y. Z. (2003). Evaluation of motion of the Duffing equation from its general properties. *Journal of Sound and Vibration, 264*, 491–497.

Cheng, Y. K., Chen, S. H., & Lau, S. L. (1991). A modified Lindstedt–Poincaré method for certain strongly non-linear oscillators. *International Journal of Nonlinear Mechanics, 26*, 367–378.

Colm, I. J., & Clark, N. J. (1988). *Forming, shaping and working of high-performance ceramics*. New York: Blackie.

Cooper, K., & Mickens, R. E. (2002). Generalized harmonic balance—numerical method for determining analytical approximations to the periodic solutions of the $x^{4/3}$ potential. *Journal of Sound and Vibration, 250*, 951–954.

Coppola, V. T., & Rand, R. H. (1990). Averaging using elliptic functions: Approximation of limit cycles. *Acta Mechanica, 81*, 125–142.

Cortopassi, C., & Englander, O. (2010). *Nonlinear springs for increasing the maximum stable deflection of MEMS electrostatic gap closing actuators*. Berkeley: University of Berkeley. http://www-basic-eecs.berkeley.edu/pister/245/project/CortopassiEnglander.pdf.

Cveticanin, L. (1992). Approximate analytical solutions to a class of nonlinear equations with complex functions. *Journal of Sound and Vibration, 157*, 289–302.

Cveticanin, L. (1993). An asymptotic solution to weak nonlinear vibrations of the rotor. *Mechanism and Machine Theory, 28*, 495–506.

Cveticanin, L. (1998). Analytical methods for solving strongly non-linear differential equations. *Journal of Sound and Vibration, 214*, 325–338.

Cveticanin, L. (2001). Analytic approach for the solution of the complex-valued strong non-linear differential equation of Duffing type. *Physica A: Statistical Mechanics and its Applications, 297*, 348–360.

Cveticanin, L. (2003). Vibrations of the system with quadratic non-linearity and a constant excitation force. *Journal of Sound and Vibration, 261*, 169–176.

Cveticanin, L. (2004). Vibrations of the non-linear oscillator with quadratic non-linearity. *Physica A: Statistical Mechanics and its Applications, 341*, 123–135.

Cveticanin, L. (2005a). The homotopy-perturbation method applied for solving complex—valued differential equations with strong cubic non-linearity. *Journal of Sound and Vibration, 285*, 1171–1179.

Cveticanin, L. (2005b). Free vibration of a Jeffcott rotor with pure cubic non-linear elastic property of the shaft. *Mechanism and Machine Theory, 40*, 1330–1344.

Cveticanin, L. (2006). Homotopy-perturbation method for pure non-linear differential equation. *Chaos, Solitons and Fractals, 30*, 1221–1230.

Cveticanin, L. (2008). Analyses of oscillators with non-polynomial damping terms. *Journal of Sound and Vibration, 317*, 866–882.

Cveticanin, L. (2009a). The approximate solving methods for the cubic Duffing equation based on the Jacobi elliptic functions. *International Journal of Nonlinear Science and Numerical Simulation, 10*, 1491–1516.

Cveticanin, L. (2009b). Oscillator with strong quadratic damping force. *Publications de L'Institut Mathematique, Nouvelle serie, 85*, 119–130.

Cveticanin, L. (2011). Oscillator with nonlinear elastic and damping force. *Computers and Mathematics with Applications, 62*, 1745–1757.
Cveticanin, L., Kalami-Yazdi, M., Saadatnia, Z., & Askari, H. (2010a). Application of Hamiltonian approach to the generalized nonlinear oscillator with fractional power. *International Journal of Nonlinear Sciences and Numerical Simulation, 11*, 997–1001.
Cveticanin, L., Kovacic, I., & Rakaric, Z. (2010b). Asymptotic methods for vibrations of the pure non-integer order oscillator. *Computers and Mathematics with Applications, 60*, 2616–2628.
Cveticanin, L., Kalami-Yazdi, M., & Askari, H. (2012). Analytical solutions for a generalized oscillator with strong nonlinear terms. *Journal of Engineering Mathematics, 77*, 211–223.
Dixon, J. C. (1996). *Tires, suspension and handling*. Warrandale: Society of Automative Engineers.
Dymnikov, S. I. (1972). Stiffness computation for rubber rings and cords. *Issues on Dynamics and Strength, 24*, 163–173. (in Russian).
Gatti, G., Kovacic, I., & Brennan, M. J. (2010). On the response of a harmonically excited two degree-of-freedom system consisting of linear and nonlinear quasi-zero stiffness oscillators. *Journal of Sound and Vibration, 329*, 823–835.
Gottlieb, H. P. W. (2003). Frequencies of oscillators with fractional-power non-linearities. *Journal of Sound and Vibration, 261*, 557–566.
Haslach, H. W. (1985). Post-buckling behavior of columns with non-linear constitutive equations. *International Journal of Non-Linear Mechanics, 20*, 53–67.
Haslach, H. W. (1992). Influence of adsorbed moisture on the elastic post-buckling behavior of columns made of non-linear hydrophilic polymers. *International Journal of Non-Linear Mechanics, 27*, 527–546.
He, J. H. (1998a). Homotopy perturbation technique. *Computational Methods in Applied Mechanics and Engineering, 178*, 257–262.
He, J. H. (1998b). An approximate solution technique depending upon an artificial parameter. *Communications in Nonlinear Science and Numerical Simulation, 3*, 92–97.
He, J. H. (2002a). Modified Lindstedt–Poincaré methods for some strongly nonlinear oscillations, Part I: Expansion of a constant. *International Journal of Non-Linear Mechanics, 37*, 309–314.
He, J. H. (2002b). Modified Lindstedt–Poincaré methods for some strongly nonlinear oscillations, Part II: A new transformation. *International Journal of Non-Linear Mechanics, 37*, 315–320.
van Horssen, W. T. (2003). On the periods of the periodic solutions of the non-linear oscillator equation $\ddot{x} + x^{1/(2n+1)} = 0$. *Journal of Sound and Vibration, 260*, 961–964.
Hu, H., & Xiong, Z.-G. (2003). Oscillations in an $\mathrm{x}^{(2m+2)/(2n+1)}$ potential. *Journal of Sound and Vibration, 259*, 977–980.
Hu, H. (2007). Solution of a mixed parity nonlinear oscillator: Harmonic balance. *Journal of Sound and Vibration, 299*, 331–338.
Ibrahim, R. A. (2008). Recent advances in nonlinear passive vibration isolators. *Journal of Sound and Vibration, 314*, 371–452.
Jones, T. B., & Nenadic, N. G. (2013). *Electromechanics and MEMS*. New York: Cambridge University Press.
Jutte, C.V. (2008). *Generalized synthesis methodology of nonlinear springs for prescribed load—displacement functions*. Ph.D. Thesis, Michigan: Mechanical Engineering, The University of Michigan.
Kanai, Y., & Yabuno, H. (2012). Creation-annihilation process of limit cycles in the Rayleigh–Duffing oscillator. *Nonlinear Dynamics, 70*, 1007–1016.
Kermani, M. M., & Dehestani, M. (2013). Solving the nonlinear equations for one-dimensional nano-sized model including Rydberg and Varshni potentials and Casimir force using the decomposition method. *Applied Mathematical Modelling, 37*, 3399–3406.
Kovacic, I. (2011). The method of multiple scales for forced oscillators with some real-power nonlinearities in the stiffness and damping force. *Chaos, Solitons & Fractals, 44*, 891–901.
Kovacic, I., Brennan, M. J., & Waters, T. P. (2008a). A study of a non-linear vibration isolator with quasi-zero stiffness characteristic. *Journal of Sound and Vibration, 315*, 700–711.

Kovacic, I., & Brennan, M. J. (2008b). On the use of two classical series expansion methods to determine the vibration of harmonically excited pure cubic oscillators. *Physics Letters A, 372*, 4028–4032.

Kovacic, I., Rakaric, Z., & Cveticanin, L. (2010). A non-simultaneous variational approach for the oscillators with fractional-order power nonlinearities. *Applied Mathematics and Computation, 217*, 3944–3954.

Krylov, N., & Bogolubov, N. (1943). *Introduction to nonlinear mechanics*. New Jersey: Princeton University Press.

Leung, A. Y. T., Guo, Z. J., & Yang, H. X. (2012). Residue harmonic balance analysis for the damped Duffing resonator driven by a van der Pol oscillator. *International Journal of Mechanical Sciences, 63*, 59–65.

Lewis, G., & Monasa, F. (1982). Large deflections of cantilever beams of non-linear materials of the Ludwick type subjected to an end moment. *International Journal of Non-Linear Mechanics, 17*, 1–6.

Liao, S. J., & Tan, Y. (2007). A general approach to obtain series solutions of nonlinear differential equations. *Studies in Applied Mathematics, 119*, 297–355.

Lo, C. C., & Gupta, S. D. (1978). Bending of a nonlinear rectangular beam in large deflection. *Journal of Applied Mechanics, ASME, 45*, 213–215.

Lou, J. J., Zhu, S. J., He, L., & He, Q. W. (2009). Experimental chaos in non-linear vibration isolation system. *Chaos, Solitons and Fractals, 40*(2009), 1367–1375.

Mickens, R. E. (1981). A uniformly valid asymptotic solution for $d^2y/dt^2 + y = a + \varepsilon y^2$. *Journal of Sound and Vibration, 76*, 150–152.

Mickens, R. E. (2001). Oscillations in an $x^{4/3}$ potential. *Journal of Sound and Vibration, 246*, 375–378.

Mickens, R. E. (2003). A combined equivalent linearization and averaging perturbation method for non-linear oscillator equations. *Journal of Sound and Vibration, 264*, 1195–1200.

Mickens, R. E. (2005). A generalized iteration procedure for calculating approximations solutions of "truly nonlinear oscillators". *Journal of Sound and Vibration, 287*, 1045–1051.

Mickens, R. E. (2006). Iteration method solutions for conservative and limit-cycle $x^{1/3}$ force oscillators. *Journal of Sound and Vibration, 292*, 964–968.

Mickens, R. E. (2010). *Truly nonlinear oscillations*. Singapore: World Scientific.

Mickens, R. E., & Oyedeji, K. (1985). Construction of approximate analytical solutions to a new class of non-linear oscillator equation. *Journal of Sound and Vibration, 102*, 579–582.

Mojahedi, M., Zand, M. M., Ahmadian, M. T., & Babaei, M. (2001). Analytic solutions to the oscillatory behavior and primary resonance of electrostatically actuated microbridges. *International Journal of Structural Stability and Dynamics, 11*, 1119–1137.

Nayfeh, A. H., & Mook, D. T. (1979). *Nonlinear oscillations*. New York: Wiley.

Nijse, G. L. P. (2001). *Linear motion systems: A modular approach for improved straightness performance*. Ph.D. Thesis, Delft: University of Technology.

Ozis, T., & Yildirm, T. A. (2007). Determination of periodic solution for a $u^{1/3}$ force by He's modified Lindstedt–Poincaré method. *Journal of Sound and Vibration, 301*, 415–419.

Patten, W. N., Sha, S., & Mo, C. (1998). A vibration model of open celled polyurethane foam automotive seat cushions. *Journal of Sound and Vibration, 217*, 145–161.

Pilipchuk, V. N. (2007). Strongly nonlinear vibration of damped oscillators with two nonsmooth limits. *Journal of Sound and Vibration, 302*, 398–402.

Pilipchuk, V. N. (2010). *Nonlinear dynamics: Between linear and impact limits*. New York: Springer.

Polo, M. F. P., Molina, M. P., & Chica, J. G. (2009). Chaotic dynamic and control for micro-electro-mechanical systems of massive storage with harmonic base excitation. *Chaos, Solitons and Fractals, 39*, 1356–1370.

Polo, M. F. P., Molina, M. P., & Chica, J. G. (2010). Self-oscillations and chaotic dynamic of a nonlinear controlled nano-oscillator. *Journal of Computational and Theoretical Nanoscience, 7*, 2463–2477.

Prathap, G., & Varadan, T. K. (1976). The inelastic large deformation of beams. *ASME Journal of Applied Mechanics, 43*, 689–690.

Ravindra, B., & Mallik, A. K. (1994). Role of nonlinear dissipation in soft Duffing oscillators. *Physical Review E, 49*, 4950–4954.

Rivin, E. I. (2003). *Passive vibration isolation.* New York: ASME Press.

Russell, D., & Rossing, T. (1998). Testing the nonlinearity of piano hammers using residual shock spectra. *Acta Acustica, 84*, 967–975.

Sanjuan, M. A. F. (1999). The effect of nonlinear damping on the universal escape oscillator. *International Journal of Bifurcation and Chaos, 9*, 735–744.

Sharma, A., Patidar, V., Purohit, G., & Sud, K. K. (2012). Effects on the bifurcation and chaos in forced Duffing oscillator due to nonlinear damping. *Communication in Nonlinear Science and Numerical Simulation, 17*, 2254–2269.

Siewe, M. S., Cao, H., & Sanjuan, M. A. F. (2009). Effect of nonlinear dissipation on the boundaries of basin of attraction in two-well Rayleigh–Duffing oscillators. *Chaos, Solitons & Fractals, 39*, 1092–1099.

de Sudipto, K., & Aluru, N. R. (2006a). Complex nonlinear oscillations in electrostatically actuated microstructures. *IEEE Journal of Microelectromechanical Systems, 5*, 355–369.

de Sudipto, K., & Aluru, N. R. (2006b). U-sequence in electrostatic micromechanical systems (MEMS). *Proceedings of the Royal Society A, 462*, 3435–3464.

Trueba, J. L., Rams, J., & Sanjuan, M. A. F. (2000). Analytical estimates of the effect of nonlinear damping in some nonlinear oscillators. *International Journal of Bifurcation and Chaos, 10*, 2257–2267.

Waluya, S. B., & van Horssen, W. T. (2003). On the periodic solutions of a generalized non-linear Van der Pol oscillator. *Journal of Sound and Vibration, 268*, 209–215.

Xing, J. T., Xiong, Y. P., & Price, W. G. (2005). Passive active vibration isolation systems to produce zero of infinite dynamic modulus: Theoretical and conceptual design strategies. *Journal of Sound and Vibration, 286*, 615–636.

Xiong, Y. P., Xing, J. T., & Price, W. G. (2005). Interactive power flow characteristics of an integrated equipment—nonlinear isolator—travelling flexible ship excited by sea waves. *Journal of Sound and Vibration, 287*, 245–276.

Yuste, S. B., & Bejarano, J. D. (1986). Construction of approximate analytical solution to a new class of a non-linear oscillator equations. *Journal of Sound and Vibration, 110*, 347–350.

Yuste, B. S., & Bejarano, J. D. (1990). Improvement of a Krylov–Bogoliubov method that uses Jacobi elliptic functions. *Journal of Sound and Vibration, 139*, 151–163.

Zhu, Q., & Ishitoby, M. (2004). Chaos and bifurcations in a nonlinear vehicle model. *Journal of Sound and Vibration, 275*, 1136–1146.

Chapter 3
Pure Nonlinear Oscillator

In this Chapter the pure nonlinear oscillator with pure nonlinearity of any rational order is considered. The mathematical model of the oscillator is

$$\ddot{x} + f(x) = 0, \tag{3.1}$$

where $f(x)$ is the pure nonlinearity given as (2.1). Substituting (2.1) into (3.1) it is

$$\ddot{x} + c_\alpha^2 x \left| x \right|^{\alpha-1} = 0, \tag{3.2}$$

with the initial conditions

$$x(0) \equiv x_0 = A, \qquad \dot{x}(0) = 0, \tag{3.3}$$

where, as it is previously mentioned, c_α^2 is a positive constant that need not to be small, and the order of nonlinearity $\alpha \in \mathbb{R}_+$ is the positive rational number written as a termination decimal or as an exact fraction, $\alpha \in \mathbb{Q}_+ = \left\{ \frac{m}{n} > 0 : m \in \mathbb{Z}, n \in \mathbb{Z}, n \neq 0 \right\}$ and $\mathbb{Z}$ is integer.

The Chapter is organized in 8 sections. In Sect. 3.1 the qualitative analysis of the oscillator (3.2) is considered and the period of vibration is determined. In Sect. 3.2 the exact solution of (3.2) for the initial conditions (3.3) in the form of the Ateb function is obtained. Due to its complexity it is not suitable for application by engineers and technicians It is the reason that the approximate solution procedures for (3.2) with (3.3) are developed. In Sects. 3.3 and 3.4 the adopted and the modified Lindstedt–Poincaré (LP) methods are given. Using the exact period of vibration, amplitude and maximal velocity of vibration the solutions in the form of the Jacobi elliptic function and trigonometric function are given (see Sects. 3.6 and 3.7, respectively). In Sect. 3.8 the pure nonlinear oscillator with linear damping is investigated. The approximate solution for this type of oscillator is determined. Some numerical examples are also considered, and the accuracy of the suggested methods is tested. The Chapter ends with a Reference list.

L. Cveticanin, *Strong Nonlinear Oscillators*, Mathematical Engineering,
DOI 10.1007/978-3-319-58826-1_3

3.1 Qualitative Analysis

For the qualitative analysis of the oscillator it is convenient to use the first integral of the system (3.2). The first integral reads

$$\dot{x}^2 + \frac{2c_\alpha^2}{\alpha+1}|x|^{\alpha+1} = K\,, \tag{3.4}$$

where K is a constant dependent on the initial conditions. Being both addends on the left nonnegative, the associated phase paths represent a generalized Lamé super-ellipse in the x - $\dot{x}$ phase plane. There is a single equilibrium point $x = \dot{x} = 0$ such that is a centre, therefore the solutions of (3.2) are periodic in time. Namely, the result of the qualitative analysis shows that the motion of the oscillator with a pure non-integer order restoring force is periodical.

Let us apply the initial conditions (3.3). The system described by (3.2) has the following energy-type first integral

$$\dot{x}^2 = \frac{2c_\alpha^2}{\alpha+1}(|A|^{\alpha+1} - |x|^{\alpha+1}). \tag{3.5}$$

The corresponding phase trajectories are bounded

$$-A \leq x \leq A, \tag{3.6}$$

$$-\sqrt{\frac{2c_\alpha^2\,|A|^{\alpha+1}}{\alpha+1}} \leq \dot{x} \leq \sqrt{\frac{2c_\alpha^2\,|A|^{\alpha+1}}{\alpha+1}}. \tag{3.7}$$

The extremum of the coordinate depends only on the initial amplitude A, while the extreme value of the velocity depends on the initial amplitude A, as well as on the internal parameters of the system: the fraction value α and the parameter c_α.

Since $\alpha > 1$, the velocity is also limited by the value $c_\alpha A$. Two different cases can be distinguished. If $\sqrt[\alpha-1]{\frac{\alpha+1}{2}} > A$, then

$$\sqrt{\frac{2c_\alpha^2\,|A|^{\alpha+1}}{\alpha+1}} \leq |\dot{x}|_{\max} < c_\alpha A. \tag{3.8}$$

In case $\sqrt[\alpha-1]{\frac{\alpha+1}{2}} < A$, one has

$$c_\alpha A < |\dot{x}|_{\max} \leq \sqrt{\frac{2c_\alpha^2\,|A|^{\alpha+1}}{\alpha+1}}. \tag{3.9}$$

The limiting phase trajectories are shown for both cases in Fig. 3.1a, b as full lines ($\alpha = 1$) and dotted line ($\alpha = 2$). In addition, the phase trajectories corresponding to $\alpha = 4/3$ (dashed line) and $\alpha = 5/3$ (dashed-dotted line) are also plotted.

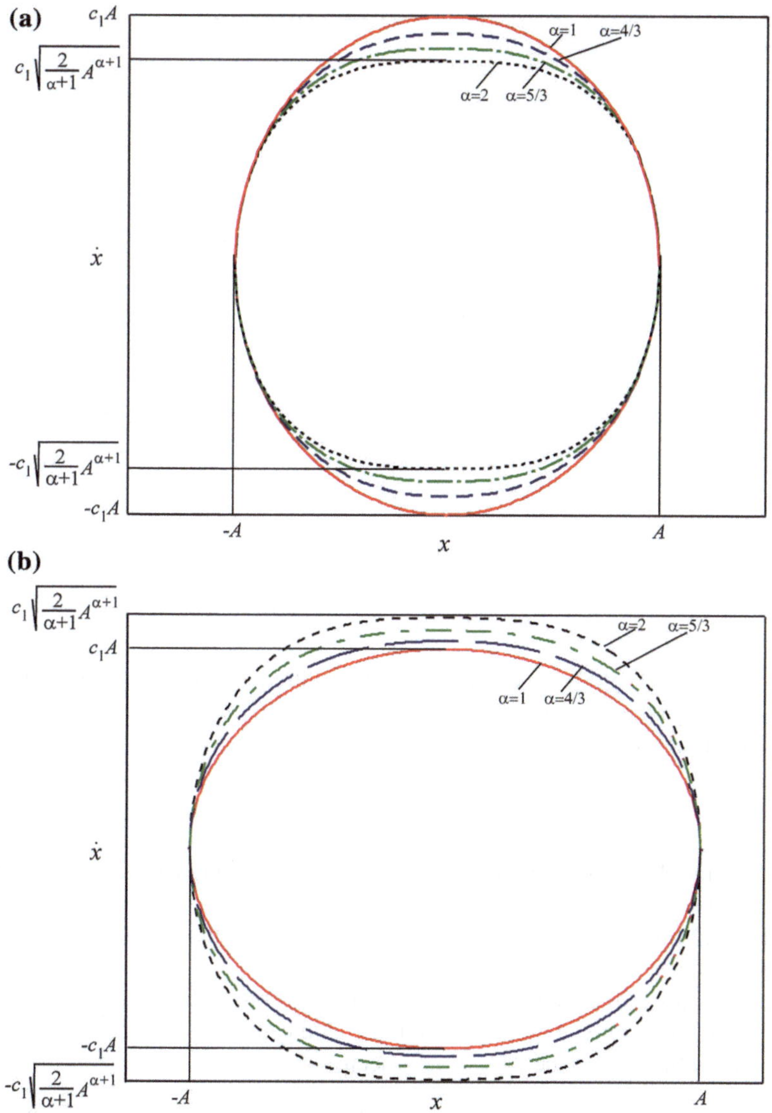

Fig. 3.1 The phase trajectories for different values of the parameter α and: **a** $A \leq 1$, **b** $A > 1$

If $\alpha \rightarrow \infty$ and $A \leq 1$, the lower limit of the velocity tends to zero, i.e. the dotted phase trajectory in Fig. 3.1a compresses along the vertical axes.

For $\alpha \rightarrow \infty$ and $A > 1$, the upper limit of the velocity tends to infinity, which means that the dotted phase trajectory in Fig. 3.1b extends along the vertical axes.

Fig. 3.2 $A^{(\alpha-1)/2} - A$ diagram for various values of the parameter α

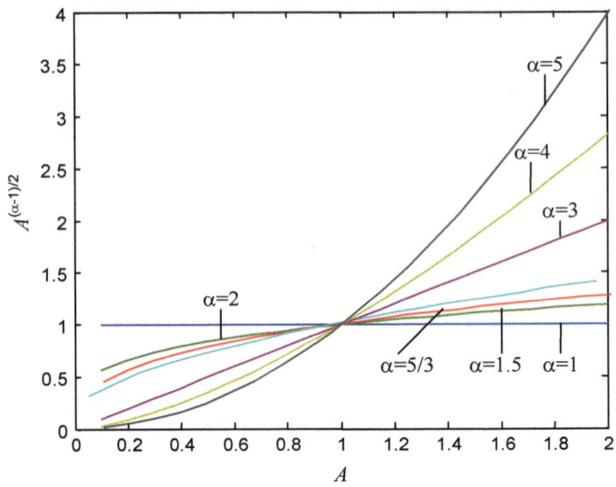

It should be noted that

$$\inf(|\dot{x}|_{\max}) \to c_\alpha A, \tag{3.10}$$

if $\alpha \to 1$ and $A \leq 1$. In other words, the dotted phase trajectory in Fig. 3.1a approaches then the outer limit depicted by a full line.

On the other hand

$$\sup(|\dot{x}|_{\max}) \longrightarrow c_\alpha A, \tag{3.11}$$

for $\alpha \to 1$ and $A > 1$. Under these conditions, the dotted phase trajectory in Fig. 3.1b becomes closer to the inner limit given by a full line.

In Fig. 3.2 the $|A|^{(\alpha-1)/2} - A$ diagram for various values of the parameter α is plotted.

The diagram proves the previously obtained results. These results are already published in Cveticanin et al. (2010). In Fig. 3.2, the curve, $|A|^{(\alpha-1)/2} - A$, is plotted for various values of α. The following is seen:

1. For all $A < 1$ and independently on α, it is $|A|^{(\alpha-1)/2} < 1$. This term makes the absolute error smaller.
2. For $A > 1$ and all values of α, it is $|A|^{(\alpha-1)/2} > 1$. This term makes the absolute error higher. The higher the A, the larger the term $|A|^{(\alpha-1)/2}$.
3. For $\alpha = 3$, the curve $|A|^{(\alpha-1)/2} - A$ transforms into a linear function which represents a 'boundary'. Namely, for $1 < \alpha < 3$ and the constant initial amplitude $A < 1$, the term $|A|^{(\alpha-1)/2}$ increases as α increases. In contrary, for $1 < \alpha < 3$ and the constant initial amplitude $A > 1$, the term $|A|^{(\alpha-1)/2}$ decreases for the increase of α. For $\alpha > 3$, the term $|A|^{(\alpha-1)/2}$ increases with α for all the values of the initial amplitude A.

3.1.1 Exact Period of Vibration

The energy-type first integral (3.5) is rewritten as

$$\frac{\dot{x}^2}{2} + \frac{c_\alpha^2}{\alpha+1}|x|^{\alpha+1} = \frac{c_\alpha^2}{\alpha+1}|A|^{\alpha+1}. \tag{3.12}$$

Analyzing (3.12) it is obvious that the both terms on the left side are positive and the motion is periodical (see Mickens 2001, Gottlieb 2003 or Belendez et al. 2007) with period

$$T_{ex} = 4\int_0^A \frac{dx}{|\dot{x}|} = 4\sqrt{\frac{\alpha+1}{2c_\alpha^2}}\int_0^A \frac{dx}{\sqrt{|A|^{\alpha+1} - |x|^{\alpha+1}}}. \tag{3.13}$$

Substituting the new variable $|x| = |A|\,|u|^{1/(\alpha+1)}$ into (3.13), the transformed version of period is

$$T_{ex} = \frac{4\,|A|^{(1-\alpha)/2}}{c_\alpha\sqrt{2(\alpha+1)}}\int_0^1 (1-|u|)^{-1/2}u^{-\alpha/(\alpha+1)}du. \tag{3.14}$$

Introducing the Euler Beta function $B(m,n)$ (see Rosenberg 1963)

$$B(m,n) = \int_0^1 (1-|u|)^{n-1}u^{m-1}du, \tag{3.15}$$

the relation (3.14) can be rewritten as follows

$$T_{ex} = \frac{4\,|A|^{(1-\alpha)/2}}{c_\alpha\sqrt{2(\alpha+1)}}B\left(\frac{1}{\alpha+1},\frac{1}{2}\right). \tag{3.16}$$

Due to

$$B(m,n) = \frac{\Gamma(m)\Gamma(n)}{\Gamma(m+n)}, \tag{3.17}$$

the exact period is

$$T_{ex} = \frac{4\,|A|^{(1-\alpha)/2}}{c_\alpha\sqrt{2(\alpha+1)}}\frac{\Gamma(\frac{1}{\alpha+1})\Gamma(\frac{1}{2})}{\Gamma(\frac{3+\alpha}{2(\alpha+1)})}, \tag{3.18}$$

where Γ is the Euler Gamma function (Gradstein and Rjizhik 1971). Analyzing the relation (3.16) i.e., (3.18) it is obvious, that the period of vibration T_{ex} depends on

Table 3.1 The values of parameter T^* for various α

α	1/3	2/3	1	4/3	5/3	2	3
T^*	5.86966	6.07863	6.28319	6.48313	6.67845	6.86926	7.41630

the Euler Beta function i.e., Euler Gamma function and is the function of fraction value α, coefficient of the rigidity c_α and initial amplitude A.

Using $\Gamma(\frac{1}{2}) = \sqrt{\pi}$, let us rewrite (3.18) into the form

$$T_{ex} = \frac{T^*}{c_\alpha \left|A\right|^{(\alpha-1)/2}}, \tag{3.19}$$

where

$$T^* = \left(\frac{4\sqrt{\pi}}{\sqrt{2(\alpha+1)}}\right) \frac{\Gamma(\frac{1}{\alpha+1})}{\Gamma(\frac{3+\alpha}{2(\alpha+1)})}. \tag{3.20}$$

In Table 3.1 the values of T^* for various orders of nonlinearity α are shown.

It is evident that for higher orders of nonlinearity the coefficient T^* is also higher. Thus, according to (3.19) and (3.20) the following is concluded:

1. For the coefficient of nonlinearity $c_\alpha = 1$ and the initial amplitude $A = 1$, the period of vibration is longer for higher orders of nonlinearity.
2. For the oscillator with fixed order of nonlinearity and fixed initial amplitude, the period of vibration increases by decreasing of the coefficient of nonlinearity c_α.
3. If the nonlinear properties of the oscillator are unchanged, the period of vibration depends on the initial amplitude for all $\alpha \neq 1$: the higher the value of the initial amplitude A, the shorter the period of vibration.
4. For the linear oscillator when $\alpha = 1$ the period of vibration is independent on the initial amplitude.

Using the frequency—period of vibration expression, $\Omega = 2\pi/T_{ex}$, and the exact period of vibration (3.18), the exact frequency of vibration follows as

$$\Omega = \frac{\sqrt{2\pi c_\alpha^2(\alpha+1)}}{2\left|A\right|^{(1-\alpha)/2}} \frac{\Gamma(\frac{3+\alpha}{2(\alpha+1)})}{\Gamma(\frac{1}{\alpha+1})}. \tag{3.21}$$

The frequency of vibration directly depends on the coefficient of nonlinearity and is the function of the initial amplitude A and order of nonlinearity α as is already published in Cveticanin (2009a).

3.2 Exact Periodical Solution

It is not hard to see that the Eq. (3.4) can be rewritten as

$$\dot{x} = \pm K\sqrt{1-\left(\frac{|x|}{A}\right)^{\alpha+1}}, \qquad K = \frac{\sqrt{2}\,|c_\alpha|A^{(\alpha+1)/2}}{\sqrt{\alpha+1}}. \tag{3.22}$$

Now, we are ready to solve (3.22) analytically. Let us choose the positive right–hand–side expression in (3.22). It will be

$$L = \int \frac{\mathrm{d}x}{\sqrt{1-\left(\frac{|x|}{A}\right)^{\alpha+1}}} = Kt + C, \tag{3.23}$$

where C denotes the integration constant. Expanding the integrand in L into a binomial series and then integrating term wise, we conclude:

$$L = \int \sum_{n=0}^{\infty} (-1)^n \binom{-\frac{1}{2}}{n} \left(\frac{|x|}{A}\right)^{(\alpha+1)n} \mathrm{d}x = A \sum_{n=0}^{\infty} \binom{-\frac{1}{2}}{n} \frac{1}{(\alpha+1)n+1} \left(\frac{|x|}{A}\right)^{(\alpha+1)n+1}.$$

Employing the Pochhammer symbol $(a)_n = a(a+1)\cdots(a+n-1),\, n \in \mathbb{N}$ *mutatis mutandis*

$$\binom{-\frac{1}{2}}{n} = (-1)^n \frac{\left(\frac{1}{2}\right)_n}{n!},$$

and

$$\frac{1}{(\alpha+1)n+1} = \frac{\frac{1}{\alpha+1}}{n+\frac{1}{\alpha+1}} = \frac{\left(\frac{1}{\alpha+1}\right)_n}{\left(\frac{\alpha+2}{\alpha+1}\right)_n},$$

hence

$$L = |x| \sum_{n=0}^{\infty} \frac{\left(\frac{1}{2}\right)_n \left(\frac{1}{\alpha+1}\right)_n}{\left(\frac{\alpha+2}{\alpha+1}\right)_n n!} \left(\frac{|x|}{A}\right)^{(\alpha+1)n} = |x|\, {}_2F_1\left[\begin{matrix}\frac{1}{2}, \frac{1}{\alpha+1} \\ \frac{\alpha+2}{\alpha+1}\end{matrix} \,\middle|\, \left(\frac{|x|}{A}\right)^{\alpha+1}\right]. \tag{3.24}$$

Now, by the well-known formula (see http: 2002a and http: 2002b) such that connects the Gaussian hypergeometric ${}_2F_1$ and the incomplete Beta function B_z:

$${}_2F_1\left[\begin{matrix}a, 1-b \\ a+1\end{matrix} \,\middle|\, z\right] = \frac{a}{z^a} B_z(a,b), \qquad |z| < 1,$$

letting

$$a = \frac{1}{\alpha+1}, \qquad b = \frac{1}{2},$$

we get

$$B_{\left(\frac{|x|}{A}\right)^{\alpha+1}}\left(\frac{1}{\alpha+1},\frac{1}{2}\right)=\sqrt{2(\alpha+1)}\,|c_\alpha|\,A^{(\alpha-1)/2}\,t+C\,. \tag{3.25}$$

Finally, the initial condition $x(0) = A$ clearly gives

$$\begin{aligned}\frac{(\alpha+1)|x|}{A}\,{}_2F_1\left[\begin{matrix}\frac{1}{\alpha+1},\ \frac{1}{2}\\ 1+\frac{1}{\alpha+1}\end{matrix}\Bigg|\left(\frac{|x|}{A}\right)^{\alpha+1}\right]&=\mathrm{B}_{\left(\frac{|x|}{A}\right)^{\alpha+1}}\left(\frac{1}{\alpha+1},\frac{1}{2}\right)\\&=\mathrm{B}\left(\frac{1}{\alpha+1},\frac{1}{2}\right)+\sqrt{2(\alpha+1)}\,|c_\alpha|\,A^{(\alpha-1)/2}\,t\,.\end{aligned} \tag{3.26}$$

This relation is the main tool in determining the explicit solution of Cauchy problem (3.2) with (3.3). Introducing the inverse incomplete Euler Beta functions which are called "sine Ateb function" *sa* and "cosine Ateb function" *ca* (see Rosenberg's paper in 1963 and Senik's exhaustive article in 1969 in which the author constructed the periodic Ateb functions), the solution of the differential equation (3.2) and its first time derivative have the form (see Cveticanin and Pogany 2012)

$$x=Aca\left(\alpha,1,t\sqrt{\frac{c_\alpha^2\,|A|^{\alpha-1}\,(\alpha+1)}{2}}+\theta\right), \tag{3.27}$$

and

$$\dot{x}=-\frac{\sqrt{2c_\alpha^2}}{\sqrt{\alpha+1}}\,A^{(\alpha+1)/2}sa\left(1,\alpha,t\sqrt{\frac{c_\alpha^2\,|A|^{\alpha-1}\,(\alpha+1)}{2}}+\theta\right), \tag{3.28}$$

where A and θ are arbitrary constants. Due to initial conditions (3.3) the phase angle θ is zero and the solution of (3.27) and its time derivative are

$$x=Aca\left(\alpha,1,\sqrt{\frac{\alpha+1}{2}}\,|c_\alpha|A^{(\alpha-1)/2}\,t\right), \tag{3.29}$$

and

$$\dot{x}=-\sqrt{\frac{2}{\alpha+1}}|c_\alpha|\,A^{(\alpha+1)/2}sa\left(1,\alpha,\sqrt{\frac{\alpha+1}{2}}\,|c_\alpha|A^{(\alpha-1)/2}\,t\right). \tag{3.30}$$

The period of the Ateb function (see Appendix A) corresponds to the period of vibration (3.16) presented in the previous Section.

To provide an insight into the type of the response described by the Ateb function derived, a Fourier series representation is used. The Fourier series is expressed as

$$\mathrm{ca}\,(\alpha, 1, t) = \sum_{N=1}^{\infty} C_{2N-1}\,(\alpha) \cos\left[(2N-1)\,\frac{2\pi}{T}t\right], \tag{3.31}$$

where C_{2N-1} are the Fourier coefficients dependent on α and T is the period. Calculation of the coefficients of the series is given in Appendix B.

3.2.1 Linear Case

For $\alpha = 1$ the Eq. (3.2) reduces the linear one

$$\ddot{x} + c_1^2 x = 0\,. \tag{3.32}$$

By virtue of the initial condition $x(0) = A$ formula (3.26) reduces to

$$|x|\,{}_2F_1\left[\begin{matrix}\frac{1}{2},\frac{1}{2}\\ \frac{3}{2}\end{matrix}\,\middle|\left(\frac{x}{A}\right)^2\right] = A \arcsin\left(\frac{x}{A}\right) = A\left(\frac{\pi}{2} + |c_1|t\right),$$

therefore

$$x(t) = \pm A\,\sin\left(\frac{\pi}{2} + |c_1|t\right) = A\,\cos\left(|c_1|t\right). \tag{3.33}$$

The solution is in the form of a trigonometric function.

3.2.2 Odd Quadratic Nonlinearity

Equation (3.2) transforms into

$$\ddot{x} + c_2^2 x\,|x| = 0\,. \tag{3.34}$$

According to (3.29), the exact solution is

$$x = A\,\mathrm{ca}\left(2, 1, \frac{\sqrt{3}\,|c_2|}{\sqrt{2}}\,A^{1/2}\,t\right), \tag{3.35}$$

where the period of the ca function is

$$\Pi_2 := \mathrm{B}\left(\frac{1}{3}, \frac{1}{2}\right) = \frac{\Gamma(\frac{1}{3})\Gamma(\frac{1}{2})}{\Gamma(\frac{5}{6})}.$$

Using the frequency of the ca function (3.21)

$$\Omega = \sqrt{\frac{3}{2}}|c_2|\, A^{1/2},$$

the period of vibration is

$$T = \frac{2\Pi_2}{\Omega} = \frac{\Gamma(\frac{1}{3})}{\Gamma(\frac{5}{6})}\frac{\sqrt{2\pi}}{\sqrt{3}\,|c_2|A^{1/2}}. \tag{3.36}$$

Comparing (3.36) with the period (3.19) for $\alpha = 2$ it is seen that the results are equal.

The solution given by Eq. (3.35) corresponding to $c_2 = 1$ and $A = 1$ can be expressed using the Fourier series expansion of the ca Ateb function (Appendix B) as

$$x\,(t) = \mathrm{ca}\left(2,\,1,\,\sqrt{\frac{3}{2}}t\right) \tag{3.37}$$

$$\approx 0.97480\cos\,(0.91468t) + 0.02572\cos\,(2.74404t) - 0.00064\cos\,(4.57341t) + 0.00014\cos\,(6.40277t)\,, \tag{3.38}$$

This implies that the content is such that the first harmonic dominates with 97% in the response, the third one takes around 2.5% and the rest of them are much smaller. The fifth harmonic has the negative coefficient.

3.2.3 *Cubic Nonlinearity*

Using http: 2002a and http: 2002c, let us recall that

$${}_2F_1\left[{\frac{1}{4},\frac{1}{2} \atop \frac{5}{4}}\Big|z\right] = \frac{1}{\sqrt[4]{z}}\,F(\arcsin\sqrt[4]{z}\,|-1) = \frac{1}{\sqrt[4]{z}}\,\mathrm{sn}^{-1}(z|-1), \quad |z| < 1\,, \tag{3.39}$$

where

$$F(z|m) = \int_0^{z=\mathrm{am}(z|m)} \frac{\mathrm{d}t}{\sqrt{1 - m\sin^2 t}}, \qquad \mathrm{sn}(z|m) = \sin\mathrm{am}(z|m)\,,$$

and F, am, sn denote the incomplete elliptic integral of the first kind, the Jacobi amplitude, the Jacobi elliptic sn function and sn^{-1} stands for the inverse Jacobi elliptic sn functions, respectively.

Thus, for $\alpha = 3$ we have (see http: 2002d)

$$|x|\,{}_2F_1\left[\begin{matrix}\frac{1}{4},\frac{1}{2}\\ \frac{5}{4}\end{matrix}\middle|\left(\frac{x}{A}\right)^4\right] = A\cdot \mathrm{sn}^{-1}\left(\frac{|x|}{A}\middle| -1\right) = C + \frac{|c_3|A^2}{\sqrt{2}}t. \tag{3.40}$$

Since the initial condition $x(0) = A$, we conclude

$$C = A\cdot \mathrm{sn}^{-1}(1|-1) = A\cdot K(-1)\,,$$

where $K(m) = F\left(\frac{\pi}{2}|m\right)$ denotes the celebrated complete elliptic integral of the first kind. The Jacobi elliptic $sn(z|m)$ has period $\omega = 4K(m)$, for now

$$4K(-1) = B\left(\frac{1}{4},\frac{1}{2}\right) = \frac{\Gamma^2\left(\frac{1}{4}\right)}{\sqrt{2\pi}} \approx 5.244116\,,$$

is the period of the function $\mathrm{sn}(z|-1)$. Employing the quarter-period transformation formula for the Jacobi amplitude (see Byrd and Friedman 1954)

$$\mathrm{am}(K(m)-z|m) = \frac{\pi}{2} - \mathrm{am}\left(\sqrt{1-m}\,z\middle|\frac{m}{m-1}\right), \qquad m\le 1\,,$$

for $m = -1$, one deduces by (3.40) that

$$\begin{aligned} x(t) &= A\cdot sn\left(K(-1)+\frac{|c_3|A}{\sqrt{2}}t\middle| -1\right) = A\cdot \sin\mathrm{am}\left(K(-1)+\frac{|c_3|A}{\sqrt{2}}t\middle| -1\right)\\ &= A\cdot\sin\left\{\frac{\pi}{2} - \mathrm{am}\left(K(-1)+\frac{|c_3|A}{\sqrt{2}}t\middle| -1\right)\right\} = A\cdot\cos\mathrm{am}\left(|c_3|A\,t\middle|\frac{1}{2}\right). \end{aligned}$$

Thus

$$x(t) = A\cdot cn\left(|c_3|A\,t\middle|\frac{1}{2}\right). \tag{3.41}$$

Here $cn(z|m) = \cos\mathrm{am}(z|m)$ denotes the Jacobi elliptic cn function (Byrd and Friedman 1954). Being $cn(0|\frac{1}{2}) = 1$, formula (3.41) has the interpolative property $x(0) = A$. It is worth to say that Lyapunov in his classical paper in 1893 introduced the Jacobi elliptic functions (cn and sn) which are the special case of Ateb cosine and Ateb sinus functions for $\alpha = 3$. The same functions are used for solving the third order nonlinear differential equation of Duffing type by Yuste and Bejarano (1990), Chen and Cheung (1996), Cveticanin (2009b), (2011), Kovacic et al. (2016) The in-built subroutine Jacobi CN $[z, k^2]$ in *Mathematica* enables the exact curve plotting of the Jacobi elliptic functions (see du Val 1973).

By using Eq. (3.2), the solution given by Eqs. (3.41) and (3.29) corresponding to $c_3 = 1$ and $A = 1$ can be expressed using the Fourier series expansion of the ca Ateb function (Appendix B) as

$$x(t) = \mathrm{ca}\left(3, 1, \sqrt{2}t\right)$$
$$\approx 0.95501\cos(0.84721t) + 0.04305\cos(2.54164t) +$$
$$0.00186\cos(4.23607t) + 0.00008\cos(5.93049t)\,. \quad (3.42)$$

The content is such that the first harmonic dominates with 95% in the response, the third one takes around 4% and the rest of them are again very small. All the Fourier coefficients are positive.

In spite of the fact that the exact solution of (3.2) for (3.3) exists, a significant number of approximative solution procedures are developed. Let us mention some of them.

3.3 Adopted Lindstedt–Poincaré Method

Equation (3.2) is not appropriate for the application of the standard perturbation techniques since it contains neither a linear term nor a small parameter. Hence, this equation should be transformed appropriately so that the Lindstedt-Poincaré (LP) method (see Cheng et al. 1991, Amore and Aranda 2005) can be applied. At this point we rewrite (3.2) in the form

$$\ddot{x} + c_1^2 x = c_1^2 x - c_\alpha^2 x\,|x|^{\alpha-1}\,, \quad (3.43)$$

as it was done by Belendez et al. (2007) for the pure nonlinear differential equation with the fractional power 1/3. For (3.43), we establish the following mapping

$$\ddot{x} + c_1^2 x = p(c_1^2 x - c_\alpha^2 x\,|x|^{\alpha-1}), \quad (3.44)$$

where $p \in [0, 1]$ is the dummy parameter. For $p = 0$, the Eq. (3.44) transforms into a linear equation that corresponds to the original one (3.2) with $\alpha = 1$

$$\ddot{x} + c_1^2 x = 0. \quad (3.45)$$

For $p = 1$, Eq. (3.44) becomes the original one (3.2). The solution in the first approximation is assumed as

$$x = A\cos(\omega t), \quad (3.46)$$

where ω is the unknown frequency that need to be determined. The parameter p is used to expand the solution x and the square of the unknown angular frequency ω as follows

$$x = x_0 + px_1 + \cdots \quad (3.47)$$

and

$$\omega^2 = c_1^2 + p\omega_1^2 + \cdots\,, \quad (3.48)$$

i.e.

$$c_1^2 = \omega^2 - p\omega_1^2 - \cdots, \tag{3.49}$$

where ω_1 is an unknown constant. Substituting (3.47) and (3.49) into (3.44) and separating the terms with the same order of the parameter p, the system of linear differential equations is obtained

$$p^0: \quad \ddot{x}_0 + \omega^2 x_0 = 0, \tag{3.50}$$
$$p^1: \quad \ddot{x}_1 + \omega^2 x_1 = \omega_1^2 x_0 + c_1^2 x_0 - c_\alpha^2 x_0 \left| x_0 \right|^{\alpha-1}, \tag{3.51}$$
$$\ldots$$

with the initial conditions

$$p^0: \quad x_0(0) = A, \qquad \dot{x}_0(0) = 0, \tag{3.52}$$
$$p^1: \quad x_1(0) = 0, \qquad \dot{x}_1(0) = 0, \tag{3.53}$$
$$\ldots$$

The exact solution of (3.50) with the conditions (3.52) is

$$x_0 = A\cos(\omega t). \tag{3.54}$$

Substituting (3.54) into (3.51), we obtain

$$\ddot{x}_1 + \omega^2 x_1 = (\omega_1^2 + c_1^2) A\cos(\omega t) - c_\alpha^2 A \left| A \right|^{\alpha-1} \cos(\omega t) \left| \cos(\omega t) \right|^{\alpha-1}. \tag{3.55}$$

Using the Fourier series expansion for a rational order trigonometric function (Mickens 2004), as well as the fact that $\cos\theta$ is an even function, leads to

$$\left| \cos\theta \right|^{\alpha-1} = \frac{a_0}{2} + \sum_{n=1}^{\infty} a_n \cos(n\theta), \tag{3.56}$$

where

$$a_n = \frac{4}{\pi} \int_0^{\pi/2} \left| \cos\theta \right|^{\alpha-1} \cos(n\theta) d\theta = \frac{4}{(2^\alpha)\alpha B(\frac{\alpha+n+1}{2}, \frac{\alpha-n+1}{2})}, \qquad n = 0, 1, 2, ..., \tag{3.57}$$

while $\theta = \omega t$ and $B(\frac{\alpha+n+1}{2}, \frac{\alpha-n+1}{2})$ is the Euler Beta function (Gradstein and Rjizhik 1971), which can be defined in terms of the Gamma function Γ

$$B(\frac{\alpha+n+1}{2}, \frac{\alpha-n+1}{2}) = \frac{\Gamma(\frac{\alpha+n+1}{2})\Gamma(\frac{\alpha-n+1}{2})}{\Gamma(\alpha+1)}. \tag{3.58}$$

Introducing (3.58) into (3.57), one obtains that the coefficients of the Fourier series are the functions of the fraction α

$$a_n = \frac{4\Gamma(\alpha+1)}{(2^\alpha)\alpha\Gamma(\frac{\alpha+n+1}{2})\Gamma(\frac{\alpha-n+1}{2})}, \qquad n = 0, 1, 2, ... \tag{3.59}$$

Using the fact that $\Gamma(\alpha+1) = \alpha\Gamma(\alpha)$ (see Abramowitz and Stegun 1979), this expression transforms to

$$a_n = \frac{4\Gamma(\alpha)}{(2^\alpha)\Gamma(\frac{\alpha+n+1}{2})\Gamma(\frac{\alpha-n+1}{2})}, \qquad n = 0, 1, 2, ... \tag{3.60}$$

Substituting (3.60) into (3.55) yields

$$\ddot{x}_1 + \omega^2 x_1 = (c_1^2 + \omega_1^2) A\cos(\omega t) - c_\alpha^2 A\,|A|^{\alpha-1}\cos(\omega t)\left[\frac{a_0}{2} + \sum_{n=1}^{\infty} a_n \cos(n\omega t)\right]. \tag{3.61}$$

Eliminating the secular terms on the right-hand side of (3.61), i.e., the terms with $\cos(\omega t)$, the correction of the frequency in the first approximation is obtained

$$\omega_1^2 = c_\alpha^2\,|A|^{\alpha-1}\,\frac{a_0 + a_2}{2} - c_1^2, \tag{3.62a}$$

where

$$a_0 = \frac{4\Gamma(\alpha)}{(2^\alpha)\frac{(\alpha-1)^2}{4}[\Gamma(\frac{\alpha-1}{2})]^2}, \qquad a_2 = \frac{4\Gamma(\alpha)}{(2^\alpha)\frac{\alpha^2-1}{4}[\Gamma(\frac{\alpha-1}{2})]^2} = \frac{\alpha-1}{\alpha+1}a_0. \tag{3.63}$$

The comparison of (3.48) and (3.62a) for $p = 1$ gives

$$\omega^2 = c_\alpha^2\,|A|^{\alpha-1}\,\frac{a_0 + a_2}{2}, \tag{3.64}$$

i.e., the corrected version of the vibration frequency in the first approximation is

$$\omega = c_\alpha\,|A|^{\frac{\alpha-1}{2}}\sqrt{\frac{\alpha a_0}{\alpha+1}}. \tag{3.65}$$

Thus, the solution in the first approximation for $\alpha > 1$ takes the form

$$x = A\cos(q c_\alpha t\,|A|^{\frac{\alpha-1}{2}}), \tag{3.66}$$

where

$$q = \sqrt{\frac{(2^{4-\alpha})\alpha\Gamma(\alpha-1)}{(\alpha-1)^2[\Gamma(\frac{\alpha-1}{2})]^2}}]. \tag{3.67}$$

For the sake of simplicity and computational reasons, let us express the Euler Gamma function as the product series (Gradstein and Rjizhik 1971) and re-express the approximate values of the frequency as

$$\omega_A = q_A c_\alpha \left|A\right|^{\frac{\alpha-1}{2}}, \tag{3.68}$$

where

$$q_A = 6.9643\frac{1}{\alpha-1}\sqrt{\frac{\alpha}{2^\alpha(\alpha+1)}}10^{-3}\sqrt{\frac{(\prod\limits_{k=0}^{10}(k+\frac{\alpha-1}{2}))^2}{\prod\limits_{k=0}^{10}(k+\alpha)}}, \tag{3.69}$$

represents the first eleven terms of the series expansion of Euler Gamma function which is enough for technical reasons.

3.4 Modified Lindstedt-Poincaré Method

The modified Lindstedt–Poincaré method, which gives uniformly valid asymptotic expansions for the periodic solutions of weakly nonlinear oscillators (Nayfeh and Mook 1979) is adopted by Cheng et al. (1991), and He (2002), so that the strongly nonlinear differential equations can be studied. The method is named the modified Lindstedt–Poincaré (MLP) one. It is also adopted to the nonlinear systems with non-polynomial elastic restoring forces (Ozis and Yildirm 2007).

To apply the MLP method, we express (3.2) in the form

$$\ddot{x} + \omega_0^2 x + px\left|x\right|^{\alpha-1} = 0, \tag{3.70}$$

where $c_\alpha^2 \equiv p$ and the introduced parameter ω_0^2 is equal to zero. This parameter is assumed to be the first term of the series expansion of the frequency ω

$$\omega^2 = \omega_0^2 + p\omega_1^2 + \cdots \tag{3.71}$$

Assuming the solution in the form (3.47), Eq. (3.70) transforms to

$$\begin{aligned}&(\ddot{x}_0 + p\ddot{x}_1 + \cdots) + (\omega^2 - p\omega_1^2 - \cdots)(x_0 + px_1 + \cdots)\\&= -p(x_0 + px_1 + \cdots)\left|x_0 + px_1 + \cdots\right|^{\alpha-1}.\end{aligned} \tag{3.72}$$

Separating the terms with the same order of the parameter p, the following system of linear differential equations is obtained

$$p^0: \quad \ddot{x}_0 + \omega^2 x_0 = 0, \tag{3.73}$$

$$p^1: \quad \ddot{x}_1 + \omega^2 x_1 = \omega_1^2 x_0 - x_0 |x_0|^{\alpha-1}, \tag{3.74}$$

$$\ldots$$

with the initial conditions (3.52), (3.53). Substituting (3.54) which is the exact solution of (3.73) into the right-hand side of (3.74) we obtain

$$\ddot{x}_1 + \omega^2 x_1 = \omega_1^2 A \cos(\omega t) - A |A|^{\alpha-1} \cos(\omega t) |\cos(\omega t)|^{\alpha-1}. \tag{3.75}$$

Using the Fourier series expansion (3.56) and substituting it into (3.75), we have

$$\ddot{x}_1 + \omega^2 x_1 = \omega_1^2 A \cos(\omega t) - A |A|^{\alpha-1} \cos(\omega t)[\frac{a_0}{2} + \sum_{n=1}^{\infty} a_n \cos(n\omega t)], \tag{3.76}$$

where the Fourier constants are given by (3.60). Eliminating the secular terms on the right-hand side of (3.76), the correction to the frequency is obtained

$$\omega_1^2 = |A|^{\alpha-1} \frac{a_0 + a_2}{2}. \tag{3.77}$$

For $p = c_\alpha^2$, $\omega_0^2 = 0$ and ω_1^2 given with (3.77), the frequency (3.71) in the first approximation corresponds to (3.64) and the solution in the first approximation to (3.66). In the first approximation the LP and the MLP method give the same results.

3.4.1 Comparison of the LP and MLP Methods

To make a comparison between the approximate frequencies (3.64) and (3.68) with numerically obtained one, the relative errors are defined, as

$$\omega_{\%error} = 100 \left| \frac{\omega - \omega_e}{\omega_e} \right| = 100 \left| \frac{q - q_e}{q_e} \right|, \tag{3.78}$$

$$\omega_{\%Aerror} = 100 \left| \frac{\omega_A - \omega_e}{\omega_e} \right| = 100 \left| \frac{q_A - q_e}{q_e} \right|, \tag{3.79}$$

where ω_e is the numerically calculated frequency by solving of the Eq. (3.2) using the Runge-Kutta solution procedure, and $q_e = (\omega_e |A|^{\frac{1-\alpha}{2}})/c_\alpha$. The absolute errors are the differences between the approximate frequencies ω and ω_A, and the numerically obtained frequency ω_e

$$\omega_{error} = |\omega - \omega_e| = c_\alpha |A|^{(\alpha-1)/2} |q - q_e|, \tag{3.80}$$

Table 3.2 The frequencies ω, ω_A and ω_e as functions of various values of the parameter α

α	ω	ω_A	ω_e
1	c_α	c_α	c_α
4/3	$0.97013c_\alpha \|A\|^{\frac{1}{6}}$	$0.95452c_\alpha \|A\|^{\frac{1}{6}}$	$0.96916c_\alpha \|A\|^{\frac{1}{6}}$
3/2	$0.95671c_\alpha \|A\|^{1/4}$	$0.93298c_\alpha \|A\|^{1/4}$	$0.95469c_\alpha \|A\|^{1/4}$
5/3	$0.94418c_\alpha \|A\|^{\frac{1}{3}}$	$0.91212c_\alpha \|A\|^{\frac{1}{3}}$	$0.94081c_\alpha \|A\|^{\frac{1}{3}}$
2	$0.92130c_\alpha \|A\|^{1/2}$	$0.87214c_\alpha \|A\|^{1/2}$	$0.91468c_\alpha \|A\|^{1/2}$

Table 3.3 The relative errors (3.78), (3.79) of the frequencies ω and ω_A with respect to the exact frequency ω_e and the absolute errors (3.80), (3.81)

α	$\omega_{\%error}$ (%)	$\omega_{\%Aerror}$ (%)	ω_{error}	ω_{Aerror}
1	0	0	0	0
4/3	0.10000	1.5106	$0.00097c_\alpha \|A\|^{\frac{1}{6}}$	$0.01464c_\alpha \|A\|^{\frac{1}{6}}$
3/2	0.21159	2.2740	$0.00202c_\alpha \|A\|^{1/4}$	$0.02171c_\alpha \|A\|^{1/4}$
5/3	0.35820	3.0495	$0.00337c_\alpha \|A\|^{\frac{1}{3}}$	$0.02869c_\alpha \|A\|^{\frac{1}{3}}$
2	0.72375	4.6508	$0.00662c_\alpha \|A\|^{1/2}$	$0.04254c_\alpha \|A\|^{1/2}$

$$\omega_{Aerror} = |\omega_A - \omega_e| = c_\alpha |A|^{(\alpha-1)/2} |q_A - q_e| . \tag{3.81}$$

In Table 3.2, the frequencies obtained by using the LP and MLP methods ω (3.64), ω_A (3.68), and the numerically calculated frequency ω_e.are presented.

In Table 3.3, the relative errors (3.78), (3.79), and absolute errors (3.80), (3.81) are given for various values of the fraction order α.

The relative error (3.78) i.e. (3.79) depends only on α and it is independent on the coefficient of nonlinearity c_α and the initial amplitude A. In contrary, the absolute error depends not only on α, but also on c_α and A.

It should be noted that although the analyses given above are carried out to the first approximation only, the analytical results are in good agreement with the corresponding exact values. However, all the procedures shown have the potential to be applied for finding higher approximations if necessary.

3.4.2 Conclusion

The adopted LP method and the adopted MLP method give the same result for motion in the first approximation. The approximate solution is the function of a cosine trigonometric function. The frequency of vibration depends on the initial amplitude, the coefficient of nonlinearity and the value of the fractional power. For technical reasons, it is recommendable to replace the Euler Gamma functions with the series of products in the frequency solution.

Analyzing the absolute error and the influence of the initial amplitude, it is concluded that the approximate solutions are valid for $1 < \alpha < 3$. The results obtained by LP method and the MLP method for the cases when the non-integer order is close to unity or to the value three agree very good with the exact analytical results when $|A| \approx 1$ and $c_\alpha^2 \leq 1$.

Otherwise, from the viewpoint of accuracy, the approximations are appropriate for the interval $1 < \alpha < 2$, when the influence of the initial amplitude is minimal. The standard LP and the MLP technique use the linear differential equation as the generating one. It is the reason that the approximation is satisfactory only for $1 < \alpha < 2$. This limitation is eliminated with the new technique introduced in the following sections.

3.5 Exact Amplitude, Period and Velocity Method

Using the results presented in the previous Sections it can be concluded that the approximately obtained solutions are valid only for:

- small initial amplitudes of vibration and
- short time of motion.

As the period and frequency of vibration depend on the initial amplitude A [see (3.18) and (3.21)], the higher its value, the larger the difference between the exact and approximate frequency of vibration. Besides, as the difference for one vibration period between the approximate T_a and the exact T_e one is $\Delta T \neq 0$, for n periods it is $n(\Delta T)$. The higher the value of ΔT and longer the time of oscillation, the difference $n(\Delta T)$ is higher. The accuracy of the approximate solution decreases in time and the solution tends to be useless for quantitative analysis.

To eliminate these lacks of the solution procedures, an improved solution methods based on the exact frequency and amplitude of vibration of the oscillator, and also on the exact values of the maximal velocity of vibration of oscillator are developed.

Namely, for the conservative oscillator (3.2) it is known that the motion is periodical with the period T (3.18) and amplitude of vibration A (3.27) which corresponds to the initial amplitude (3.3), i.e.,

$$x(0) = x(T) = \ldots = x(nT) = A. \tag{3.82}$$

Using the expression (3.28), the maximal velocity of vibration has to be determined as

$$v_{\max} = \frac{\sqrt{2c_\alpha^2}}{\sqrt{\alpha+1}} A^{(\alpha+1)/2}, \tag{3.83}$$

which exists for

$$x_V = 0. \tag{3.84}$$

Finally, the suggested approximate solution of (3.2) has to satisfy the following conditions:

- initial conditions (3.3),
- maximal amplitude of vibration $x_{\max} = A$,
- frequency of vibration (3.106)
- maximal velocity (3.83) of vibration $\dot{x}_{\max} = v_{\max}$.

These four criteria will be useful for the choice of the most appropriate approximate solution of (3.2). There are many functions which approximately satisfy the given requirements. In this book two of them will be considered: the Jacobi elliptic function (Appendix D) and the trigonometric function.

3.6 Solution in the Form of Jacobi Elliptic Function

Let us assume the approximate solution of (3.2) in general (for any value of α) as the Jacobi elliptic function (Byrd and Friedman 1954)

$$x = Acn(\Omega_1 t, k^2), \tag{3.85}$$

where Ω_1 and k^2 are the unknown frequency and modulus of the Jacobi elliptic function. The suggested approximate solution satisfies the given initial conditions and the requirement of the exact amplitude, i.e. the first two required criteria from the previous section. Our task is to determine the unknown parameters of the function: Ω_1 and k^2.

As it is shown in Appendix D, the relation between the frequency of the Jacobi elliptic function Ω_1 and the period of vibration T_{ex} is

$$\Omega_1 = \frac{4K(k^2)}{T_{ex}}, \tag{3.86}$$

where $4K(k^2)$ is the period of the Jacobi elliptic function and $K(k^2)$ is the complete elliptic integral of the first kind. Substituting the exact period of vibration (3.18) into the relation (3.86), we obtain the frequency of the Jacobi elliptic function

$$\Omega_1 = \Omega_1^* A^{(\alpha-1)/2} \sqrt{c_\alpha^2}, \tag{3.87}$$

where

$$\Omega_1^* = \frac{2K(k^2)}{\pi}\Omega_\alpha, \qquad \Omega_\alpha = \frac{\sqrt{2\pi(\alpha+1)}}{2} \frac{\Gamma(\frac{3+\alpha}{2(\alpha+1)})}{\Gamma(\frac{1}{\alpha+1})}. \tag{3.88}$$

It is evident that the frequency of the Jacobi elliptic function depends on the modulus k^2 which has to be determined.

The modulus k^2 is calculated for the maximal vibration velocity (3.83) when (3.84). Namely, using the condition (3.84) and the relation (3.85), it is obtained that the maximal vibration velocity is for

$$t_V = \frac{(2n-1)K(k^2)}{\Omega_1}, \tag{3.89}$$

where $n = 1, 2, 3, \ldots$The first time derivative of the approximate solution (3.85) is

$$\dot{x} = -A\Omega_1 sn(\Omega_1 t, k^2) dn(\Omega_1 t, k^2). \tag{3.90}$$

Substituting (3.89) into (3.90) it is

$$\dot{x}(t_V) = -A\Omega_1 sn((2n-1)K(k^2), k^2) dn((2n-1)K(k^2), k^2), \tag{3.91}$$

where sn and dn are Jacobi elliptic functions (see Appendix D) which satisfy the relations

$$sn((2n-1)K(k^2), k^2) = \pm 1, \qquad dn((2n-1)K(k^2), k^2) = k' = \sqrt{1-k^2}, \tag{3.92}$$

i.e.,

$$sn((2n-1)K(k^2), k^2) dn((2n-1)K(k^2), k^2) = \pm k', \tag{3.93}$$

and k' is the complementary modulus of the Jacobi elliptic function (Byrd and Friedman 1954). Due to (3.91), (3.93) and (3.87) with (3.88), the maximal value of the first time derivative is

$$\dot{x}_{\max} \equiv \dot{x}(t_V) = A\Omega_1 k' = \frac{2K(k^2)k'}{\pi} \Omega_\alpha A^{(\alpha+1)/2} \sqrt{c_\alpha^2}. \tag{3.94}$$

Equating the relation (3.94) with the maximal velocity of vibration (3.83), it follows

$$k'K(k^2) = \frac{\sqrt{\pi}}{(\alpha+1)} \frac{\Gamma(\frac{1}{\alpha+1})}{\Gamma(\frac{3+\alpha}{2(\alpha+1)})}. \tag{3.95}$$

The solution of (3.95) gives the value of the modulus of the Jacobi elliptic function k^2. For that certain value of k^2, the frequency of the Jacobi elliptic function (3.87) is calculated. Substituting the corresponding value of k^2 and (3.87) into (3.85) and (3.90), the approximate solution of (3.2) and its first time derivative are obtained.

Analyzing the relation (3.95) it is obvious that the modulus of the Jacobi elliptic function depends only on the order of the nonlinearity and is $k^2(\alpha)$. Solving the nonlinear algebraic equation (3.95) for various values of α, the modulus k^2 is calculated and given in Table 3.4. In the Table 3.4 also the constants (3.88) of the frequency of vibration Ω_α and of the Jacobi elliptic function Ω_1^* for certain value of α are given. Knowing the order of nonlinearity α and taking the correspondent modulus of the

Table 3.4 Values of k^2, Ω_α and Ω_1^* for various values of α

α	Ω_α	k^2	Ω_1^*
0.0	1.110720	−1.807855	0.843971
0.2	1.086125	−1.062291	0.898978
0.4	1.062822	−0.622992	0.938194
0.6	1.040748	−0.338896	0.966232
0.8	1.019832	−0.065976	0.986138
1.0	1.000000	0.000000	1.000000
1.2	0.981179	0.107572	1.009292
1.4	0.963300	0.191252	1.015083
1.6	0.946298	0.257978	1.018169
1.8	0.930110	0.312288	1.019137
2.0	0.914681	0.357261	1.018444
4.0	0.792402	0.575240	0.970417
6.0	0.708220	0.652256	0.906434
8.0	0.645978	0.691091	0.848161
10	0.597617	0.714386	0.797865
20	0.455377	0.760706	0.630870
30	0.382158	0.775993	0.536665
40	0.335713	0.575240	0.474776
50	0.302909	0.788140	0.430235
60	0.278155	0.791164	0.396230
70	0.258622	0.793320	0.369179
80	0.242698	0.794935	0.346998
90	0.229395	0.796191	0.328384
100	0.218063	0.797194	0.312474

Jacobi elliptic function k^2 and the constants Ω_α and Ω_1^*, the frequency of the Jacobi elliptic function (3.87) is calculated.

Analyzing the data in the Table 3.4 it is evident:

(1) For the linear case, when $\alpha = 1$, the both constants: Ω_α and Ω_1^* are equal to 1.
(2) The value of Ω_α decreases from 1 by increasing of the order of nonlinearity α, and increases with decreasing of the order of nonlinearity α.

Considering the relation (3.95), the following is concluded:

(1) For the linear oscillator, when $\alpha = 1$, the right hand side of the Eq. (3.95) is $\pi/2$ and according to (3.95) the corresponding value of the modulus of the Jacobi elliptic function is $k^2 = 0$. Then, the *cn* Jacobi elliptic function transforms into the cosine trigonometric function.
(2) For $0 < \alpha < 1$ the modulus of the Jacobi elliptic function is according to (3.95) negative. For $k^2 < 0$, the *cn* Jacobi elliptic function transforms into $cd \equiv cn/dn$ Jacobi elliptic function

$$cn(\Omega_1 t, -k^2) = cd\left(\Omega_1 t\sqrt{1+|k^2|}, \frac{|k^2|}{1+|k^2|}\right). \tag{3.96}$$

(3) If α tends to zero the right hand side of the equation tends to 2, and $k^2 = -1.807854731$. Using the transformation (3.96) the corresponding modulus of the cd Jacobi elliptic function tends to $k^2 = 0.643\,86$.
(4) Introducing the new parameter $s = 1/(\alpha+1)$ which tends to zero when α tends to infinity, the right hand side of the relation (3.95) is

$$\lim_{\alpha\to\infty}\frac{\sqrt{\pi}}{(\alpha+1)}\frac{\Gamma(\frac{1}{\alpha+1})}{\Gamma(\frac{3+\alpha}{2(\alpha+1)})} = \lim_{s\to 0}\sqrt{\pi}\frac{s\Gamma(s)}{\Gamma(s+\frac{1}{2})} = \lim_{s\to 0}\sqrt{\pi}\frac{\Gamma(s+1)}{\Gamma(s+\frac{1}{2})} = \sqrt{\pi}\frac{\Gamma(1)}{\Gamma(\frac{1}{2})} = 1. \tag{3.97}$$

Then, the value of the modulus of the Jacobi elliptic function tends to a constant value: $k^2 = 0.806192642$.
(5) For $\alpha = 3$, the modulus of the elliptic function is $k^2 = 1/2$ and the frequency of the Jacobi function is $\Omega_1 = A\sqrt{c_\alpha^2}$. Thereby, the pure cubic differential equation

$$\ddot{x} + c_\alpha^2 x^3 = 0, \tag{3.98}$$

for initial conditions (3.3), has the exact solution

$$x = Acn(At\sqrt{c_\alpha^2}, 1/2), \tag{3.99}$$

with constant modulus of the Jacobi elliptic function.

3.6.1 Example

Let us consider the example of an oscillator with the non-integer order of nonlinearity ($\alpha = 5/3$) and with parameter value $c_{5/3}^2 = 1$:

$$\ddot{x} + x\,|x|^{2/3} = 0. \tag{3.100}$$

The initial conditions (3.3) are

$$x(0) = A = 2, \qquad \dot{x}(0) = 0. \tag{3.101}$$

From the Table 3.4 and (3.95) the modulus of Jacobi elliptic function is $k^2 = 0.277285$ and the corresponding complete elliptic integral of the first kind $K(k^2) = 1.700838$. Introducing the initial amplitude (3.101), the solution of (3.100) is due to (3.85) and (3.86)

$$x = 2cn(1.2835t, 0.277285), \tag{3.102}$$

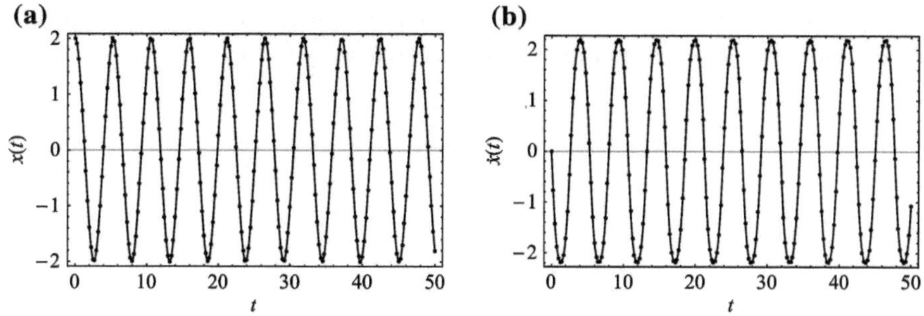

Fig. 3.3 **a** $x - t$ diagrams obtained analytically (*full line*) and numerically (*dotted-line*), **b** $\dot{x} - t$ diagrams obtained analytically (*full line*) and numerically (*dotted-line*), for α=5/3

and its first time derivative

$$\dot{x} = -2.567sn(1.2835t, 0.277285)dn(1.2835t, 0.277285). \tag{3.103}$$

Applying the Runge–Kutta solving procedure, the Eq. (3.98) is solved numerically. In Fig. 3.3 the obtained solution and its time derivative are compared with the analytically obtained ones, (3.102) and (3.103).

(a) $x - t$ diagrams obtained analytically (full line) and numerically (dotted-line), (b) $\dot{x} - t$ diagrams obtained analytically (full line) and numerically (dotted-line), for $\alpha = 5/3$.

Analyzing the results it is evident that the expression for x (3.102) has to be assumed as the approximate solution of (3.2), as it is very close the numerical one. The same is valid for the corresponding first time derivatives.

3.7 Solution in the Form of a Trigonometric Function

The method described in the previous section assumes the solution in the form of Jacobi elliptic function. In spite of the fact that the approximation to the exact solution is excellent, the practical application of this special function is not easy: It is the reason that a method which uses the solution in the form of a well known trigonometric function is developed. The solution is based on the exact value of the period of vibration and constant amplitude of vibration. The approximate solution of (3.2) and its first time derivative are assumed in the form of trigonometric functions

$$x = A\cos(\omega t), \qquad \dot{x} = -A\omega\sin(\omega t), \tag{3.104}$$

where A is the initial amplitude and ω is the angular frequency of vibration. The relation (3.104) is an approximation to the exact solution of (3.2) which corresponds to a truncated Fourier expansion where only the first term is retained. Based on the exact period of vibration (3.18) and the assumption of periodical harmonic function (3.104) the explicit expression for the angular frequency follows

$$\omega = \frac{2\pi}{T_{ex}} = \frac{\pi\sqrt{2c_\alpha^2(\alpha+1)}}{2\,|A|^{(1-\alpha)/2}} \frac{\Gamma(\frac{3+\alpha}{2(\alpha+1)})}{\Gamma(\frac{1}{\alpha+1})\Gamma(\frac{1}{2})}, \tag{3.105}$$

i.e., for (3.18)

$$\omega = \sqrt{\frac{c_\alpha^2(\alpha+1)}{2}} \frac{\sqrt{\pi}\Gamma(\frac{3+\alpha}{2(\alpha+1)})}{\Gamma(\frac{1}{\alpha+1})} |A|^{(\alpha-1)/2}. \tag{3.106}$$

The oscillations are periodical and have the constant amplitude A. The angular frequency of vibration depends not only on the fraction value α but also on the initial amplitude A, as it is well known for nonlinear oscillators.

3.7.1 Example

1. Let us apply the previously mentioned approximate solution procedure for the oscillator (3.100) where $\alpha = 5/3$. For the initial conditions (3.100) when $A = 2$, and the parameter of nonlinearity is $c_{5/3}^2 = 1$, the approximate solution is

$$x = 2\cos(1.1853t), \tag{3.107}$$

and its first time derivative

$$\dot{x} = -2.3706\sin(1.1853t). \tag{3.108}$$

In Fig. 3.4, the analytical solution and its time derivative are compared with the numerical solution of (3.100) and its time derivative. It is evident that the analytical

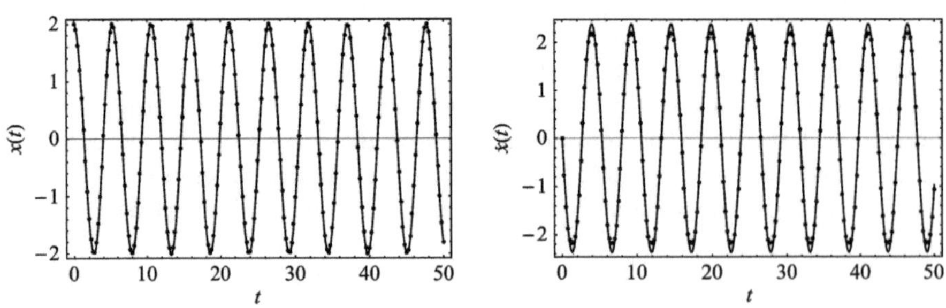

Fig. 3.4 **a** $x - t$ diagrams obtained analytically (*full line*) and numerically (*dotted-line*), **b** $\dot{x} - t$ diagrams obtained analytically (*full line*) and numerically (*dotted-line*), for $\alpha = 5/3$

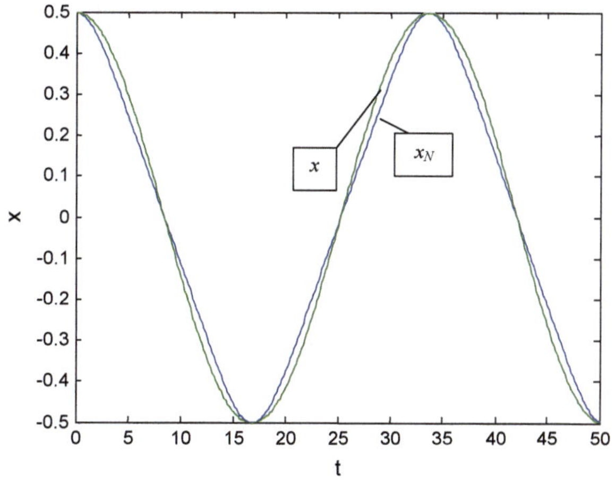

Fig. 3.5 $x - t$ and $x_N - t$ diagrams for $\alpha = 5$

solution is on the top of the numerical solution, but the first time derivatives differ for a bit.

Finally, it can be concluded that the Jacobi elliptic function (3.102) gives the more appropriate approximation than the trigonometric function (3.107).

2. For the pure fifth order differential equation

$$\ddot{x} + c_5^2 x\,|x|^4 = 0, \tag{3.109}$$

the approximate analytical solution is

$$x = A\cos(0.74683 c_5 t\,|A|^2). \tag{3.110}$$

In spite of the fact that the amplitude and the period of vibration of the oscillator are exactly determined by the suggested analytical procedure, the shape of the approximate solution (3.110) differs from the exact numerical one of (3.109) (see Fig. 3.5. for $A = 0.5$ and $c_5 = 1$).

This clearly shows that for the higher-order of nonlinearity the solution in the form of trigonometric function differs from the exact one.

3.7.2 *Conclusion*

Based on the obtained results the following advantages of the suggested method are evident:

(1) The choice of the approximate solution is directed to those periodical functions which satisfy not only the initial conditions, but also the exact values of the

amplitude and period of vibration. Besides, the maximal value of the first time derivative of the function has to be equal or close to the maximal value of the vibration velocity.

(2) The better choice of the function for the asymptotic solution gives the smaller difference between the exact and asymptotic solution inside a period of vibration.
(3) There is no error accumulation in the approximate solution, as the period of vibration and the amplitude are fixed and are exact values. As the period of vibration is exactly known and included into the approximate solution, there is no difference in the period between the exact and approximate solution. This property excludes one of the most important errors in the approximate solution: the error caused by the difference between the exact and approximate periods. Due to this fact for our method there is no superposition of the period errors as is usual in other approximate methods. This fault causes the approximate solution to be quite incorrect and useless for long time consideration.
(4) The previous mentioned property of the solution gives the possibility for the long time analyses of the motion.
(5) Only one period of vibration is enough to be analyzed. Namely, the relation between the approximate and the exact solution in the first period of vibration is quite the same as for the second, third or even n-th period of vibration. It means that it is enough to analyze only the first period of vibration, as the approximate solution for the n-th period of vibration is equal to the first.
(6) Based on the previous property, the prediction of any period of vibration is possible.
(7) The suggested procedure does not require averaging, series expansions or simplification of the differential equation and is much simpler than the previously published one.

Additionally, it can be concluded:

(1) The *cn* Jacobi elliptic function with its frequency and modulus, which are determined to give the exact period of vibration and to satisfy the condition of equality of the maximal value of its first time derivative with the velocity of vibration, is discovered as a very accurate approximate solution of the pure nonlinear differential equation.
(2) The approximation of the oscillations in the form of the trigonometric function is very good for $1 \leq \alpha \leq 3$. Namely, difference between the shapes of the approximate and exact time history diagrams is negligible.

3.8 Pure Nonlinear Oscillator with Linear Damping

The model of a pure nonlinear oscillator with linear viscous damping is

$$\ddot{x} + \kappa\dot{x} + c_{\alpha}^{2} x\, |x|^{\alpha-1} = 0, \tag{3.111}$$

where κ is the damping coefficient and the initial conditions are given with (3.3). In spite of the fact that the relation seems to be simple, to determine its solution is not easy. To explain the physical sense of the problem the energy function E of the oscillator is introduced

$$E = \frac{\dot{x}^2}{2} + \frac{c_\alpha^2 x^{i+1}}{\alpha+1}, \tag{3.112}$$

which gives the rewritten form of the Eq. (3.111)

$$\frac{dE}{dt} = -2\Phi, \tag{3.113}$$

where Φ is the dissipation function

$$\Phi = \kappa \frac{\dot{x}^2}{2}. \tag{3.114}$$

Equation (3.113) is the velocity of the change of the total mechanical energy. Namely, for $\kappa > 0$ the dissipation function is positive and the total mechanical energy of the system decreases. The velocity of energy dissipation is higher for higher values of damping coefficients κ (see (3.114)). Furthermore, the amplitude of vibration decreases, too, and the period of vibration is not a constant value. Due to these facts the approximate solution of (3.111) is assumed as a product of a time variable amplitude function and a *ca* Ateb function (given in Appendix A),

$$x = A\exp(-\delta t) ca(\alpha, 1, \psi(t)), \tag{3.115}$$

where δ is an unknown constant value, A and θ are arbitrary constants, $\omega(t)$ is the time variable frequency and $\psi(t)$ is an unknown time functions which satisfies the relation

$$\psi(t) = \theta + \int \omega(t)dt. \tag{3.116}$$

Using the derivatives of the Ateb functions (see Appendix A and Rosenberg 1963), the first and the second time derivatives of the solution (3.115) follow as

$$\dot{x} = -\delta A\exp(-\delta t) ca(\alpha, 1, \psi(t)) - \frac{2A\omega(t)}{\alpha+1}\exp(-\delta t) sa(1, \alpha, \psi(t)), \tag{3.117}$$

$$\ddot{x} = \exp(-\delta t)(\delta^2 A ca(\alpha, 1, \psi(t)) + \frac{4\omega(t)}{\alpha+1}\delta A sa(1, \alpha, \psi(t))$$
$$-\frac{2A\dot{\omega}(t)}{\alpha+1} sa(1, \alpha, \psi(t)) - \frac{2A\omega^2(t)}{\alpha+1} ca^\alpha(\alpha, 1, \psi(t))). \tag{3.118}$$

Substituting (3.115), (3.117) and (3.118) into (3.111) and using the notation $ca(\alpha, 1, \psi(t)) \equiv ca$, $sa(1, \alpha, \psi(t)) \equiv sa$, $\omega(t) \equiv \omega$ we have

$$0 = \delta^2 Aca + \frac{4\omega}{\alpha+1}\delta Asa - \frac{2A\dot{\omega}}{\alpha+1}sa - \frac{2A\omega^2}{\alpha+1}ca^{\alpha}$$
$$-\delta\kappa Aca - \kappa\frac{2A\omega}{\alpha+1}sa + c_{\alpha}^2 Aca\,|A\exp(-\delta t)ca|^{\alpha-1}. \tag{3.119}$$

Now, according to the harmonic balance procedure the terms with the same order of the Ateb function have to be equated. Separating the terms with functions sa and $ca\,|ca|^{\alpha-1}$, the following two equations are obtained

$$sa: \quad 2\omega\delta - \dot{\omega} - \kappa\omega = 0,$$
$$ca\,|ca|^{\alpha-1}: \quad -\frac{2\omega^2}{\alpha+1} + c_{\alpha}^2\,|A\exp(-\delta t)|^{\alpha-1} = 0. \tag{3.120}$$

Solving the relations (3.120), we have

$$\omega^2 = \frac{\alpha+1}{2}c_{\alpha}^2\,|A\exp(-\delta t)|^{\alpha-1}, \quad \dot{\omega} = -\frac{1}{2}\delta(\alpha-1)\omega,$$
$$\delta = \frac{2\kappa}{3+\alpha} = const. \tag{3.121}$$

Substituting (3.121) into (3.119) it is obvious that (3.115) approximately satisfies (3.111). Namely, for (3.121) the terms with the function ca in (3.119) are not zero, but have the value ω_1^2 which depends on the order of the nonlinearity and on the damping coefficient, i.e.,

$$ca: \quad \omega_1^2 = \frac{2\kappa^2(1+\alpha)}{(3+\alpha)^2} \neq 0. \tag{3.122}$$

It is of interest to analyze the value of the parameter (3.122). We rewrite (3.122) into $\omega_1^2 = (p\kappa)^2$ where

$$p = \frac{\sqrt{2(1+\alpha)}}{3+\alpha} = const. \tag{3.123}$$

For $\alpha \in [0, \infty)$, the parameter p^2 is in the interval $[2/9, 0)$. For $\kappa < 1$ independently on the order of nonlinearity α the coefficient $(p\kappa)^2 \ll \kappa$. Using this result and the relation (3.122) it is obvious that the solution (3.115) with (3.121) is the exact solution of a differential equation

$$\ddot{x} + \kappa\dot{x} + (p\kappa)^2 x + c_{\alpha}^2 x\,|x|^{\alpha-1} = 0, \tag{3.124}$$

which represents the extension of the Eq. (3.111) with an additional linear term with a small coefficient $(p\kappa)^2$. Comparing (3.111) with (3.124) and using the parameter value for $(p\kappa)^2 \ll \kappa$, it can be concluded that the Eq. (3.111) is very close to (3.124). Based on this result we state that the solution

$$x = A(\exp(-\frac{2\kappa t}{3+\alpha}))ca[\alpha, 1, \sqrt{\frac{\alpha+1}{2}c_\alpha^2}\frac{(3+\alpha)A^{(\alpha-1)/2}}{\kappa(\alpha-1)}(1-\exp(-\kappa t\frac{\alpha-1}{3+\alpha})) + \theta], \tag{3.125}$$

is the appropriate approximate solution of (3.111).

3.8.1 Parameter Analysis

Analyzing the solution (3.125) it is obvious that the amplitude of vibration has the tendency of decrease and it is

$$A_d = A(\exp(-\frac{2\kappa t}{3+\alpha})). \tag{3.126}$$

The amplitude decrease depends not only on the coefficient of damping κ but also on the order of nonlinearity α and initial amplitude A. The rewritten form of (3.126)

$$\frac{A_d}{A} = \exp(-\frac{2\kappa t}{3+\alpha}), \tag{3.127}$$

is convenient for investigation of the amplitude ratio variation. For the linear damped oscillator, when $\alpha = 1$, the relation (3.127) transforms into the well known one: $A_d/A = \exp(-\kappa t/2)$. Besides, the higher the order of nonlinearity ($\alpha > 1$), the argument in the exponential function (3.127) is smaller and the amplitude decrease is slower than for the linear case. For $\alpha \to \infty$, the amplitude of vibration tends to the constant initial amplitude A. Opposite, the smaller the order of nonlinearity ($\alpha < 1$), the amplitude of vibration decreases faster than for the linear oscillator. For $\alpha \to 0$, the amplitude decrease is according to $A_d/A = \exp(-2\kappa t/3)$.

Analyzing the function

$$\omega(t) = A^{(\alpha-1)/2}\sqrt{\frac{\alpha+1}{2}c_\alpha^2}\left|\exp(-\frac{\kappa(\alpha-1)}{3+\alpha}t)\right|, \tag{3.128}$$

it is evident that the frequency of vibration is a time variable function and strongly depends on the order of nonlinearity: for $\alpha > 1$, the frequency of vibration decreases in time, while for $\alpha < 1$ the frequency of vibration increases in time. For $\alpha \to 0$, the frequency behaves as $\omega \sim \exp(\kappa t/3)$ and for $\alpha \to \infty$, it is $\omega \sim \exp(-\kappa t)$. For the linear case ($\alpha = 1$), as is well known, the frequency of vibration is a constant value.

The period of vibration is also a time variable function

$$T = \bar{T}\left|\exp(\frac{\kappa(\alpha-1)}{3+\alpha}t)\right|, \tag{3.129}$$

where $\bar{T}$ is the period of vibration for the undamped oscillator

$$\bar{T} = \frac{2\Pi}{A^{(\alpha-1)/2}\sqrt{\frac{\alpha+1}{2}c_\alpha^2}}, \tag{3.130}$$

and 2Π is the period of the Ateb function (see Appendix A). The motion is quasi-periodic with time variable amplitude and period of vibration.

Examples

To prove the previously mentioned result about the approximate solution, let us consider two examples of viscous damped pure nonlinear oscillators: one, where the nonlinearity is of order $\alpha = 1/3$ and the second, where the nonlinearity is of cubic order ($\alpha = 3$).

For the parameter values $c_\alpha^2 = 1$ and $\kappa = 0.1$ the differential equations are

$$\ddot{x} + 0.1\dot{x} + x\,|x|^{-2/3} = 0, \tag{3.131}$$

and

$$\ddot{x} + 0.1\dot{x} + x^3 = 0. \tag{3.132}$$

The approximate solution of (3.131) and (3.132) is, respectively, for $\alpha = 1/3$

$$x = A(\exp(-0.06t))ca[\frac{1}{3}, 1, -40.825A^{-1/3}(1 - \exp(0.02t)) + \theta], \tag{3.133}$$

and for $\alpha = 3$

$$x = A(\exp(-0.03333t))ca[3, 1, 42.426A(1 - \exp(-0.03333t)) + \theta]. \tag{3.134}$$

The numerical solutions of (3.131) and (3.132) are obtained by using the Runge-Kutta procedure. In Fig. 3.6 the analytical and numerical results calculated for the initial conditions $x(0) = 1$ and $\dot{x}(0) = 0$ are compared. In Fig. 3.6a the analytical solution (3.133) and and numerical solution of (3.131), where $\alpha = 1/3$, is shown. In Fig. 3.6b the analytical solution (3.134) is compared with numerical solution of (3.132) for $\alpha = 3$.Comparing the analytical and numerical solutions it is evident that they are in good agreement. From Fig. 3.6 it is evident that the amplitude decreases faster for $\alpha = 1/3$ than for $\alpha = 3$, as was previously discussed. The frequency of vibration is higher for $\alpha = 1/3$ than for $\alpha = 3$. Besides, for $\alpha = 1/3$ the period of vibration decreases in time, while for $\alpha = 3$ it increases.

In Fig. 3.7 the $x - t$ curves for various initial amplitude A (1.2; 1; 0.8) and for: (a) $\alpha = 1/3$ (3.133) and (b) $\alpha = 3$ (3.134) are plotted.

The influence of the initial amplitude is evident: For $\alpha = 1/3$, the higher the initial amplitude the period of vibration is longer. For $\alpha = 3$, it is seen that the period of vibration is shorter for higher initial amplitudes than for the smaller ones.

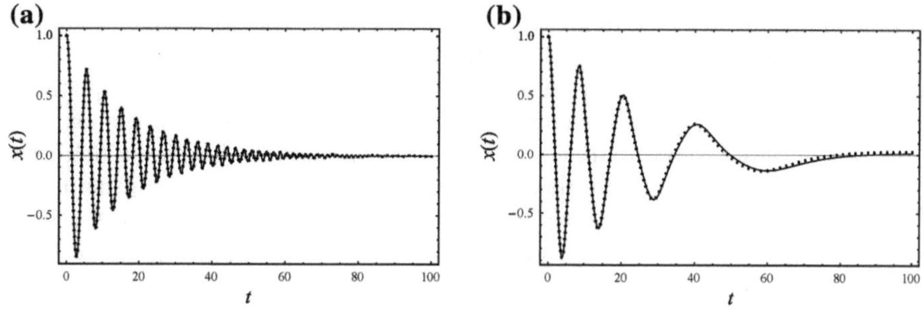

Fig. 3.6 $x - t$ diagrams obtained analytically (*full line*) and numerically (*dotted line*) for: **a** $\alpha = 1/3$ and **b** $\alpha = 3$.

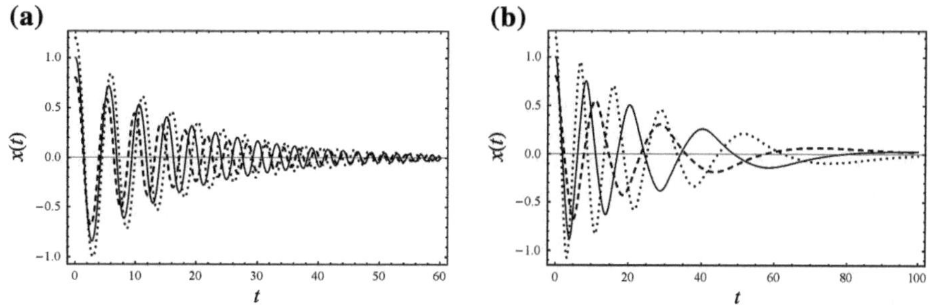

Fig. 3.7 $x - t$ diagrams for $A = 1.2$ (*dotted line*), $A = 1$ (*full line*) and $A = 0.8$ (*dashed line*) for: **a** $\alpha = 1/3$ and **b** $\alpha = 3$

3.8.2 Conclusion

Due to the previous consideration the following conclusions about thc influence of the viscous damping on the motion of the pure nonlinear oscillators are evident:

1. The approximate solution of the differential equation for the pure nonlinear oscillator with viscous damping is assumed in the form of the Ateb function and is proved that this solution is close to the numerical one. Analyzing the approximate solution the comment and suggestion about the interaction of the damping parameter and order of nonlinearity is possible to be given.
2. Due to viscous damping the amplitude of vibration decreases for all types of pure nonlinear oscillators. The amplitude decrease depends on the initial amplitude, damping coefficient and order of nonlinearity of the oscillator. The amplitude decrease is faster for the nonlinear oscillator with order of nonlinearity smaller than 1 in comparison to the linear damped oscillator. Opposite, the amplitude of linear oscillator decreases faster than of the oscillator which order of nonlinearity is higher than 1. It suggests the following: The appropriate choice of the material with elastic properties with nonlinearity order smaller than 1 will give the faster

elimination of the vibrations of the oscillator with viscous damping than the material with linear or nonlinear elastic properties of order higher than 1.

3. As it is known, for the linear oscillator with viscous damping the quasi-period of vibration is constant and depends on the damping coefficient. For the nonlinear oscillator with viscous damping the period of vibration is varying in time. The period variation of vibration depend on the damping coefficient, initial amplitude and order of nonlinearity. If the order of nonlinearity is higher than 1, the period of vibration increases in time and is longer than for the linear oscillator. If the order of nonlinearity is smaller than 1 the period of vibration decreases and the period of vibration is always shorter than for the linear case.

References

Abramowitz, M., & Stegun, I. A. (1979). *Handbook of mathematical functions with formulas, graphs and mathematical tables*. Moscow: Nauka. (in Russian).

Amore, P., & Aranda, A. (2005). Improved Lindstedt–Poincaré method for the solution of nonlinear problems. *Journal of Sound and Vibration*, *283*, 1115–1136.

Belendez, A., Pascual, C., Gallego, S., Ortuño, M., & Neipp, V. (2007). Application of a modified He's homotopy perturbation method to obtain higher-order approximations of an $x^{1/3}$ force nonlinear oscillator. *Physics Letters A*, *371*, 421–426.

Byrd, P. F., & Friedman, M. D. (1954). *Handbook of elliptic integrals for engineers and physicists*. Berlin: Springer Verlag.

Chen, S. H., & Cheung, Y. K. (1996). An elliptic perturbation method for certain strongly non-linear oscillators. *Journal of Sound and Vibration*, *192*, 453–464.

Cheng, Y. K., Chen, S. H., & Lau, S. L. (1991). A modified Lindstedt–Poincaré method for certain strongly non-linear oscillators. *International Journal of Nonlinear Mechanics*, *26*, 367–378.

Cveticanin, L. (2009a). Oscillator with fraction order restoring force. *Journal of Sound and Vibration*, *320*, 1064–1077.

Cveticanin, L. (2009b). The approximate solving methods for the cubic Duffing equation based on the Jacobi elliptic functions. *International Journal of Nonlinear Sciences and Numerical, Simulation*, *10*, 1491–1516.

Cveticanin, L. (2011). Analysis Technique for the various forms of the duffing equation, sect. 4. In I. Kovacic & M. J. Brennan (Eds.), *Duffing equation: Nonlinear oscillators and their behaviour*. London: Wiley.

Cveticanin, L., & Pogany, T. (2012). Oscillator with a sum of non-integer order non-linearities. *Journal of Applied Mathematics*, *649050*, 20 pp.

Cveticanin, L., Kovacic, I., & Rakaric, Z. (2010). Asymptotic methods for vibrations of the pure non-integer order oscillator. *Computers and Mathematics with Applications*, *60*, 2616–2628.

Gottlieb, H. P. W. (2003). Frequencies of oscillators with fractional-power non-linearities. *Journal of Sound and Vibration*, *261*, 557–566.

Gradstein, I. S., & Rjizhik, I. M. (1971). *Table of integrals, series and products*. Moscow: Nauka. (in Russian).

He, J.-H. (2002). Modified Lindstedt–Poincaré methods for some strongly non-linear oscillations Part I: Expansion of a constant. *International Journal of Nonlinear Mechanics*, *37*, 309–314.

Kovacic, I., Cveticanin, L., Zukovic, M., & Rakaric, Z. (2016). Jacobi elliptic functions: A review of nonlinear oscillatory application problems. *Journal of Sound and Vibration*, *380*, 1–36.

Mickens, R. E. (2001). Oscillations in an $x^{4/3}$ potential. *Journal of Sound and Vibration*, *246*, 375–378.

Mickens, R. E. (2004). *Mathematical methods for the natural and engineering sciences*. Hackensack, NJ: World Scientific.

Nayfeh, A. H., & Mook, D. T. (1979). *Nonlinear oscillations*. New York: Wiley.

Ozis, T., & Yildirm, T. A. (2007). Determination of periodic solution for a $u^{1/3}$ force by He's modified Lindstedt–Poincaré method. *Journal of Sound and Vibration*, *301*, 415–419.

Rosenberg, R. M. (1963). The Ateb(h)-functions and their properties. *Quarterly of Applied Mathematics*, *21*, 37–47.

Senik, P. M. (1969). Inversion of the incomplete Beta-function,.*Ukrains'kyi Matematychnyi Zhurnal*,*21*, 325–333 and *Ukrainian Mathematical Journal*,*21*, 271–278.

du Val, P. (1973). *Elliptic functions and elliptic curves*. London mathematical society lecture notes series 9. Cambridge: Cambridge University Press.

Yuste, S. B., & Bejarano, J. D. (1990). Improvement of a Krylov–Bogolubov method that uses Jacobi elliptic functions. *Journal of Sound and Vibration*, *139*, 151–163.

http://functions.wolfram.com/GammaBetaErf/Beta3/26/01/02/0001 (2002a)

http://functions.wolfram.com/HypergeometricFunctions/Hypergeometric2F1/03/09/19/02/0017 (2002b).

http://functions.wolfram.com/EllipticIntegrals/EllipticF/16/01/02/0001 (2002c).

http://functions.wolfram.com/EllipticFunctions/JacobiAmplitude/16/01/01/0001 (2002d).

Chapter 4
Free Vibrations

In this chapter the free vibration of the oscillator with pure nonlinearity (2.1) and small additional forces is investigated. Using the mathematical model of the pure nonlinear oscillator (3.2), the system with additional small nonlinearities is

$$\ddot{x} + c_\alpha^2 x\,|x|^{\alpha-1} = \varepsilon f(x, \dot{x}), \tag{4.1}$$

with the initial conditions (3.3), where εf is a small function which depends on x and $\dot{x}$.

In spite of the fact that a numerous methods are developed for analytic solving of strongly nonlinear differential equations describing the vibrations of the oscillator, the asymptotic approaches still need to be improved. Namely, all of the suggested asymptotic solving procedures have beside their advantages also some disadvantages.

One of the conceptually simplest analytic approximate procedure is the method of harmonic balance, described by Sarma and Rao (1997), which leads to algebraic equations; however the obtained approximative results may be inconsistent. Sensoy and Huseyin (1998) improved the method and developed the intrinsic harmonic balancing technique (IHB). Wu and Lim (2004) combined the linearization and the harmonic balance method and the result was the absence of sets of equations with complex nonlinearities (as is the case for the classical harmonic balance method). In Belhaq and Lakrad (2000a) the harmonic balance method involving the Jacobian elliptic function is applied. Margallo and Bejarano (1987) used the generalized Fourier series assumption in combination with the harmonic balance method to find an approximate solution of the Eq. (4.1). In the paper of Qaisi (1996) the autonomous Duffing equation is transformed into a non-autonomous one, transforming the independent time variable into an oscillating time variable function. The transformed differential equation becomes well-conditioned for a solution by the power series method. The methods proposed by He (2000) and Hu (2004) are identical and represent a classical perturbation technique. Instead of expansion of the frequency, as it is done in the normal classical perturbation method, the square frequency is given as a series of small parameter. It is shown that for this differential equation the 'innovative' technique works even for the large parameter. Cheung et al. (1991) and also He

L. Cveticanin, *Strong Nonlinear Oscillators*, Mathematical Engineering,
DOI 10.1007/978-3-319-58826-1_4

(2001a, 2002a, b), and Liu (2005) proposed a modified Lindstedt-Poincaré method, Lim and Wu (2002) presented a modified Mickens (2001, 2010) procedure for certain nonlinear oscillators. Other analytical methods such as energy method (He 2003a) and bookkeeping parameter method (He 2001b) are also introduced. In the papers of Belhaq and Lakrad (2000b), and Lakrad and Belhaq (2002) the multiple scales method, developed for the systems with small nonlinearities, is extended to the case of strongly nonlinear systems. The solution is expressed in terms of Jacobian elliptic function. Yuste and Bejarano (1990) modified the Krylov and Bogolubov (1943) method to solve strongly nonlinear systems by using elliptic functions. The same equation was studied by Coppola and Rand (1990) using an averaging method and elliptic functions. Belhaq and Lakrad (2000c) applied the averaging method combined formally with the Jacobian elliptic functions to determine an approximative solution.

In this chapter the generalization of the averaging procedure for the pure nonlinear differential equations with additional small deflection and velocity function (4.1) is presented. The method is based on the exact or approximate solution of the pure nonlinear oscillator, as is presented in the previous chapter. The solution is assumed in the same form but with time variable parameters: the periodic Ateb function with time variable amplitude and frequency, the trigonometric function with time variable amplitude and phase, and the Jacobi elliptic function with time variable amplitude, frequency and modulus. The averaging procedure is over the period of the corresponding functions. Averaging method yields a lowest order approximation conveniently, but higher order calculations become lengthy and complicated.

In this chapter the pure nonlinear oscillator with linear damping and small additional function is also considered

$$\ddot{x} + \kappa\dot{x} + c_\alpha^2 x\,|x|^{\alpha-1} = \varepsilon f(x, \dot{x}), \tag{4.2}$$

where κ is the damping coefficient. Based on the generating solution of vibrations of a pure nonlinear oscillator with linear damping, when $\varepsilon = 0$, the trial solution for the oscillator with additional small function (4.2) is considered. The parameters of the solution are assumed to be time variable. The modified averaging solution procedure is presented. The special attention is given to the pure nonlinear and linear damped-Van der Pol oscillator. The interaction of the viscous and Van der Pol damping on the motion of the pure nonlinear oscillator is investigated. The boundary for the limit cycle motion depending on the order of nonlinearity is analyzed.

All of the previously mentioned methods are based on the perturbation of the nonlinear system and certain difficulties appear. As the unperturbed system is already nonlinear the perturbation schemes themselves are difficult to be implemented. The perturbation procedure requires the existence of a small parameter which is not always the case. To eliminate these disadvantages Liao (1995), He (1999, 2000, 2003b), and Liao and Tan (2007) adopted the homotopy technique which is widely applied in differential topology for obtaining the approximate analytic solution of the second order strong nonlinear differential equation. The embedded artificial parameter $p \in [0, 1]$ is introduced and the homotopy transformation of the differential equation is done. The introduced approximate solution represents the solution of the linear differential equation and the suggested method gives the approximate solution by improving the parameters of this analytic solution. The basic idea of the method is discussed in the paper of He (2004). The method does not require an additional small parameter and it is the main advantage of the this method in comparison to the usual perturbation procedure. Cveticanin (2006) extended the homotopy method by improving the parameters of the solution determined for the pure nonlinear differential equation. Namely, the additional terms which are not the dominant one have to be used for improving the solution parameters. Belendez et al. (2007) applied the suggested method for the fractional order nonlinear differential equations. In this chapter the homotopy perturbation method is also presented.

4.1 Homotopy-Perturbation Technique

The homotopy method presented in this section is applicable for strong nonlinear differential equations and also pure nonlinear differential equations. Analyzing the differential equation (4.1) it is evident that beside the function $c_\alpha^2 x\,|x|^{\alpha-1}$ another function $\varepsilon f(x, \dot{x})$ exists which needs not to be linear and also small. The first part of nonlinearity ($c_\alpha^2 x\,|x|^{\alpha-1}$) forms a differential equation which closed form or an approximate analytic solution is known (see Chap. 3). The initial approximation of the problem (4.1) is assumed in the form of that solution. Due to the homotopy technique embedding parameter $p \in [0, 1]$ is introduced and the mapping of a function $x(t)$ into $X(t, p)$ is done. $X(t, p)$ is developed into series of p and the perturbation procedure for parameter p is applied. The solution $x(t)$ of (4.1) corresponds to the solution $X(t, p)$ for $p = 1$. In this section the solution in the first approximation is considered. The differential equation is of oscillatory type and the solution is periodical. Specially, the differential equations with strong cubic and strong quadratic nonlinearities are considered. In the paper numerical examples are solved applying the suggested procedure. The analytically obtained results are compared with numerically obtained ones.

For $\varepsilon f(x, \dot{x}) = 0$ the differential equation (4.1) simplifies to (3.2) which for the initial conditions (3.3) has the explicit or approximate analytic solution $x^0(t)$ (see Chap. 3). We assume the initial approximate solution $x_0(t)$ of (4.1), which transforms to $x^0(t)$ when the nonlinearity $\varepsilon f(x, \dot{x})$ is zero.

By introducing an embedding parameter p with values in the interval [0, 1], a transformation of the variable $x(t)$ to $X(t, p)$ is done. Equation (4.1) is transformed to

$$(1-p)([\ddot{X} + c_\alpha^2 X\,|X|^{\alpha-1}] - [\ddot{x}_0 + c_\alpha^2 x_0\,|x_0|^{\alpha-1}]) + p[\ddot{X} + F\left(X, \dot{X}\right)] = 0, \quad (4.3)$$

with the initial conditions

$$X(0, p) = A, \qquad \dot{X}(0, p) = 0, \quad (4.4)$$

where

$$F\left(X,\dot{X}\right)=c_{\alpha}^{2}X\left|X\right|^{\alpha-1}-\varepsilon f(X,\dot{X}). \tag{4.5}$$

Substituting $p=0$ in (4.3) we find

$$\ddot{X}+c_{\alpha}^{2}X\left|X\right|^{\alpha-1}=0, \tag{4.6}$$

where the differential equation (4.6) has the exact or approximate solution

$$X(t,0)=x^{0}(t). \tag{4.7}$$

When $p=1$ the equation has the same form as the original equation

$$\ddot{X}+c_{\alpha}^{2}X\left|X\right|^{\alpha-1}=\varepsilon f(X,\dot{X}), \tag{4.8}$$

and the solution is

$$X(t,1)=x(t). \tag{4.9}$$

It can be concluded that the differential equation (4.3) differs for different values of p and the solution $X(t,p)$ of (4.3) depends on p. The process of change of p from zero to unity is the process of continual change of solution from (4.7) to (4.9).

For $X(t,p)$, which is the solution of (4.3) in the whole domain $p\in[0,1]$, and is smooth enough to have the kth-order partial derivatives with respect to p at $p=0$, we obtain

$$x_{k}(t)=\left(\frac{\partial^{k}X(t,p)}{\partial p^{k}}\right)_{p=0}, \qquad k=1,2,3,... \tag{4.10}$$

Then, the Maclaurin's series of $X(t,p)$ is as follows

$$X(t,p)=x_{0}(t)+\sum_{k=1}^{\infty}\left(\frac{x_{k}(t)}{k!}\right)p^{k}. \tag{4.11}$$

Substituting (4.11) into (4.3) and separating the terms with the same order of the parameter p a system of linear differential equations is obtained. For p^{1} the first order deformation equation is

$$\ddot{x}_{1}+q(t)x_{1}=F^{*}(t), \tag{4.12}$$

with initial conditions

$$x_{1}(0)=0, \qquad \dot{x}_{1}(0)=0, \tag{4.13}$$

where

$$q(t) = \frac{\partial}{\partial X}\left(c_\alpha^2 X\,|X|^{\alpha-1}\right)_{x_0} = \alpha c_\alpha^2 x_0^{\alpha-1},$$
$$F^*(t) = -[\ddot{x}_0 + c_\alpha^2 x_0\,|x_0|^{\alpha-1} - \varepsilon f(x_0,\dot{x}_0)]. \tag{4.14}$$

For the case when the differential equation (4.1) describes the oscillatory motion the solution $x_0(t)$ is periodical and the Eq. (4.12) represents a linear non-homogenous second order differential equation with periodic coefficients.

As the homogenous part of the Eq. (4.12) has two linear nonvanishing independent solutions $x_{11}(t)$ and $x_{12}(t)$, the fundamental set of solutions is $x_1 = C_1 x_{11}(t) + C_2 x_{12}(t)$ where C_1 and C_2 are constants. Based on this solution and applying the method of variation of constants the solution of differential equation (??) is (see Kamke 1959)

$$x_1 = x_{12}(t)\int \frac{f^*(t)x_{11}(t)}{W}dt - x_{11}\int \frac{f^*(t)x_{12}}{W}dt + C_1x_{11} + C_2x_{12}, \tag{4.15}$$

where $W(t) = x_{11}(x_{12})' - (x_{11})'x_{12}$. Unfortunately, to find the closed form solutions $x_{11}(t)$ and $x_{12}(t)$ is not always possible. Different approximate methods for solving the Hill's equation (4.12) dependently on the type of the functions $q\,(t)$ and $f^*\,(t)$ are applied. In Chap. 3 it is shown that the solution of (3.2) with (3.3) is the function of the periodical Ateb function or of the Jacobi elliptic function or of a trigonometric function. Let us consider the case when the solution of (3.2) with (3.3) is assumed as a function of a Jacobi elliptic function cn (see Appendix D and Byrd and Friedman 1954)

$$x^0 = f^*(cn), \tag{4.16}$$

where $cn \equiv cn(\omega t, k^2)$, ω is frequency and k^2 is the modulus of the Jacobi elliptic function. Based on (4.16) the initial approximate solution $x_0(t)$ is formed as

$$x_0 = f^*(cn_1), \tag{4.17}$$

where $cn_1 \equiv cn(\omega_1 t, k_1^2)$, $\omega_1 = \lambda\omega$, $k_1^2 = \beta k^2$, λ and β are correction factors of frequency and modulus of Jacobi elliptic function, respectively, which depend on parameter p. For $p = 0$ the correction factors are $\lambda = \beta = 1$ and the frequency and modulus of Jacobi function are $\omega_1 = \omega$ and $k_1^2 = k^2$. For p^1 the first order deformation equation (4.12) with (4.13) is

$$\ddot{x}_1 + q[f^*(cn_1)]x_1 = F^*[f^*(cn_1)], \tag{4.18}$$

where $q[f^*(cn_1)]$ and $F^*[f^*(cn_1)]$ are functions of Jacobi elliptic function and ω_1, k_1^2 i.e., α and β are unknown values which have to be determined. To find the closed form analytic solution for (4.18) is usually impossible. Approximate solving procedure is applied. In the paper of Cveticanin (2000) an approximate analytic method for solving of the differential equation where the parametric excitation is a sn Jacobi elliptic function is shown. If the modulus of Jacobi function tends to zero the Jacobi function transforms to the trigonometric function. For the special case the Hill's equation transforms to a Mathieu–Hill equation and one of the suggested approximate solving methods like method of multiple scales Nayfeh and Mook (1979), Bogolubov and Mitropolski (1974), etc., may be applied.

The most common method for solving (4.18) is the harmonic balance procedure. Approximate solution is assumed as a series of Jacobi elliptic function

$$x_1 = \sum_{i=1} K_i (cn_1)^i, \tag{4.19}$$

which satisfies the initial conditions (4.13)

$$\sum_{i=1} K_i [(cn_1)^i]_{t=0} = 0, \qquad \sum_{i=1} K_i [\frac{d}{dt}[(cn_1)^i]_{t=0} = 0. \tag{4.20}$$

Substituting (4.19) into (4.18), equating the terms with the same order of Jacobi elliptic function cn_1 and using relations (4.20) we obtain the constants K_i, frequency ω_1 and modulus k_1^2, i.e., the correction coefficients α and β. Due to (4.9), (4.11), (4.17), (4.19) and (4.20) solution in the first approximation is $x = x_0 + x_1$.

4.1.1 Duffing Oscillator with a Quadratic Term

As a special case, let us assume a pure cubic nonlinear differential equation where a quadratic term also exists

$$\ddot{x} + c_3^2 x^3 + c_2^2 x^2 = 0. \tag{4.21}$$

It is supposed that the cubic nonlinearity is dominant, i.e., $c_3 A^3 > c_2 A^2$. The basic equation ($c_2^2 = 0$) is pure cubic. The exact closed form solution of the pure cubic nonlinear differential equation is

$$x^0(t) = Acn(\omega t, k^2), \tag{4.22}$$

where the frequency ω and modulus k^2 are

$$\omega^2 = c_3^2 A^2, \qquad k^2 = \frac{1}{2}. \tag{4.23}$$

The initial introduced approximate solution of (4.21) is assumed in the form (4.22)

$$x_0 = Acn(\omega_1 t, k_1^2) \equiv Acn_1, \tag{4.24}$$

where ω_1 and k_1^2 depend on the parameter p and have to be determined. Applying the homotopy mapping transformation the first order deformation equation is

$$\ddot{x}_1 + 3c_3x_0^2x_1 = -(\ddot{x}_0 + c_3x_0^3 + c_2x_0^2), \tag{4.25}$$

i.e.,

$$\ddot{x}_1 + 3c_3^2A^2cn_1^2x_1 = -[-A\omega_1^2cn_1(1-2k_1^2+2k_1^2cn_1^2)+c_3^2A^3cn_1^3+c_2^2A^2cn_1^2]. \tag{4.26}$$

Solution of (4.26) is assumed as sum of a constant and a linear term of elliptic function cn_1

$$x_1 = K_1 + K_2cn_1. \tag{4.27}$$

Applying harmonic balance method and separating the terms with the same order of elliptic function cn_1 in the Eq. (4.26) the following system of algebraic equations is obtained

$$\begin{aligned}(K_1 + A)\omega_1^2(1 - 2k_1^2) &= 0,\\ 3c_3^2A^2K_0 + c_2^2A^2 &= 0,\\ 3c_3^2A^2K_1 + c_3^2A^3 - 2(K_1 + A)k_1^2\omega_1^2 &= 0.\end{aligned} \tag{4.28}$$

Due to initial conditions (4.13) the relation for K_0 and K_1 is

$$K_0 + K_1 = 0. \tag{4.29}$$

Solving Eqs. (4.28) and (4.29) it follows

$$\begin{aligned}&\omega_1^2 = c_3^2A^2\frac{A + \left(\frac{c_2}{c_3}\right)^2}{A + \frac{1}{3}\left(\frac{c_2}{c_3}\right)^2}, \qquad \lambda = 1 + \frac{\frac{2}{3}\left(\frac{c_2}{c_3}\right)^2}{A + \frac{1}{3}\left(\frac{c_2}{c_3}\right)^2},\\ &k_1^2 = k^2 = \frac{1}{2}, \qquad K_0 = -K_1 = -\frac{1}{3}\left(\frac{c_2}{c_3}\right)^2.\end{aligned} \tag{4.30}$$

Solution in first approximation of the Eq. (4.21) is

$$x = (A + K_1)cn\left(\omega_1 t, \frac{1}{2}\right) - K_0. \tag{4.31}$$

It is of special interest to discuss the validity of the so obtained solutions. Analyzing relations (4.30) it is obvious that the coefficient c_2^2 has no influence on modulus of Jacobi function. Frequency and argument of Jacobi function depend on coefficient ratio c_2/c_3. Coefficients K_0 and K_1 are also functions of ratio c_2/c_3. The accuracy of the approximate solution (4.31) depends on the ratio of coefficients of nonlinearities c_2/c_3. For smaller ratio $c_2/c_3 \ll 1$ the difference between exact solution and

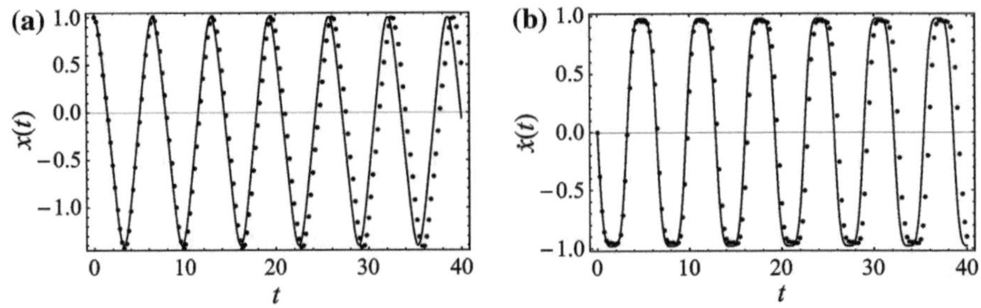

Fig. 4.1 **a** $x - t$ and **b** $\dot{x} - t$ diagrams obtained analytically (*full line*) and numerically (*dotted line*)

approximate solution is negligible. For higher values of the ratio c_2/c_3 the difference is significant and the solution in the first approximation is not acceptable.

For numerical values $A = 1, c_2^2 = 0.6, c_3^2 = 1$, using the aforementioned relations (4.30), the solution of (4.21) in the first approximation is

$$x = 1.2cn(1.1547t, 0.5) - 0.2. \tag{4.32}$$

The corresponding time derivative (see Appendix D) is

$$\dot{x} = -1.3856sn(1.1547t, 0.5)dn(1.1547t, 0.5). \tag{4.33}$$

In Fig. 4.1 the analytical solution (4.32) and the numerical solution of (4.21), obtained applying the Runge–Kutta procedure, and the corresponding first time derivatives, are plotted. Analyzing Fig. 4.1 it is evident that there is a difference between analytical and numerical solution. The error is higher for longer time interval. Besides, in Fig. 4.1b it is shown that the maximal velocity of vibration obtained analytically differs from the numerically calculated one. In spite of that, the solutions obtained by application of the homotopy method are suitable for application as give a good qualitative description of vibrations.

4.1.2 Conclusion

The homotopy method suggested in this section is applicable for analytical solving of any pure nonlinear oscillator with additional nonlinearities which need not to be weak. The approximate solution is based on the solution of the pure nonlinear differential equation, which is the part of differential equation (4.1). The solution in the form of the Jacobi elliptic function seems to be suitable for further calculation.

Namely, in the first approximation the homotopy method, transforms the strong nonlinear differential equation to a linear parametrically excited equation for which numerous approximate analytic solving methods already exist. Method of harmonic balance seems to be the most appropriate and simplest. The solution of the differential equation of oscillatory motion is assumed in the form of series of Jacobi elliptic function. The parameters of Jacobi elliptic function are obtained solving the first order deformation equation. The mathematical calculation of the approximate solution is simpler than for the classic perturbation procedure with Jacobi elliptic function. The interesting features of the proposed method are except its simplicity also its excellent accuracy in a wide range of values of oscillation amplitude. The solution obtained using the suggested method has a very high accuracy comparing with the exact numerical solution even for long time period. The suggested method is found to work extremely well in the example where the nonlinearities of even and odd order exist.

4.2 Averaging Solution Procedure

The averaging solution procedure of (4.1) is based on the well known Krylov–Bogolubov method of time variable amplitude and phase (1943), which is developed for the linear oscillator with small additional nonlinearity. Namely, for $\varepsilon = 0$, the Eq. (4.1) simplifies to the pure nonlinear second order differential equation (3.2) considered in Chap. 3. Equation (3.2) is the generating one for (4.1), while the exact (3.27) and approximate solutions (3.85) and (3.104) are the corresponding generating solutions. The trial solution of (4.1) is assumed in the form of the generating solution but with time variable amplitude $a(t)$ and phase angle $\theta(t)$, i.e. frequency $\dot{\psi}(t)$. Introducing the trial solution and its time derivatives into (4.1), the second order differential equation is rewritten into two coupled first order nonlinear differential equations

$$\dot{a} = f_1(a, \psi), \qquad \dot{\theta} = f_2(a, \psi), \tag{4.34}$$

where f_1 and f_2 are functions of a and ψ. Our task is to solve the system of differential equations (4.34). To find the analytic solution of (4.34) is an extreme heavy task. This is the reason that some simplification to the problem is introduced. As the right-hand-side function depend on the periodical function (Ateb periodical function, Jacobi elliptic function or trigonometric function), the averaging over the period T is done. The averaged differential equations are obtained

$$\dot{a} = \frac{1}{T}\int_0^T f_1(a, \psi)d\psi = f_1^*(a), \qquad \dot{\theta} = \frac{1}{T}\int_0^T f_2(a, \psi)d\psi = f_2^*(a). \tag{4.35}$$

The Eq. $(4.35)_1$ is an uncoupled single variable first order nonlinear differential equation. Solving the differential equation $(4.35)_1$ the time variable amplitude $a(t)$ is obtained. Substituting $a(t)$ into $(4.35)_2$ and integrating, the $\theta(t)$ variable is determined. Introducing $a(t)$ and $\theta(t)$ into the trial solution, the averaged approximate solution of (4.1) is obtained.

In the following sections the here explained averaging procedure will be applied for the trial solution in the form of a periodical Ateb function, a Jacobi elliptic function and a trigonometric function.

4.2.1 Solution in the Form of an Ateb Function

In this section the trial solution of (4.1) and its first time derivative are assumed in the same form as the generating solution (3.27) of the pure nonlinear differential equation (3.2) and its first time derivative (3.28), but with time variable parameters, i.e.,

$$x = a(t)ca\,(\alpha, 1, \psi(t)) \qquad \dot{\psi}(t) := \sqrt{\frac{c_\alpha^2(\alpha+1)}{2}}\,[a(t)]^{(\alpha-1)/2} + \dot{\theta}(t)\,, \tag{4.36}$$

and

$$\dot{x} = -\frac{\sqrt{2c_\alpha^2}}{\sqrt{\alpha+1}}\,a^{(\alpha+1)/2}\,sa(1, \alpha, \psi(t))\,, \tag{4.37}$$

where the amplitude a and the phase angle ψ are time dependent. Comparing (4.37) with the first time derivative of (4.36)

$$\dot{x} = \dot{a}(t)ca\,(\alpha, 1, \psi(t)) - \frac{\sqrt{2c_\alpha^2}}{\sqrt{\alpha+1}}\,a^{(\alpha+1)/2}\,sa(1, \alpha, \psi(t)) - a\dot{\theta}\sqrt{\frac{2c_\alpha^2}{\alpha+1}}\,sa(1, \alpha, \psi), \tag{4.38}$$

it is obvious that they are equal if

$$\dot{a}\,ca(\alpha, 1, \psi) - \frac{2}{\alpha+1}a\dot{\theta}\,sa(1, \alpha, \psi) = 0\,. \tag{4.39}$$

Substituting (4.36), (4.37) and the second time derivative

$$\ddot{x} = -\frac{\sqrt{2c_\alpha^2}}{\sqrt{\alpha+1}}\left[\frac{\alpha+1}{2}a^{(\alpha-1)/2}\,\dot{a}sa(1, \alpha, \psi(t)) + a^{(\alpha+1)/2}\dot{\psi}(t)\,ca^{\alpha}(1, \alpha, \psi(t))\right], \tag{4.40}$$

into (4.1), and after some simplification, we have

$$\sqrt{\frac{c_\alpha^2(\alpha+1)}{2}}a^{(\alpha-1)/2}\,\dot{a}sa(1,\alpha,\psi(t))+\sqrt{\frac{2c_\alpha^2}{\alpha+1}}a^{(\alpha+1)/2}\dot{\theta}(t)\,ca^{\alpha}(1,\alpha,\psi(t))$$

$$=\ -\varepsilon f\left(a(t)ca\left(\alpha,1,\psi(t)\right),-\sqrt{\frac{2c_\alpha^2}{\alpha+1}}\,a^{(\alpha+1)/2}\,sa(1,\alpha,\psi(t))\right). \tag{4.41}$$

The Eqs. (4.39) and (4.41) represent the rewritten version of the Eq. (4.1) in the new variables $a(t)$ and $\theta(t)$. Applying the relation between ca and sa periodical Ateb functions (A7) given in the Appendix A, Eqs. (4.39) and (4.41) are modified as

$$a^{(\alpha-1)/2}\dot{a}=-\varepsilon f\sqrt{\frac{2}{c_\alpha^2(\alpha+1)}}sa,$$

$$a^{(\alpha+1)/2}\dot{\theta}=-\varepsilon f\sqrt{\frac{\alpha+1}{2c_\alpha^2}}ca, \tag{4.42}$$

where $\varepsilon f\ \equiv\ \varepsilon f\left(a(t)ca\left(\alpha,1,\psi(t)\right),-\sqrt{\frac{2c_\alpha^2}{\alpha+1}}\,a^{(\alpha+1)/2}\,sa(1,\alpha,\psi(t))\right)$. Our task is to solve the system of coupled nonlinear first order differential equations (4.42). According to the averaging procedure, the right-hand side terms are averaged over the period of the Ateb functions $2\Pi_\alpha$ and the averaged differential equations of motion are

$$a^{(\alpha-1)/2}\dot{a}=-\frac{\varepsilon}{2\Pi_\alpha}\sqrt{\frac{2}{c_\alpha^2(\alpha+1)}}\int_0^{2\Pi_\alpha}fsad\psi,$$

$$a^{(\alpha+1)/2}\dot{\theta}=-\frac{\varepsilon}{2\Pi_\alpha}\sqrt{\frac{\alpha+1}{2c_\alpha^2}}\int_0^{2\Pi_\alpha}fcad\psi, \tag{4.43}$$

where Π_α is given with (A.5) in the Appendix A. First, solving $(4.43)_1$ the variable $a(t)$ is obtained. Then, substituting $a(t)$ into $(4.43)_2$ and integrating, the $\theta(t)$ relation is obtained. Introducing the variables $a(t)$ and $\theta(t)$ into (4.36) the approximate averaged solution of the differential equation (4.1) is determined.

Let us apply the suggested solution procedure for some special cases.

Small Nonlinear Deflection Functions

The suggested solution procedure is applied for the oscillator where the additional small force depends only on the variable x, i.e., is the sum of polynomial deflection functions. The model of the oscillator is (see Mickens 2002, Cveticanin and Pogany 2012)

$$\ddot{x}+c_\alpha^2x|x|^{\alpha-1}=-\sum_{j\in\Lambda}c_j^2x|x|^{j-1}, \tag{4.44}$$

where $c_\alpha^2x|x|^{\alpha-1}$ is the so-called 'dominant term'. Thus, the terms in the polynomial on the right side of the relation (4.44) satisfy the relation

$$c_\alpha^2 A|A|^{\alpha-1} >> c_j^2 A|A|^{j-1}, \tag{4.45}$$

for all $j \in \Lambda$ and $\Lambda \subset \mathbb{R}_+$ and some set of indices. For condition (4.45) the terms $c_j^2 x|x|^{j-1}$ are significantly smaller than the dominant one $c_\alpha^2 x|x|^{\alpha-1}$ and for all values of x we have

$$c_\alpha^2 x|x|^{\alpha-1} >> c_j^2 x|x|^{j-1}. \tag{4.46}$$

The right-hand-side expression in (4.44) is the perturbation term which represents a small value. Due to (4.45) we assume the solution of (4.44) as the perturbed version of the solution of the differential equation (3.2).

Remark 1 Usually, in the literature it is stated that if the right–hand–side coefficients are small ($c_j^2 << 1$) in comparison to c_α^2 the solution of the differential equation (4.44) differs from the solution of the differential equation (3.2) just for a bit. Unfortunately, this conclusion is not valid for strong nonlinear differential equations where the initial conditions have a significant influence. Then, the limitation (4.45) has to be introduced and satisfied.

According to the previous solution procedure, the averaged differential equations of motion (4.43) become

$$\dot{a} = \frac{\sqrt{2}\, a^{(1-\alpha)/2}}{2\Pi_\alpha \sqrt{c_\alpha^2(\alpha+1)}} \int_0^{2\Pi_\alpha} \sum_{j\in\Lambda} c_j^2\, a^j\, ca^j(\alpha, 1, \psi) sa(1, \alpha, \psi) d\psi, \tag{4.47}$$

$$\dot{\theta} = \frac{\sqrt{\alpha+1}\, a^{(1-\alpha)/2}}{2\Pi_\alpha \sqrt{2c_\alpha^2}} \int_0^{2\Pi_\alpha} \sum_{j\in\Lambda} c_j^2\, a^{j-1}\, ca^{j+1}(\alpha, 1, \psi) d\psi\,. \tag{4.48}$$

The averaging procedure immediately gives

$$\dot{a} = \frac{a^{(1-\alpha)/2}}{\sqrt{2c_\alpha^2(\alpha+1)}\Pi_\alpha} \sum_{j\in\Lambda} c_j^2\, a^j \int_0^{2\Pi_\alpha} sa(1, \alpha, \psi) ca^j(\alpha, 1, \psi) d\psi = 0\,. \tag{4.49}$$

By the initial condition $x(0) = A$, we deduce the claim

$$a(t) = A\,. \tag{4.50}$$

Now, it remains to solve (4.48). Because $ca(\alpha,\, 1,\, z)$ is even and periodic, by averaging the right-hand-side expression in (4.48), we conclude

$$\dot{\theta} = \frac{\sqrt{\alpha+1}\, a^{(1-\alpha)/2}}{2\sqrt{2c_\alpha^2}\, \Pi_\alpha} \sum_{j\in\Lambda} c_j^2\, a^{j-1} \int_0^{2\Pi_\alpha} ca^{j+1}(\alpha, 1, \psi)\, d\psi\,. \tag{4.51}$$

To find the value of the inner–most integral we apply a result by Drogomirecka (1997), reads as follows

$$\int_0^{2\Pi_\alpha} sa^p(n,m,z)\,ca^q(m,n,z)\,dz \tag{4.52}$$
$$= \frac{1}{2}\left(1+(-1)^p+(-1)^q+(-1)^{p+q}\right)B\left(\frac{p+1}{n+1},\frac{q+1}{m+1}\right),$$

for all $p, q \in \{\frac{r}{s} : r \in \mathbb{Z},\ s = 2k-1, k \in \mathbb{N}\}$, i.e. the powers are rational, having integer numerator and odd positive denominator. Hence

$$\dot{\theta} = \frac{\sqrt{\alpha+1}\,a^{(1-\alpha)/2}}{\sqrt{2c_\alpha^2}\,\Pi_\alpha}\sum_{j\in\Lambda_{\text{odd}}} c_j^2 a^{j-1} B\left(\frac{j+2}{\alpha+1},\frac{1}{2}\right), \tag{4.53}$$

so, according to $a(t) \equiv A$, we get

$$\theta(t) = \left\{\frac{\sqrt{\alpha+1}}{\sqrt{2c_\alpha^2}\,\pi}\sum_{j\in\Lambda_{\text{odd}}} c_j^2\, A^{j-(\alpha+1)/2}\, B\left(\frac{j+2}{\alpha+1},\frac{1}{2}\right)\right\} t\,. \tag{4.54}$$

Thus, the exact averaged perturbed solution of the Cauchy problem (4.44) with the initial conditions (3.3), reads

$$x(t) = A\cdot ca\left(\alpha, 1, \sqrt{\frac{c_\alpha^2(\alpha+1)}{2}}\,A^{(\alpha-1)/2}\left\{1+\frac{c^*}{c_\alpha^2\,\pi}\right\} t\right), \tag{4.55}$$

with the correction constant which describes the influence of the right-hand-side perturbation terms

$$c^* = \sum_{j\in\Lambda_{\text{odd}}} c_j^2\, A^{j-\alpha} B\left(\frac{j+2}{\alpha+1},\frac{1}{2}\right). \tag{4.56}$$

Correction constant depends not only on the coefficients c_j^2 and c_α^2, but also on the initial amplitude and the order of the binoms in polynomial.

Differential Equation with a Linear Dominant Term

Let us consider the case when the dominant term is linear and the second order nonlinear Cauchy problem is

$$\ddot{x} + c_1^2 x = -\sum_{j\in\Lambda} c_j^2 x|x|^{j-1}, \qquad x(0) = A, \quad \dot{x}(0) = 0\,. \tag{4.57}$$

Let us mention that this system one reduces to (3.32) associated with the linear case $\alpha = 1$. The general solution of (3.32) can be taken in the form (3.33)

$$x = a\cos(c_1 t + \theta)\,, \tag{4.58}$$

where a and θ stand for the related integration constants. Assuming that the right-hand side perturbation terms are small ones ($c_j^2|A|^{j-1} \ll c_1^2$), it is supposed that the vibrations of the perturbed system are close to that of the linear system. So, we are looking for an approximate solution of (4.57) in the form of (4.58), but with time variable amplitude and phase

$$x = a(t)\cos\psi(t) =: a\cos\psi\,, \tag{4.59}$$

where $\psi = \psi(t) = c_1 t + \theta(t)$. First time derivative of x is

$$\dot{x} = -ac_1 \sin\psi, \tag{4.60}$$

for

$$\dot{a}\cos\psi - a\dot{\theta}\sin\psi = 0\,. \tag{4.61}$$

Substituting (4.59) and the time derivative of (4.60) into (4.57), we obtain

$$\dot{a}\sin\psi + a\dot{\theta}\cos\psi = \frac{1}{c_1}\sum_{j\in\Lambda} c_j^2 a^j \cos^j\psi =: \Phi\,. \tag{4.62}$$

Hence, the second order nonlinear differential equation (4.57) is rewritten into a system of two coupled first order differential equations (4.61) and (4.62). Solving this system with respect to $\dot{\theta}$ and $\dot{a}$, we get

$$\dot{\psi} = c_1 + \frac{\Phi}{a}\cdot\cos\psi\,, \qquad \dot{a} = \Phi\cdot\sin\psi\,. \tag{4.63}$$

Due to the suggested procedure averaging of the Eq. (4.63) over the period of vibration equal to $T_1 = 2\pi/c_1$ is done. Assuming that $\Re\{j\} > -1$, the averaged differential equations are

$$\begin{aligned}
\dot{a} &= 0,\\
\dot{\theta} &= \frac{1}{2\pi}\sum_{j\in\Lambda} c_j^2 a^{j-1}\int_0^{2\pi/c_1}\cos^{j+1}(c_1 t + \theta)\,\mathrm{d}t\\
&= \frac{1}{2c_1\sqrt{\pi}}\sum_{j\in\Lambda} c_j^2\,\frac{a^{j-1}(1+(-1)^{j+1})\Gamma\left(\frac{j+2}{2}\right)}{\Gamma\left(\frac{j+3}{2}\right)}\,.
\end{aligned} \tag{4.64}$$

For the initial conditions (3.3) the solutions of the averaged differential equation are

$$a(t) = A = const., \tag{4.65}$$

$$\psi(t) = \left(c_1 + \frac{1}{c_1\sqrt{\pi}} \sum_{j\in\Lambda_{\text{odd}}} c_j^2 \frac{A^{j-1}\Gamma\left(\frac{j+2}{2}\right)}{\Gamma\left(\frac{j+3}{2}\right)} \right) t, \tag{4.66}$$

where Λ_{odd} denotes the set of indices which behave like odd integers and $\psi(t)$ takes its final form.

Having in mind the initial conditions (3.3) the solution of Eq. (4.57) in the first approximation is obtained as

$$x(t) = A\cos\left\{\left(c_1 + \frac{c^{**}}{c_1}\right)t\right\}, \tag{4.67}$$

with the correction term

$$c^{**} = \frac{1}{\sqrt{\pi}} \sum_{j\in\Lambda} c_j^2 \frac{A^{j-1}\Gamma\left(\frac{j+2}{2}\right)}{\Gamma\left(\frac{j+3}{2}\right)}. \tag{4.68}$$

which depends on the coefficients of nonlinearity, initial amplitude and the order of nonlinearity.

Two numerical examples are solved.

1. We consider an oscillator for which the mathematical model and initial conditions follow as

$$\ddot{x} + x = -c_{5/3}^2\, x|x|^{2/3}, \qquad x(0) = 0.1, \quad \dot{x}(0) = 0. \tag{4.69}$$

For the case when $c_{5/3}^2 = 0.01$ the related first approximation solution is

$$x_A = 0.1\,\cos\left[\left(1 + 0.01\cdot 0.1^{2/3}\frac{\Gamma\left(\frac{11}{6}\right)}{\sqrt{\pi}\,\Gamma\left(\frac{7}{3}\right)}\right)t\right] \approx 0.1\,\cos(1.00096\,t), \tag{4.70}$$

and for the case when $c_{5/3}^2 = 1$ the related first approximation solution is

$$x_{A1} = 0.1\,\cos(1.096\,t). \tag{4.71}$$

Using the Runge–Kutta solving procedure the Eq. (4.69) for certain parameter values is solved numerically. In Fig. 4.2. the approximate analytic solutions, x_A (4.70) and x_{A1} (4.71), and the numerical ones, x_N and x_{N1} for (4.69) with $c_{5/3}^2 = 0.01$ and $c_{5/3}^2 = 1$, respectively, are plotted.

It can be seen that the difference is negligible. In spite of the fact that the coefficient of the nonlinear term $c_{5/3}^2$ is of the same order as of the linear term c_1^2 the averaging solving procedure gives very accurate results due to the fact that the relation (4.45) is satisfied, i.e. $0.1 > 0.1^{5/3} \equiv 0.021544$.

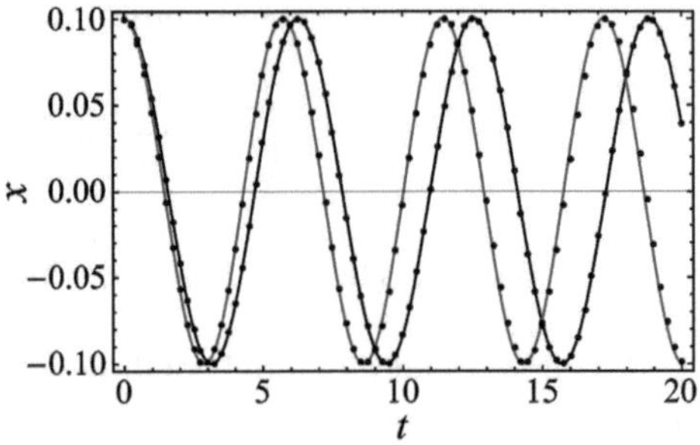

Fig. 4.2 $x - t$ diagrams obtained numerically (*dotted line*) and analytically for $c_{5/3}^2 = 0.01$ (*black line*) and $c_{5/3}^2 = 1$ (*gray line*)

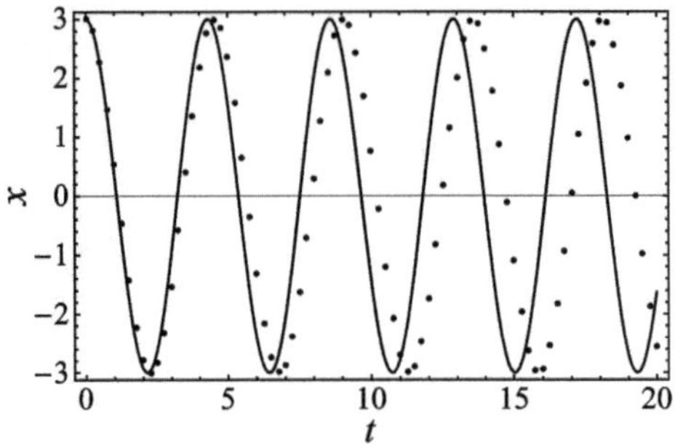

Fig. 4.3 $x - t$ diagrams obtained analytically(*full line*) and numerically (*dotted line*)

2. For the oscillator and initial conditions

$$\ddot{x} + x = -0.5x|x|^{2/3}, \qquad x(0) = 3.0, \quad \dot{x}(0) = 0, \tag{4.72}$$

the analytical approximate solution is

$$x_A = 3\cos(1.46343t)\,. \tag{4.73}$$

In Fig. 4.3. the analytical solution x_A (4.73) and the numerically obtained one x_N for (4.72) are plotted. Numerical solutions are obtained by applying of the Runge–Kutta procedure.

It can be seen that the difference is significant. Namely, the suggested criteria (4.45) is not satisfied: $3 \not\gg 0.5 \cdot 3^{5/3} \equiv 3.12$. The criteria that the coefficient c_1^2 is higher than $c_{5/3}^2$ is not valid for application of the approximate solving procedures in the strong nonlinear oscillators.

Remark 2 The approximate averaging method is suitable for application if the perturbation of the oscillatory motion is small and the criteria (4.45) i.e. (4.46) is satisfied (see also Cveticanin 2009). If the coefficient of perturbation is small it does not mean that the method is applicable for the strong nonlinear systems. The suggested solving method does not require the existence of the small parameter but the condition (4.45) i.e. (4.46) to be fulfilled.

4.2.2 Solution in the Form of the Jacobi Elliptic Function

If the motion of the oscillator (4.1) is in the neighborhood of (3.2), the trial solution of (4.1) has to be close to the generating solution of (3.2). If the generating solution has the form of a Jacobi elliptic function (3.85) and also its first time derivative (3.90), the trial solution and its time derivative are the same functions but with time variable amplitude and phase, i.e.

$$x = a(t)cn(\psi(t), k^2), \tag{4.74}$$
$$\dot{x} = -a(t)\Omega_1(t)sn(\psi(t), k^2)dn(\psi(t), k^2), \tag{4.75}$$

where

$$\dot{\psi}(t) \equiv \dot{\psi} = \Omega_1(t) + \dot{\theta}(t), \qquad a(t) \equiv a, \qquad \theta(t) = \theta,$$
$$\Omega_1 = \Omega_1^* a^{(\alpha-1)/2}\sqrt{c_\alpha^2}. \tag{4.76}$$

Let us assume the first time derivative of (4.74)

$$\dot{x} = \dot{a}cn(\psi, k^2) - a(\Omega_1 + \dot{\theta})sn(\psi, k^2)dn(\psi, k^2). \tag{4.77}$$

Comparing (4.75) and (4.77) it is

$$\dot{a}cn(\psi, k^2) - a\dot{\theta}sn(\psi, k^2)dn(\psi, k^2) = 0. \tag{4.78}$$

The time derivative of (4.75) is

$$\begin{aligned}\ddot{x} = &-a(\Omega_1 + \dot{\theta})\Omega_1 cn(\psi, k^2)(1 - 2k^2 + 2k^2 cn^2(\psi, k^2)) \\ &-(\dot{a}\Omega_1 + a\dot{\Omega}_1)sn(\psi, k^2)dn(\psi, k^2).\end{aligned} \tag{4.79}$$

Substituting (4.74), (4.75) and (4.79) into (4.1) and using the relation

$$- a(\Omega_1 + \dot{\theta})\Omega_1 cn(1 - 2k^2 + 2k^2 cn^2) + c_\alpha^2 a^\alpha cn^\alpha \approx 0, \tag{4.80}$$

we have

$$\begin{aligned}&-a\dot{\theta}\Omega_1 cn(1 - 2k^2 + 2k^2 cn^2) - \dot{a}(1 + \frac{\alpha - 1}{2}\Omega_1^* a^{(\alpha-1)/2}\sqrt{c_\alpha^2})sndn \\ &= \varepsilon f\,(acn, -a\Omega_1 sndn)\,.\end{aligned} \tag{4.81}$$

Equations (4.78) and (4.81) represent the transformation of the second order differential equation (4.1) into two first order differential equations, i.e.

$$a\dot{\theta}\Omega_1\left(\frac{\alpha+1}{2}sn^2dn^2+cn^2(1-2k^2+2k^2cn^2)\right)=-\varepsilon f\,(acn,-a\Omega_1sndn)\,cn, \tag{4.82}$$

$$\dot{a}\Omega_1\left(\frac{\alpha+1}{2}sn^2dn^2+cn^2(1-2k^2+2k^2cn^2)\right)=-\varepsilon f\,(acn,-a\Omega_1sndn)\,sndn. \tag{4.83}$$

For simplification, the averaging over the period of the Jacobi elliptic function $4K(k)$ is done. The averaged differential equations (4.82) and (4.83) are

$$a\dot{\theta}\Omega_1=-\frac{\varepsilon}{r}\int_0^{4K(k)}f\,(acn,-a\Omega_1sndn)\,cn\,d\psi, \tag{4.84}$$

$$\dot{a}=-\frac{\varepsilon}{r}\int_0^{4K(k)}f\,(acn,-a\Omega_1sndn)\,sndn\,d\psi, \tag{4.85}$$

where

$$r=\frac{\alpha+1}{2}(1-k^2+C_2(2k^2-1)-k^2C_4)+C_2(1-2k^2)+2k^2C_4, \tag{4.86}$$

and

$$C_2=\int_0^{4K(k)}cn^2d\psi=\frac{1}{k^2}\left(\frac{E}{K}-1+k^2\right)\approx\frac{1}{2}+\frac{k^2}{16}+..., \tag{4.87}$$

$$C_4=\int_0^{4K(k)}cn^4d\psi=\frac{1}{3k^2}\left[2(2k^2-1)C_2+(1-k^2)\right]\approx\frac{3}{8}+\frac{k^2}{32}+.... \tag{4.88}$$

$E\equiv E(k^2)$ is the complete elliptic integral of the second kind (Byrd and Friedman 1954). The constant r depends on the order of the pure nonlinear oscillator. Solving (4.85) the averaged amplitude $a(t)$ function is obtained. Substituting $a(t)$ into (4.84) and integrating the phase angle $\psi(t)$ functions is determined. Based on these functions the averaged solution (4.74) and its first time derivative (4.75) follow.

Oscillator with Nonlinear Elastic Force

As a special case, let us assume the oscillator with additional nonlinear elastic force

$$\ddot{x}+c_\alpha^2x\,|x|^{\alpha-1}=\varepsilon f(x). \tag{4.89}$$

For the case when the additional function depends only on x, Eqs. (4.84) and (4.85) transform into

$$\dot{a} = 0, \qquad a\dot{\theta}\Omega_1 = -\frac{\varepsilon}{r} \int_0^{4K(k)} f\,(acn)\,cn\,d\psi. \tag{4.90}$$

Applying the initial conditions $a(0) = A$ and $\theta(0) = 0$, the solutions of (4.90) are

$$a = A = const., \qquad \theta = \theta^* t, \tag{4.91}$$

where

$$\theta^* = -\frac{\varepsilon F(A)}{Ar\Omega_1(A)} = const., \tag{4.92}$$

and

$$F(A) = \int_0^{4K(k)} f\,(Acn)\,cn\,d\psi = const., \qquad \Omega_1(A) = \Omega_1^* A^{(\alpha-1)/2}\sqrt{c_\alpha^2} = const. \tag{4.93}$$

For (4.76) the phase function $\psi(t)$ is

$$\psi(t) = \omega^* t,$$

where

$$\omega^* = \Omega_1(A) + \theta^* = const. \tag{4.94}$$

The approximate averaged solution (4.74) is

$$x = Acn(\omega^* t, k^2). \tag{4.95}$$

The amplitude of vibration is constant and is equal to the initial value, while the frequency of vibration is corrected with the constant term θ^* which depends on the properties of the additional small perturbation function εf. According to (4.75) and to the suggested procedure, the assumed averaged first time derivative of the solution is

$$\dot{x} = -A\Omega_1(A) sn(\omega^* t, k^2) dn(\omega^* t, k^2). \tag{4.96}$$

Deriving the relation (4.95) we have

$$\dot{x} = -A\omega^* sn(\omega^* t, k^2) dn(\omega^* t, k^2). \tag{4.97}$$

Comparing (4.96) and (4.97) it is obvious that the amplitudes differ for θ^*. Namely, the relation (4.97) is the improved version for the velocity of vibration. Using this relation the correction of the modulus of the Jacobi elliptic function k^* is possible to be calculated. The first time derivative of (4.97) gives the maximal velocity which is according to (3.94)

$$\dot{x}_{\max} = A\omega^*(k^*)', \tag{4.98}$$

i.e.

$$\dot{x}_{\max} = (k^*)' \left(\frac{2K(k^{*2})}{\pi} \Omega_\alpha A^{(\alpha+1)/2} \sqrt{c_\alpha^2} - \frac{\varepsilon F(A)}{r^* \Omega_1(A)} \right), \tag{4.99}$$

for

$$t_V = \frac{K(k^*)}{\omega^*}. \tag{4.100}$$

The relation r^* is the function of the corrected modulus of the Jacobi elliptic function

$$r^* = \frac{\alpha+1}{2}(1 - k^{*2} + C_2(2k^{*2} - 1) - k^{*2}C_4) + C_2(1 - 2k^{*2}) + 2k^{*2}C_4.$$

As the oscillator (4.89) is a conservative one, the total mechanical energy is constant, i.e.,

$$\frac{\dot{x}^2}{2} + \frac{c_\alpha^2}{\alpha+1} |x|^{\alpha+1} + \varepsilon F_1(x) = \frac{c_\alpha^2}{\alpha+1} |A|^{\alpha+1} + \varepsilon F_1(A), \tag{4.101}$$

where

$$\varepsilon F_1(x) = -\int \varepsilon f(x) dx. \tag{4.102}$$

For $x_V = 0$, the maximal velocity is

$$v_{\max} = \sqrt{\frac{2c_\alpha^2}{\alpha+1} |A|^{\alpha+1} + 2\varepsilon F_1(A)}. \tag{4.103}$$

Equating the relations (4.98) and (4.103), neglecting the terms with the second and higher order of small parameter $O(\varepsilon^2)$, we have

$$(k^*)'^2 \left(K^2(k^{*2}) - \frac{\varepsilon \pi^2 F(A)}{2r^* A^\alpha \Omega_\alpha^2 c_\alpha^2} \right) = \frac{\pi^2}{2\Omega_\alpha^2} \left(\frac{1}{\alpha+1} + \frac{\varepsilon F_1(A)}{c_\alpha^2 |A|^{\alpha+1}} \right). \tag{4.104}$$

From (4.104) it can be seen that the corrected version of the modulus k depends not only on the order of nonlinearity of the pure nonlinear oscillator, but also on the initial amplitude and the properties of the oscillator due to existence of the additional small deflection function in the oscillator. Finally, the corrected solution and its first derivative of (4.89) is

$$x = Acn(\omega^* t, k^{*2}), \tag{4.105}$$

$$\dot{x} = -A\omega^* sn(\omega^* t, k^{*2}) dn(\omega^* t, k^{*2}). \tag{4.106}$$

Example A numerical example is considered. The mathematical model of a strong nonlinear oscillator with additional linear deflection function is described as

$$\ddot{x} + x\,|x|^{2/3} = -0.1x, \tag{4.107}$$

with initial conditions

$$x(0) = A = 2, \qquad \dot{x}(0) = 0. \tag{4.108}$$

According to (4.84) and (4.85) the Eq. (4.107) is rewritten into two first order averaged differential equations

$$\dot{A} = 0, \qquad \dot{\theta}\Omega_1(A) = 5.4016 \times 10^{-2}. \tag{4.109}$$

For the initial conditions (4.108), the parameters of the solution in the form of the Jacobi elliptic function are

$$A = 2, \quad \omega^* = 1.3256, \quad T = 5.1323, \quad k^2 = 0.296331, \quad \theta^* = 4.2085 \times 10^{-2}, \quad \omega = 1.2835. \tag{4.110}$$

The solution in the first approximation is (4.95)

$$x = 2cn(1.3256t, 0.296331), \tag{4.111}$$

and the first time derivative according to (4.96)

$$\dot{x} = -2.5670sn(1.3256t, 0.296331)dn(1.3256t, 0.296331). \tag{4.112}$$

The improved version (4.97) of the first time derivative is

$$\dot{x} = -2.6512sn(1.3256t, 0.296331)dn(1.3256t, 0.296331). \tag{4.113}$$

Using the process of correction of the modulus of the Jacobi elliptic function, the corrected solution (4.105) and its first time derivative (4.106) follow as

$$x = 2cn(1.3527t, 0.277285), \tag{4.114}$$

and

$$\dot{x} = -2.7054sn(1.3527t, 0.277285)dn(1.3527t, 0.277285). \tag{4.115}$$

Using the Runge–Kutta solving procedure, the numerical solution of (4.107) for the initial conditions (4.108) is calculated. This result and its first time derivative are used for comparison with the analytically obtained results. Namely, in Fig. 4.4 the $x - t$ (4.111) and $\dot{x} - t$ (4.112) diagrams are compared with numerical ones. In Fig. 4.5 the $x - t$ (4.111) and the adopted version of the first derivative $\dot{x} - t$ (4.113)

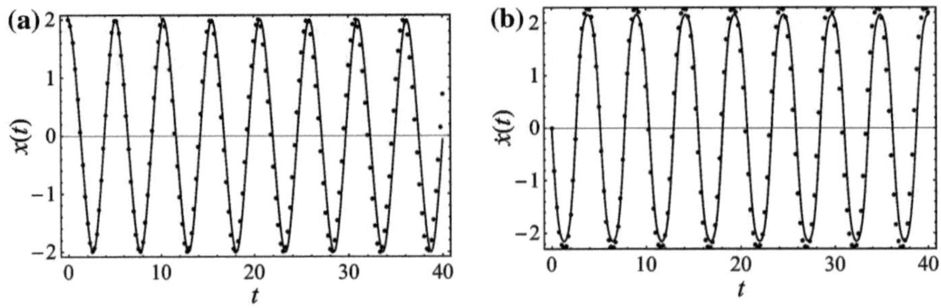

Fig. 4.4 **a** $x - t$ and **b** $\dot{x} - t$ diagrams obtained by averaging (*full line*) and numerically (*dotted line*)

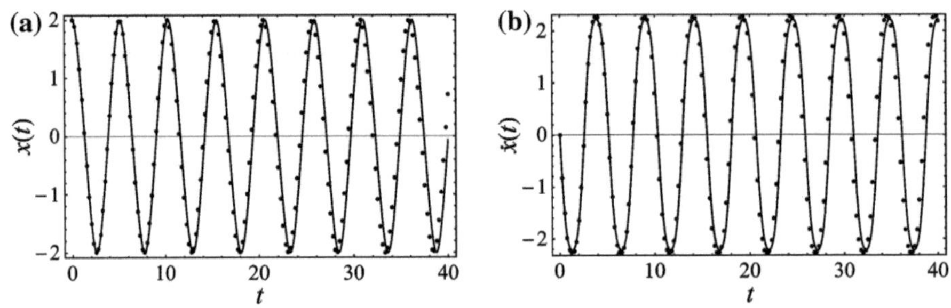

Fig. 4.5 **a** $x-t$ obtained by averaging (*full line*) and numerically (*dotted line*) and **b** $\dot{x}-t$ diagrams obtained numerically (*dotted line*) and with improvement (*full line*)

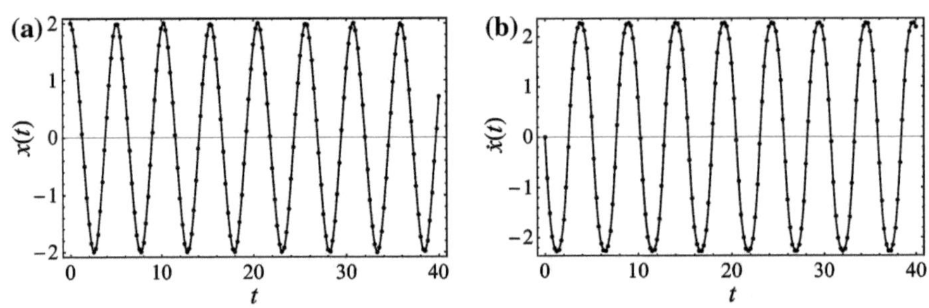

Fig. 4.6 **a** $x - t$ and **b** $\dot{x} - t$ diagrams with correction (*full line*) and numerical ones (*dotted line*)

is compared with numerical ones. Finally, in Fig. 4.6, the solution with modified modulus of the Jacobi elliptic function (4.114) and its time derivative (4.115), and also the numerically obtained curves are plotted. It can be concluded that the solution with the modified modulus of elliptic function gives very accurate results not only in a short but also long time interval.

The $x - t$ diagrams in Fig. 4.6a, i.e., Fig. 4.5a are appropriate for short time, but due to disagreement of the exact and approximate period of vibration the error increases in time. The $\dot{x} - t$ curve shown in Fig. 4.5b is more accurate than in Fig. 4.4b. The adopted version of the first time derivative gives better results.

Due to complexity of calculation of the corrected modulus of the Jacobi elliptic function, it is suggested to apply the adopted version (4.111) with the derivative (4.113).

4.2.3 Solution in the Form of a Trigonometric Function

In spite of the fact that the Jacobi elliptic function gives very accurate solution of the differential equation (4.1), the calculation is very complex for practical use. For simplification it is appropriate to assume the trial solution of (4.1) in the from of a trigonometric function. Namely, for the generating the Eq. (3.2), when $\varepsilon = 0$, the approximate solution is (3.104) with (3.106). Then the trial solution of (4.1) and its first time derivative are assumed in the form of the generating solution with variable amplitude $a(t)$, frequency $\omega(A)$ and phase $\theta(t)$

$$x = a(t)\cos\psi(t) \equiv a\cos\psi, \tag{4.116}$$

$$\dot{x} = -a(t)\omega(a)\sin\psi(t) \equiv -a\omega(a)\sin\psi, \tag{4.117}$$

with

$$\dot{a}\cos\psi(t) - a\dot{\theta}\sin\psi = 0, \tag{4.118}$$

and

$$\psi \equiv \psi(t) = \int_t \omega(a)dt + \theta(t), \tag{4.119}$$

where $a \equiv a(t)$, $\psi \equiv \psi(t)$, $\theta \equiv \theta(t)$ and $\omega \equiv \omega(a)$. Substituting the time derivative of (4.117) and the solution (4.116) into (4.1) we obtain

$$\dot{a}\omega\sin\psi + a\dot{a}\omega'\sin\psi + a\omega\dot{\theta}\cos\psi = -\varepsilon f(a\cos\psi, -a\omega\sin\psi), \tag{4.120}$$

where $\omega' \equiv d\omega/da$ as the frequency $\omega(a)$ depends on the amplitude. The transformation of (4.118) and (4.120) gives the two first order differential equations corresponding to (4.1)

$$\dot{a}(\omega + a\omega'\sin^2\psi) = -\varepsilon f(a\cos\psi, -a\omega\sin\psi)\sin\psi, \tag{4.121}$$

$$a\omega\dot{\theta} + a\dot{a}\omega'\sin\psi\cos\psi = -\varepsilon f(a\cos\psi, -a\omega\sin\psi)\cos\psi, \tag{4.122}$$

i.e., using (3.106)

$$\dot{a}(1+\frac{\alpha-1}{2}\sin^2\psi)=-\frac{1}{q\,|a|^{(\alpha-1)/2}}\varepsilon f(a\cos\psi,-q\,|a|^{(\alpha+1)/2}\sin\psi)\sin\psi, \tag{4.123}$$

$$a\dot{\theta}+\dot{a}\frac{\alpha-1}{4}\sin 2\psi=-\frac{1}{q\,|a|^{(\alpha-1)/2}}\varepsilon f(a\cos\psi,-aq\,|a|^{(\alpha-1)/2}\sin\psi)\cos\psi, \tag{4.124}$$

where

$$\omega\equiv\omega(a)=q\,|a|^{(\alpha-1)/2}, \qquad q=\frac{\sqrt{(\alpha+1)c_\alpha^2}}{\sqrt{2}}\frac{\sqrt{\pi}\Gamma(\frac{3+\alpha}{2(\alpha+1)})}{\Gamma(\frac{1}{\alpha+1})}. \tag{4.125}$$

Averaging the Eqs. (4.123) and (4.124) over the period 2π, yields

$$\dot{a}=-\frac{2\varepsilon}{\pi(\alpha+3)\omega(a)}\int_0^{2\pi}f(a\cos\psi,-a\omega(a)\sin\psi)\sin\psi d\psi, \tag{4.126}$$

$$\dot{\psi}=\omega(a)-\frac{\varepsilon}{2\pi a\omega(a)}\int_0^{2\pi}f(a\cos\psi,-a\omega(a)\sin\psi)\cos\psi d\psi. \tag{4.127}$$

The differential equations (4.126) and (4.127) are solved for the initial values

$$a(0)=A, \qquad \psi(0)=0. \tag{4.128}$$

The averaged approximate solutions depend on the type of the additional function εf. Let us consider the oscillator with small linear viscous damping.

Oscillator with Small Linear Damping

The differential equation of the oscillator with linear damping is

$$\ddot{x}+c_\alpha^2 x\,|x|^{\alpha-1}=-\varepsilon\dot{x}, \tag{4.129}$$

where $\varepsilon << 1$ is the damping coefficient. The corresponding averaged first order Eqs. (4.126) and (4.127) become

$$\dot{a}=-\frac{2\varepsilon a}{\pi(\alpha+3)}\int_0^{2\pi}\sin^2\psi d\psi=-\frac{2\varepsilon a}{\alpha+3}, \tag{4.130}$$

$$\dot{\psi}=q\,|a|^{(\alpha-1)/2}-\frac{\varepsilon}{4\pi}\int_0^{2\pi}\sin 2\psi d\psi=q\,|a|^{(\alpha-1)/2}. \tag{4.131}$$

Integrating (4.130) and (4.131) for the initial values (4.128) we obtain

$$a = A\exp\left(-\frac{2\varepsilon t}{\alpha+3}\right), \quad \psi = -\frac{\alpha+3}{\varepsilon(\alpha-1)} q\,|A|^{(\alpha-1)/2}\left[\exp\left(-\frac{\varepsilon(\alpha-1)t}{\alpha+3}\right)-1\right], \tag{4.132}$$

and the approximate solution of (4.129) yields

$$x = A\exp(-\frac{2\varepsilon t}{\alpha+3})\cos\left(\frac{\alpha+3}{\varepsilon(\alpha-1)} q\,|A|^{(\alpha-1)/2}\left[1-\exp\left(-\frac{\varepsilon(\alpha-1)t}{\alpha+3}\right)\right]\right). \tag{4.133}$$

Analyzing (4.133), it can be concluded:

1. The fraction deflection function has the significant influence on the amplitude decrease: the higher the values of parameter α, the slower the amplitude decreases. For $\alpha = 1$ the amplitude-time decrease is $A\exp(-\frac{\varepsilon t}{2})$ and for $\alpha \to \infty$ the amplitude tends to the initial constant value A independently on the value of the small parameter ε.

2. The frequency of vibration of the linear damped oscillator depends on the initial amplitude A, fraction value α and the damping coefficient ε (see (4.132)). For small damping coefficient ε the frequency (4.132) tends to (3.106) and the period of vibration to T_{ex} (3.18). Namely, using the series expansion of exponential function and using only the first two terms the function ψ simplifies to

$$\psi \approx qt\,|A|^{(\alpha-1)/2}. \tag{4.134}$$

3. For $\varepsilon \to 0$, when the damping is omitted, the amplitude (4.132) is $a = A$ and the angle $\psi \to \omega(A)t$ as

$$\lim_{\varepsilon\to 0}\frac{\alpha+3}{\varepsilon(\alpha-1)}\left[\exp\left(-\frac{\varepsilon(\alpha-1)t}{\alpha+3}\right)-1\right] = -t. \tag{4.135}$$

4. For the linear oscillator, when $\alpha = 1$, the angle ψ is

$$\lim_{\alpha\to 1}\frac{\alpha+3}{\varepsilon(\alpha-1)}\left[\exp\left(-\frac{\varepsilon(\alpha-1)t}{\alpha+3}\right)-1\right] = -t, \tag{4.136}$$

and the solution in the first approximation becomes

$$x = A\exp\left(-\frac{\varepsilon t}{2}\right)\cos(\omega t). \tag{4.137}$$

Comparing (4.137) with the exact solution for the linear damped system, it is obvious that the approximation (4.137) is valid only for $\varepsilon \ll 1$ when the damping influence on the period of vibration can be neglected.

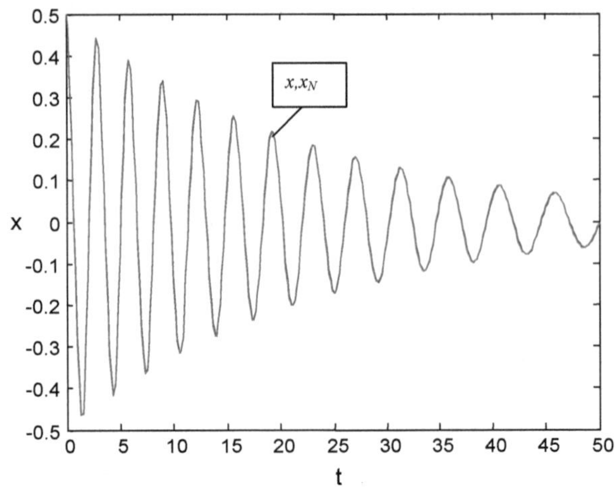

Fig. 4.7 Analytical, $x - t$, and numerical, $x_N - t$, solutions for $\alpha = 5/3$

5. For $\alpha = \frac{5}{3}$, the first order approximation is

$$x = A \exp\left(-\frac{3\varepsilon t}{7}\right) \cos\left(\frac{6.5857}{\varepsilon} q \left|A\right|^{1/3}\left[1 - \exp\left(-\frac{\varepsilon t}{7}\right)\right]\right). \tag{4.138}$$

In Fig. 4.7 the analytical (4.138) and numerical solution of

$$\ddot{x} + c_{5/3}^2 x \left|x\right|^{2/3} = -\varepsilon \dot{x}, \tag{4.139}$$

for $c_{5/3} = 3$, $\varepsilon = 0.1$, $A = 0.5$ are plotted. In spite of the long time interval the difference between approximate x and the exact numerical solution x_N is negligible.

6. For $\alpha = 2$ and

$$\ddot{x} + c_1^2 x \left|x\right| = -\varepsilon \dot{x}, \tag{4.140}$$

the vibration of the linear damped pure quadratic oscillator is approximately

$$x = A \exp\left(-\frac{2\varepsilon t}{5}\right) \cos\left(\frac{4.5734}{\varepsilon} c_1 \left|A\right|^{1/2}\left[1 - \exp\left(-\frac{\varepsilon t}{5}\right)\right]\right). \tag{4.141}$$

The approximate solution x (4.141) is compared with numerical one x_N for (4.140) and plotted in Fig. 4.8.

The parameters are $c_2 = 3$, $\varepsilon = 0.1$ and the initial amplitude $A = 0.5$. The example shows that the suggested procedure is applicable independently on the fact that the order of nonlinearity is even.

7. Yuste and Bejarano (1990) considered the linear damped pure cubic oscillator ($\alpha = 3$) described with the model

$$\ddot{x} + c_3^2 x \left|x\right|^2 = -\varepsilon \dot{x}, \tag{4.142}$$

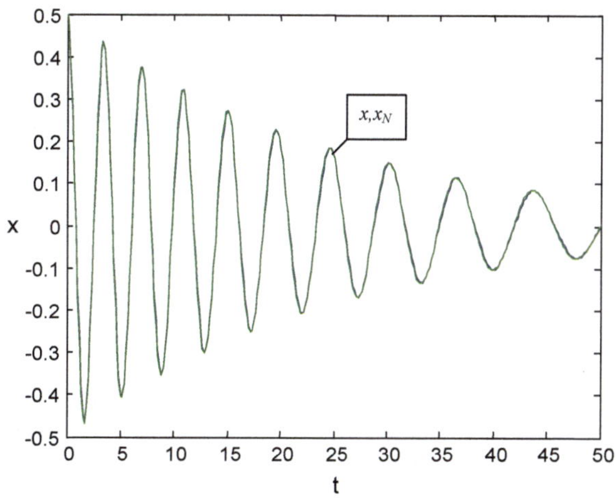

Fig. 4.8 Analytical, $x - t$, and numerical, $x_N - t$, solutions for $\alpha = 2$

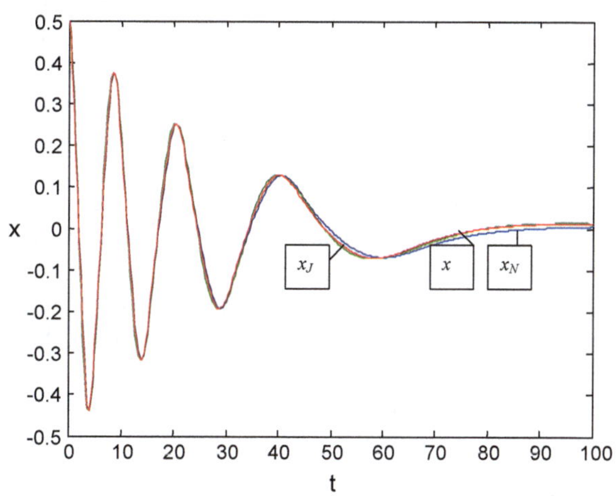

Fig. 4.9 Analytical solutions $x - t$ and $x_J - t$, and the numerical $x_N - t$ solutions for $\alpha = 3$

where ε is a small parameter. Using the solution procedure given in the previous section, the approximate solution in the form of the Jacobi elliptic function is obtained

$$x_J = A\exp\left(-\frac{\varepsilon t}{3}\right) cn\left(\frac{3c_3\,|A|}{\varepsilon}\left[1-\exp\left(-\frac{\varepsilon t}{3}\right)\right], 1/2\right). \tag{4.143}$$

The solution (4.95) of (4.142) in the first approximation is

$$x = A\exp\left(-\frac{\varepsilon t}{3}\right)\cos\left(\frac{2.54163c_3\,|A|}{\varepsilon}\left[1-\exp\left(-\frac{\varepsilon t}{3}\right)\right]\right). \tag{4.144}$$

Comparing the obtained result x (4.144), the solution with Jacobi elliptic function x_J (4.143) and the numeric solution of (4.142), calculated by Runge–Kutta method, for $A = 0.5$, $c_3 = 2$ and $\varepsilon = 0.1$ (see Fig. 4.9), it can be concluded that the both approximations are with high accuracy.

The approximation with the Jacobi elliptic function is more appropriate than the trigonometric solution, but the calculating procedure for trigonometric function is quite easy.

4.2.4 Conclusion

The averaging procedure for solving pure nonlinear ordinary second order differential equation, with additional small terms which are functions of the displacement and velocity, and describe the free vibration of the oscillator, presented in this chapter gives very accurate results in comparison to the numerical ones. The method based on the trial solution in the form of the Ateb periodical function with time variable amplitude and phase gives the best approximation not only of the solution but also of its first time derivative in comparison to the exact ones. The solution in the form of the Jacobi elliptic function with time variable amplitude and frequency and also corrected value of the modulus of the function based on the requirement for the maximal velocity of vibration of the conservative oscillator represents the appropriate solution which is suitable for motion analysis. The averaging procedure seems to be simple, but the parameter calculation is complex. Nevertheless, the analytical solutions show the influence of certain parameters on the vibration properties of the oscillator. Finally, the solution of the equation in the form of a trigonometric function with time variable amplitude and phase reads the most simple model for averaging and discussion, but the accuracy of the results is the smallest.

For the nonlinear conservative oscillator the averaging procedure gives the amplitude of vibration which corresponds to the initial one, and the frequency of vibration depends not only on the initial amplitude but also on the coefficients and orders of the nonlinearity.

4.3 Hamiltonian Approach Solution Procedure

In this section the Hamiltonian approach solution procedure (He 2010) is applied for solving vibration of conservative oscillators with the sum of integer and/or non-integer order strong nonlinearities. Solution of the nonlinear equation is assumed in the form of a trigonometric function with unknown frequency. The characteristic frequency equation is obtained based on the hypothesis that the derivative of the Hamiltonian in amplitude of vibration is zero. The accuracy of the approximate solution is treated with two different approaches: comparing the analytical value for period of vibration with the period obtained numerically and using the error estimation method based on the ratio between the averaged residual function and the total constant energy of system (Navarro and Cveticanin 2016). Two examples are treated: an oscillator with a strong nonlinear term and an oscillator where the nonlinearity is of polynomial type.

4.3.1 Approximate Frequency of Vibration

The oscillator whose nonlinearity is of polynomial type with orders which are rational numbers is

$$\ddot{x} + \sum_{k=1}^{p} c_k^2 x \left|x\right|^{\alpha_k - 1} = 0, \tag{4.145}$$

with initial conditions

$$x(0) = A, \; \dot{x}(0) = 0, \tag{4.146}$$

where c_k^2 are positive small or large constants and the order of nonlinearities α_k are rational numbers (integers or non-integers).

Equation (4.145) describes a conservative oscillator with kinetic energy

$$K = \frac{\dot{x}^2}{2}, \tag{4.147}$$

and potential energy

$$F(x) = \sum_{k=1}^{p} \frac{c_k^2}{\alpha_k + 1} x^{\alpha_k + 1}. \tag{4.148}$$

The total energy of the oscillator corresponds to its Hamiltonian

$$H = \frac{1}{2}\dot{x}^2 + F(x) = \frac{1}{2}\dot{x}^2 + \sum_{k=1}^{p} \frac{c_k^2}{\alpha_k + 1} x^{\alpha_k + 1}. \tag{4.149}$$

For the conservative oscillator the total energy keeps unchanged during the oscillation. It implies that the Hamiltonian is constant, i.e., $H = H_0 = const.$, and

$$\frac{1}{2}\dot{x}^2 + \sum_{k=1}^{p} \frac{c_k^2}{\alpha_k + 1} x^{\alpha_k + 1} - H_0 = 0, \tag{4.150}$$

where due to initial conditions (4.146)

$$H_0 = \sum_{k=1}^{p} \frac{c_k^2}{\alpha_k + 1} A^{\alpha_k + 1}. \tag{4.151}$$

Very often it is impossible to obtain the closed form analytic solution of (4.150). Trial solution is assumed in the form

$$x(t) = A\cos(\omega t), \tag{4.152}$$

where ω is the unknown frequency of vibration and A is the amplitude of the oscillator. Substituting (4.152) into (4.149), we have

$$\tilde{H} = \frac{1}{2}A^2\omega^2\sin^2(\omega t) + \sum_{k=1}^{p}\frac{c_k^2}{\alpha_k+1}A^{\alpha_k+1}\left[\sqrt{\cos^2(\omega t)}\right]^{\alpha_k+1}. \tag{4.153}$$

Usually, the frequency ω is determined from the derivative of (4.150)

$$\frac{\partial \tilde{H}}{\partial A} = 0. \tag{4.154}$$

Unfortunately, the obtained result is far from being accurate. To overcome this problem, He (2010) developed the so called Hamiltonian approach. He introduced a new function

$$\bar{H} = \int_0^{T/4} \tilde{H} dt, \tag{4.155}$$

where T is the period of vibration. After integration it is $\bar{H} = \tilde{H}T/4$, i.e.,

$$\tilde{H} = 4\frac{\partial \bar{H}}{\partial T}. \tag{4.156}$$

According to (4.154) it is

$$\frac{\partial}{\partial A}\left(\frac{\partial \bar{H}}{\partial T}\right) = \frac{\partial}{\partial A}\left(\frac{\partial \bar{H}}{\partial \omega^{-1}}\right) = \frac{\partial}{\partial A}\left(\frac{\partial \bar{H}}{\partial \omega}\right) = 0. \tag{4.157}$$

Solving the algebraic equation (4.157), the approximate frequency of a conservative nonlinear oscillator is obtained. Thus, for

$$\bar{H}(x, \dot{x}) = \int_0^{T/4}\left(\frac{1}{2}\dot{x}^2 + \sum_{k=1}^{p}\frac{c_k^2}{\alpha_k+1}x^{\alpha_k+1}\right)dt, \tag{4.158}$$

and trial function (4.152), we can express $\bar{H}$ as

$$\bar{H}(A,\omega) = \int_0^{T/4}\left\{\frac{1}{2}A^2\omega^2\sin^2(\omega t) + \sum_{k=1}^{p}\frac{c_k^2}{\alpha_k+1}A^{\alpha_k+1}\left|\cos(\omega t)\right|^{\alpha_k+1}\right\}dt. \tag{4.159}$$

Substituting the variable $\psi = \omega t$ and the period of vibration $T = 2\pi/\omega$, we have

$$\bar{H}(A,\omega) = \int_0^{\pi/2} \left\{ \frac{1}{2} A^2 \omega \sin^2(\psi) + \omega^{-1} \sum_{k=1}^{p} \frac{c_k^2}{\alpha_k + 1} A^{\alpha_k+1} \left|\cos(\psi)\right|^{\alpha_k+1} \right\} d\psi. \tag{4.160}$$

Using the integration rule (see Askey and Roy 2010)

$$\int_0^{\pi/2} \sin^{\mu-1}(\psi) \cos^{\nu-1}(\psi) d\psi = \frac{1}{2} \mathrm{B}\left(\frac{\mu}{2}, \frac{\nu}{2}\right), \tag{4.161}$$

where $\Re\mu > 0$, $\Re\nu > 0$ are real positive parts, and B is the Euler beta function, the integrals in (4.160) are

$$\int_0^{\pi/2} \sin^2(\psi) d\psi = \frac{\pi}{4}, \quad \int_0^{\pi/2} \cos^{\alpha_k+1}(\psi) d\psi = \frac{1}{2} \mathrm{B}\left(\frac{1}{2}, \frac{\alpha_k+2}{2}\right). \tag{4.162}$$

Substituting (4.162) into (4.160) it follows

$$\bar{H}(A,\omega) = \frac{1}{2} A^2 \omega \frac{\pi}{4} + \omega^{-1} \sum_{k=1}^{p} \frac{c_k^2}{\alpha_k + 1} A^{\alpha_k+1} \frac{1}{2} \mathrm{B}\left(\frac{1}{2}, \frac{\alpha_k+2}{2}\right). \tag{4.163}$$

Expression (4.157) yields

$$\frac{\partial}{\partial A}\left(\frac{\partial \bar{H}}{\partial \omega}\right) \equiv A \frac{\pi}{4} - \omega^{-2} \sum_{k=1}^{p} c_k^2 A^{\alpha_k} \frac{1}{2} \mathrm{B}\left(\frac{1}{2}, \frac{\alpha_k+2}{2}\right) = 0. \tag{4.164}$$

Solving the algebraic equation (4.164), the approximate frequency of vibration is obtained

$$\omega = \sqrt{\frac{2}{\pi}} \left\{ \sum_{k=1}^{p} c_{\alpha_k}^2 A^{\alpha_k-1} \mathrm{B}\left(\frac{1}{2}, \frac{\alpha_k+2}{2}\right) \right\}^{1/2}. \tag{4.165}$$

Frequency of vibration depends on the coefficients of rigidity of the oscillator $c_{\alpha_k}^2$, on orders of nonlinearity and initial amplitude of vibration.

Using the mathematical relation between beta B and gamma Γ special functions

$$\mathrm{B}(m,n) = \frac{\Gamma(m)\Gamma(n)}{\Gamma(m+n)}, \tag{4.166}$$

and $\Gamma(1/2) = \sqrt{\pi}$, the expression (4.165) for frequency is rewritten as

$$\omega = \frac{\sqrt{2}}{\pi^{1/4}} \left\{ \sum_{k=1}^{p} c_k^2 A^{\alpha_k-1} \frac{\Gamma\left(1 + \frac{\alpha_k}{2}\right)}{\Gamma\left(\frac{\alpha_k+3}{2}\right)} \right\}^{1/2}. \tag{4.167}$$

The approximate solution of the Eq. (4.145) is

$$x(t) = A\cos\left(\frac{\sqrt{2}}{\pi^{1/4}}\left\{\sum_{k=1}^{p} c_k^2 A^{\alpha_k-1}\frac{\Gamma\left(1+\frac{\alpha_k}{2}\right)}{\Gamma\left(\frac{\alpha_k+3}{2}\right)}\right\}^{1/2} t\right), \tag{4.168}$$

and the approximate period of vibration is

$$T = \frac{2\pi}{\omega} = \sqrt{2}\pi^{5/4}\left\{\sum_{k=1}^{p} c_k^2 A^{\alpha_k-1}\frac{\Gamma\left(1+\frac{\alpha_k}{2}\right)}{\Gamma\left(\frac{\alpha_k+3}{2}\right)}\right\}^{-1/2}. \tag{4.169}$$

Comparing the approximate frequency of vibration (4.167) with the result obtained by Cveticanin et al. (2010)

$$\omega = \left\{\frac{\sum_{k=1}^{p} c_k^2 a_k A^{\alpha_k-1}}{a_0}\right\}^{1/2} \tag{4.170}$$

it can be seen that the form is similar.

4.3.2 Error Estimation

Substituting (4.152) into (4.301) we have

$$\tilde{H} = \frac{1}{2}A^2\omega^2\sin^2(\omega t) + \sum_{k=1}^{p}\frac{c_k^2}{\alpha_k+1}A^{\alpha_k+1}\left|\cos(\omega t)\right|^{\alpha_k+1}, \tag{4.171}$$

where the frequency of vibration ω is expressed with (4.167) and $\tilde{H}$ is the Hamiltonian calculated for the approximate solution. It is obvious that, for the approximate solution (4.152), the Hamiltonian (4.171) differs from the total energy H_0. The difference between $\tilde{H}$ and H_0 gives us the residual

$$R(t) = \tilde{H} - H_0 = \frac{1}{2}A^2\omega^2\sin^2(\omega t) + \sum_{k=1}^{p}\frac{c_k^2}{\alpha_k+1}A^{\alpha_k+1}\left|\cos(\omega t)\right|^{\alpha_k+1} - \sum_{k=1}^{p}\frac{c_k^2}{\alpha_k+1}A^{\alpha_k+1}. \tag{4.172}$$

It is clear that the residual varies in time. Relation (4.172) gives us the difference between the calculated energy based on the approximate solution for every special value of time and the total energy of the oscillator.

To determine the average value, the error is estimated as the integral of the square of difference between calculated energy and the total energy, i.e., integral of the square of the residual function over the period of vibration. For estimation the following functional is introduced

$$\Delta_2 = \frac{1}{T}\int_0^T R(t)^2 dt = \frac{1}{T}\int_0^T (\tilde{H} - H_0)^2 dt. \tag{4.173}$$

Substituting (4.172) into (4.173) it is

$$\Delta_2 = \frac{4}{T}\int_0^{T/4}\left[\frac{1}{2}A^2\omega^2\sin^2(\omega t) + \sum_{k=1}^{p}\frac{c_k^2}{\alpha_k+1}A^{\alpha_k+1}\left|\cos(\omega t)\right|^{\alpha_k+1} - \sum_{k=1}^{p}\frac{c_k^2}{\alpha_k+1}A^{\alpha_k+1}\right]^2 dt, \tag{4.174}$$

i.e., for $\psi = \omega t$ and $T = 2\pi/\omega$

$$\Delta_2 = \frac{2}{\pi}\int_0^{\pi/2}\left[\frac{1}{2}A^2\omega^2\sin^2(\psi) + \sum_{k=1}^{p}\frac{c_k^2}{\alpha_k+1}A^{\alpha_k+1}\left|\cos(\psi)\right|^{\alpha_k+1} - \sum_{k=1}^{p}\frac{c_k^2}{\alpha_k+1}A^{\alpha_k+1}\right]^2 d\psi. \tag{4.175}$$

The obtained solution is the function of the initial amplitude, orders of nonlinearity and coefficients of nonlinearity: $\Delta_2 = \Delta_2(A, c_k, \alpha_k)$. Relative error is calculated as the ratio between the square root of the averaged residual function and the initial constant energy function H_0

$$\Delta = \frac{\sqrt{\Delta_2}}{H_0}. \tag{4.176}$$

The relative error is suitable to present in the percent form: $\Delta_{\%} = (\frac{\sqrt{\Delta_2}}{H_0})100\%$.

4.3.3 Comparison Between Approximate and Exact Solutions

One Conservative Term

For the case of one conservative term, we have $p = 1$ and the Eq. (4.145) becomes

$$\ddot{x} + c_1^2 x\left|x\right|^{\alpha_1 - 1} = 0. \tag{4.177}$$

According to (4.168), the approximate solution is

$$x(t) = A\cos(\omega t), \tag{4.178}$$

where

$$\omega^2 = c_1^2 \left|A\right|^{\alpha_1 - 1} K_{\alpha 1}, \tag{4.179}$$

and

$$K_{\alpha 1} = \frac{2}{\pi} B\left(\frac{1}{2}, \frac{\alpha_1 + 2}{2}\right). \tag{4.180}$$

The approximate period of vibration is

$$T = \frac{K_1}{c_1 \left|A\right|^{(\alpha_1 - 1)/2}}, \tag{4.181}$$

where

$$K_1 = \sqrt{2}\pi^{5/4} \sqrt{\frac{\Gamma\left(\frac{\alpha_1+3}{2}\right)}{\Gamma\left(1 + \frac{\alpha_1}{2}\right)}}. \tag{4.182}$$

The exact period of vibration of (4.177), according to Cveticanin (2009), Cveticanin and Pogany (2012), is expressed as

$$T_{ex} = \frac{K_2}{c_1 \left|A\right|^{(\alpha_1 - 1)/2}}, \tag{4.183}$$

where

$$K_2 = \frac{2\sqrt{2\pi}}{\sqrt{\alpha_1 + 1}} \frac{\Gamma\left(\frac{1}{\alpha_1+1}\right)}{\Gamma\left(\frac{3+\alpha_1}{2(\alpha_1+1)}\right)}. \tag{4.184}$$

The parameters K_1 and K_2, for various values of α_1, are calculated by Eqs. (4.182) and (4.184), respectively. In Fig. 4.10, K_1 and K_2 versus the power α_1 are plotted. It

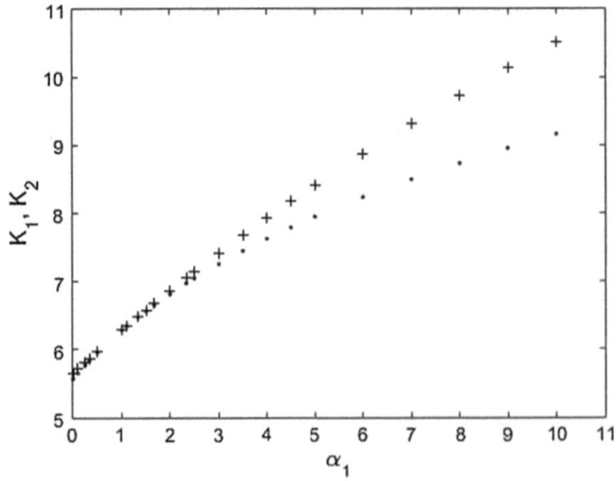

Fig. 4.10 Parameters K_1 $(\cdot)$ and K_2 (+) versus the order α_1

Table 4.1 Values of the parameters K_1, K_2 and relative error (Eq. 4.185) for some values of the α_1

α_1	K_1	K_2	Error %
0	5.56833	5.65685	1.5649
1/10	5.65200	5.72105	1.2069
1/4	5.77138	5.81679	0.7807
1/3	5.83481	5.86966	0.5938
1/2	5.95613	5.97467	0.3103
1	6.28319	6.28319	0.0000
1.1	6.34304	6.34366	0.0098
4/3	6.47666	6.48313	0.0997
3/2	6.56743	6.58136	0.2117
5/3	6.65468	6.67845	0.3560
2	6.81978	6.86926	0.7203
7/3	6.97393	7.05571	1.1591
5/2	7.04739	7.14736	1.3987
3	7.25520	7.41630	2.1723
7/2	7.44676	7.67673	2.9956
4	7.62475	7.92929	3.8407
9/2	7.79118	8.17456	4.6899
5	7.94767	8.41309	5.5321
6	8.23567	8.87180	7.1703
7	8.49642	9.30874	8.7265
8	8.73525	9.72663	10.1925
9	8.95601	10.12768	11.5690
10	9.16160	10.51373	12.8606

can be seen that for $\alpha_1 \leq 3$ there is a good agreement between the parameters. The higher the value of α_1 the difference increases.

For comparison of T_{ex} (4.183) with T (4.181) the percentage error is calculated as

$$\left| \frac{T_{ex} - T}{T_{ex}} \right| \cdot 100\% = \left| 1 - \frac{K_1}{K_2} \right| \cdot 100\%. \tag{4.185}$$

In Table 4.1, values of K_1 and K_2 and also the relative error (4.185) for various values of α_1 are presented.

The error of the approximate period of vibration in comparison to the exact one significantly increases with order of nonlinearity α_1. The conclusion is that the method is suitable up to $\alpha_1 \approx 5$.

Error estimation Hamiltonian for the Eq. (4.177) has the form

$$H = \frac{1}{2}\dot{x}^2 + \frac{1}{\alpha_1 + 1} c_1^2 \left| x \right|^{\alpha_1 + 1}. \tag{4.186}$$

For the initial conditions $x(0) = A$ and $\dot{x}(0) = 0$, the constant of the system is

$$H_0 = \frac{1}{\alpha_1 + 1} c_1^2 A^{\alpha_1+1}. \tag{4.187}$$

Substituting the approximate solution (4.178) into (4.186) and using the relation (4.173), the error function is according to (4.175)

$$\Delta_2 = \frac{2}{\pi} \int_0^{\pi/2} \left[\frac{1}{2} A^2 \omega^2 \sin^2(\psi) + \frac{c_1^2}{\alpha_1 + 1} A^{\alpha_1+1} \left|\cos(\psi)\right|^{\alpha_1+1} - \frac{c_1^2}{\alpha_1 + 1} A^{\alpha_1+1} \right]^2 d\psi. \tag{4.188}$$

After some calculation it is

$$\Delta_2 = c_1^4 A^{2(\alpha_1+1)} K_3, \tag{4.189}$$

where

$$K_3 = \frac{1 - 2C_1 + C_2}{(\alpha_1 + 1)^2} + \frac{K_{\alpha 1}^2 S_4}{4} + \frac{K_{\alpha 1}(C_1 - C_3 - S_2)}{\alpha_1 + 1}, \tag{4.190}$$

with

$$C_1 = \frac{2}{\pi} \int_0^{\pi/2} \cos^{\alpha_1+1}(\psi) d\psi = \frac{1}{\pi} B\left(\frac{1}{2}, \frac{\alpha_1 + 2}{2}\right),$$

$$C_2 = \frac{2}{\pi} \int_0^{\pi/2} \cos^{2(\alpha_1+1)}(\psi) d\psi = \frac{1}{\pi} B\left(\frac{1}{2}, \frac{2\alpha_1 + 3}{2}\right),$$

$$C_3 = \frac{2}{\pi} \int_0^{\pi/2} \cos^{\alpha_1+3}(\psi) d\psi = \frac{1}{\pi} B\left(\frac{1}{2}, \frac{\alpha_1 + 4}{2}\right),$$

$$S_2 = \frac{2}{\pi} \int_0^{\pi/2} \sin^2(\psi) d\psi = \frac{1}{\pi} B\left(\frac{1}{2}, \frac{3}{2}\right) = \frac{1}{2},$$

$$S_4 = \frac{2}{\pi} \int_0^{\pi/2} \sin^4(\psi) d\psi = \frac{1}{\pi} B\left(\frac{1}{2}, \frac{5}{2}\right) = \frac{3}{8}.$$

According to (4.176) and (4.189), the relative error (4.176) is

$$\Delta = \frac{\sqrt{\Delta_2}}{H_0} = (\alpha_1 + 1)\sqrt{K_3}. \tag{4.191}$$

In Table 4.2, the value of relative error Δ for various values of α_1 are shown.

Table 4.2 Values of relative error for various values of α_1

α_1	0	1/2	1	3/2	2	5/2	3	7/2	4	6	8	10
Δ	0.11096	0.05957	0	0.06079	0.12047	0.17823	0.23385	0.28736	0.33886	0.52752	0.69433	0.84502

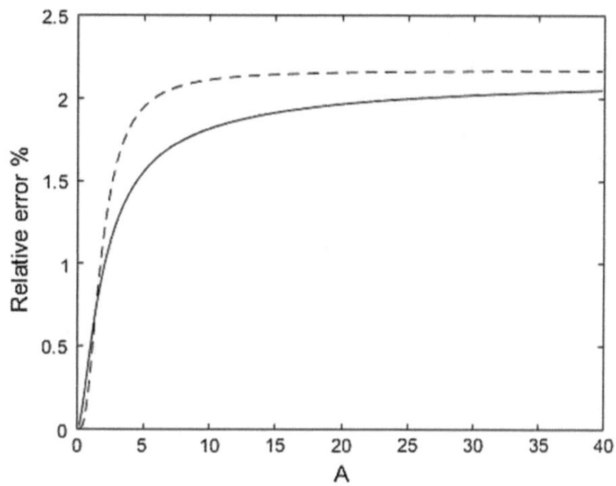

Fig. 4.11 Relative error (%) versus amplitude A for $c_1^2 = c_2^2 = c_3^2 = c_4^2 = 1$ (*solid line*) and $c_1^2 = c_4^2 = 1$, $c_2^2 = c_3^2 = 0$ (*dashed line*)

It is evident that the error is zero for the linear oscillator ($\alpha_1 = 1$) when Hamiltonian is calculated according to the exact solution of the Eq. (4.177). Besides, the error increases for higher and smaller orders of nonlinearity than $\alpha_1 = 1$.

Several Conservative Terms

Let us consider an oscillator described with the equation with non-integer order polynomial nonlinearity

$$\ddot{x} + c_1^2 x + c_2^2 x\,|x|^{2/3} + c_3^2 x\,|x|^{4/3} + c_4^2 x^3 = 0. \tag{4.192}$$

Using the suggested procedure and the expression (4.167), the approximate frequency of vibration is expressed by

$$\omega = \sqrt{c_1^2 + 0.89147 c_2^2\,|A|^{2/3} + 0.81171 c_3^2\,|A|^{4/3} + 0.75 c_4^2 A^2}. \tag{4.193}$$

In this section, we compare the approximate period (4.169) with the numerically obtained one. The 'exact' numerical period of vibration, T_n is calculated using the numerical solution of (4.192) with initial conditions $x(0) = A$, $\dot{x}(0) = 0$ by the Runge–Kutta method. Figure 4.11 illustrates the variation of the relative error in percentage versus the oscillation amplitude for two set of parameters: $c_1^2 = c_2^2 = c_3^2 = c_4^2 = 1$ (solid line) and $c_1^2 = c_4^2 = 1$, $c_2^2 = c_3^2 = 0$ (dashed line).

The vibration period is computed for time interval $0 \leqslant t \leqslant T$ with time–step of $\Delta t = 10^{-6} T$, where T is the approximate period that is a function of the amplitude

A varying in the interval $0 \leqslant A \leqslant 40$ with increment $\Delta A = 0.1$. For $A = 40$, it is observed that the relative errors converge, in both cases of the set of parameters, to small values, i. e., the maximal error for the period of vibration is approximately 2.050% for $c_1^2 = c_2^2 = c_3^2 = c_4^2 = 1$ and the error tends to 2.169% for $c_1^2 = c_4^2 = 1, c_2^2 = c_3^2 = 0$. Figure 4.11 shows that coefficients of rational terms (c_2^2 and c_3^2) in the Eq. (4.192) produce higher relative error for initial amplitude up to $A = 1.3$, approximately . After this point, the cubic nonlinearity order is dominant. Nevertheless, for both sets of parameters the errors are bounded. It can be inferred that the analytically calculated period of vibration presents high accuracy irrespectively of the value of the vibration amplitude. Namely, the error is smaller than 2.2%, even for the amplitude $A = 40$.

Error estimation Hamiltonian for the Eq. (4.192) has the form

$$H = \frac{1}{2}\dot{x}^2 + \frac{1}{2}c_1^2 x^2 + \frac{3}{8}c_2^2 x^2 \left|x\right|^{2/3} + \frac{3}{10}c_3^2 x^2 \left|x\right|^{4/3} + \frac{1}{4}c_4^2 x^4. \tag{4.194}$$

For the initial conditions $x(0) = A$ and $\dot{x}(0) = 0$, the constant of the system is

$$H_0 = \frac{1}{2}c_1^2 A^2 + \frac{3}{8}c_2^2 A^{8/3} + \frac{3}{10}c_3^2 A^{10/3} + \frac{1}{4}c_4^2 A^4. \tag{4.195}$$

Substituting the approximate solution (4.178) into (4.194) and using the relation (4.173), the error function is according to (4.175)

$$\Delta_2 = \frac{2}{\pi}\int_0^{\pi/2} [\frac{1}{2}A^2\omega^2 \sin^2(\psi) + \frac{1}{2}c_1^2 A^2 \cos^2(\psi) + \frac{3}{8}c_2^2 A^{8/3}\cos^2(\psi)\left|\cos(\psi)\right|^{2/3}$$
$$+\frac{3}{10}c_3^2 A^{10/3}\cos^2(\psi)\left|\cos(\psi)\right|^{4/3} + \frac{1}{4}c_4^2 A^4 \cos^4(\psi) - H_0]^2 d\psi. \tag{4.196}$$

i.e.,

$$\Delta_2 = \frac{2}{\pi}\{a_1^2 S_4 + a_2^2 C_4 + a_3^2 C_3 + 2a_1a_2S_3 + 2a_1a_3S_{2/3} + 2a_2a_3C_{4/3}$$
$$+ a_4^2 C_{8/3} + a_5^2 C_8 + H_0^2\frac{\pi}{2} + 2a_4a_5C_{6/3} - 2H_0a_4C_{10} - 2H_0a_5C_4$$
$$+ 2[a_1a_4C_{14} + a_2a_4C_{13} + a_3a_4C_{12}$$
$$+ 2[a_1a_5S_1 + a_2a_5C_6 + a_3a_5C_0]$$
$$- 2[a_1H_0S_2 + a_2H_0C_2 + a_3H_0C_{2/3}]\} \tag{4.197}$$

where

$$a_1 = 0.5A^2c_1^2 + 0.44574c_2^2A^{8/3} + 0.405\,86c_3^2A^{10/3} + 0.375c_4^2A^4,$$

$$a_2 = \frac{1}{2}c_1^2A^2, \quad a_3 = \frac{3}{8}c_2^2A^{8/3}, \quad a_4 = \frac{3}{10}c_3^2A^{10/3}, \quad a_5 = \frac{1}{4}c_4^2A^4,$$

and

$$S_4 = \int_0^{\pi/2} \sin^4(\psi)d\psi, \quad S_{2/3} = \int_0^{\pi/2} \sin^2(\psi)\cos^2(\psi)\,|\cos(\psi)|^{2/3}\,d\psi,$$

$$C_{12} = \int_0^{\pi/2} \cos^4(\psi)\,|\cos(\psi)|^{2/3}\,|\cos(\psi)|^{4/3}\,d\psi,$$

$$C_{13} = \int_0^{\pi/2} \cos^4(\psi)\,|\cos(\psi)|^{4/3}\,d\psi, \quad C_{14} = \int_0^{\pi/2} \sin^2(\psi)\cos^2(\psi)\,|\cos(\psi)|^{4/3}\,d\psi,$$

$$C_4 = \int_0^{\pi/2} \cos^4(\psi)d\psi, \quad C_{4/3} = \int_0^{\pi/2} \cos^4(\psi)\,|\cos(\psi)|^{2/3}\,d\psi,$$

$$C_3 = \int_0^{\pi/2} (\cos^2(\psi)\,|\cos(\psi)|^{2/3})^2 d\psi, \quad C_{8/3} = \int_0^{\pi/2} (\cos^2(\psi)\,|\cos(\psi)|^{4/3})^2 d\psi,$$

$$S_3 = \int_0^{\pi/2} \sin^2(\psi)\cos^2(\psi)d\psi, \quad C_8 = \int_0^{\pi/2} \cos^8(\psi)d\psi,$$

$$S_2 = \int_0^{\pi/2} \sin^2(\psi)d\psi, \quad C_{6/3} = \int_0^{\pi/2} \cos^6(\psi)\,|\cos(\psi)|^{4/3}\,d\psi,$$

$$C_2 = \int_0^{\pi/2} \cos^2(\psi)d\psi, \quad C_{2/3} = \int_0^{\pi/2} \cos^2(\psi)\,|\cos(\psi)|^{2/3}\,d\psi,$$

$$S_1 = \int_0^{\pi/2} \sin^2(\psi)\cos^4(\psi)d\psi, \quad C_6 = \int_0^{\pi/2} \cos^6(\psi)d\psi,$$

$$C_0 = \int_0^{\pi/2} \cos^6(\psi)\,|\cos(\psi)|^{2/3}\,d\psi, \quad C_{10} = \int_0^{\pi/2} \cos^2(\psi)\,|\cos(\psi)|^{4/3}\,d\psi.$$

After some calculation and applying the relations (6.71) and (??), the expression (??) for the error is obtained. In Table 4.3, values of error Δ for $c_1^2 = c_2^2 = c_3^2 = c_4^2 = 1$ and various values of initial amplitude A are shown. It is evident that the error is increasing with the increase of A, but for high values of the amplitude it tends to a constant value $\Delta \approx 0.227$.

Table 4.3 Relative error for $c_1^2 = c_2^2 = c_3^2 = c_4^2 = 1$ and various A

A	0.5	1	5	10	20	50
Δ	0.05739	0.09579	0.18807	0.20813	0.21914	0.22652

Table 4.4 Relative error for $c_1^2 = c_4^2 = 1$, $c_2^2 = c_3^2 = 0$ and various A

A	0.5	1	5	10	20	50
Δ	0.02598	0.07795	0.21653	0.22927	0.23269	0.23367

In Table 4.4, the relative error for $c_1^2 = c_4^2 = 1$, $c_2^2 = c_3^2 = 0$ and various values of A is presented. Increasing the amplitude of vibration, the error tends to the already given constant limit value $\Delta = 0.23385$ (see Table 4.2).

Analyzing the data in Tables 4.3 and 4.4, it can be concluded that for $A > 3$ the error rapidly grows. Besides, for $A < 3$ the accuracy is higher for the Duffing type oscillator than for the oscillator with fraction order nonlinearity. Nevertheless, for both oscillators, the error is bounded even for high initial amplitude.

4.3.4 Conclusions

Based on the obtained results we have concluded the following:

1. In the case of oscillators with one nonlinear conservative term of the order α up to 5.0, the approximate period of vibration gives accurate results when compared with those obtained numerically. For higher values of α the period of vibration diverges.

2. Error in the analytically calculated period of vibration depends on the parameters of the oscillator, order of nonlinearity and initial amplitude. It is interesting to note that for extremely high values of amplitude of vibrations its variation has a negligible influence on the error between asymptotic and exact period of vibration for a certain oscillator. The boundary value of the error depends on the properties of the oscillator if the amplitude of vibration is extremely high.

3. Using a trial solution written as a first order trigonometric function, the average relative error shows that the shape of approximate solution changes very quickly when the order of nonlinearity is higher or smaller than the linear, producing high error values. For the pure nonlinear conservative oscillators the relative error has a limit (maximal) value (see Table 4.2). For the pure quadratic oscillator it is 0.12047 and for the cubic oscillator 0.23385.

4.4 Oscillator with Linear Damping

Let us consider the pure nonlinear oscillator with significant linear damping and an additional small function εf, where $\varepsilon < 1$ is a small parameter, modeled with the Eq. (4.2). Based on the solution (3.115) of the generating solution to (4.2) and the averaging procedure given in this chapter, the approximate solution of the problem is obtained. Differential equation (4.2) represents the perturbed version of (3.111), where the perturbed terms are small in comparison to the right-hand side ones. It gives us the opportunity to assume the solution of (4.2) and its first time derivative in the form (3.115) and (3.117) but with time variable amplitude $a(t)$, frequency $\omega(t)$ and phase $\theta(t)$, i.e.,

$$x = a(t)\exp(-\delta t)ca[\alpha, 1, \psi(t)], \tag{4.198}$$

$$\dot{x} = -\delta a(t)\exp(-\delta t)ca[\alpha, 1, \psi(t)] - \frac{2}{\alpha+1}a(t)\omega(t)\exp(-\delta t)sa[1, \alpha, \psi(t)], \tag{4.199}$$

where

$$\omega(t) = (a(t))^{(\alpha-1)/2}\sqrt{\frac{\alpha+1}{2}c_\alpha^2}\,|\exp(-\delta t)|^{(\alpha-1)/2}, \tag{4.200}$$

$$\psi(t) = \theta(t) + \int \omega(t)dt. \tag{4.201}$$

Comparing (4.199) with the first time derivative of (4.198)

$$\dot{x} = a(t)\exp(-\delta t)ca[\alpha, 1, \psi(t)] - \delta a(t)\exp(-\delta t)ca[\alpha, 1, \psi(t)] - \frac{2}{\alpha+1}a(t)\left[\dot{\theta}(t) + \omega(t)\right](\exp(-\delta t)sa[1, \alpha, \psi(t)], \tag{4.202}$$

it is obvious that the relations (4.199) and (4.202) are equal for

$$\dot{a}ca - \frac{2a\dot{\theta}}{\alpha+1}sa = 0, \tag{4.203}$$

where

$$a \equiv a(t), \quad \theta \equiv \theta(t), \quad \psi \equiv \psi(t), \quad ca[\alpha, 1, \psi] \equiv ca, \quad sa[1, \alpha, \psi] \equiv sa. \tag{4.204}$$

The second time derivative of (4.199) is

$$\ddot{x} = \exp(-\delta t)[(\delta^2 a - \delta\dot{a})ca + \frac{2}{\alpha+1}\delta a(\omega+\dot{\theta})sa$$
$$-\frac{2}{\alpha+1}(\dot{a}\omega + a\dot{\omega} - a\omega\delta)sa - \frac{2}{\alpha+1}a\omega(\omega+\dot{\theta})ca^{\alpha}]. \quad (4.205)$$

Substituting (4.198), (4.199) and (4.205) into (4.2) and using the relations (4.200) and (3.119) we have

$$-\dot{a}(\omega sa + \delta ca) + \frac{2}{\alpha+1}a\dot{\theta}(\delta sa - \omega ca^{\alpha})$$
$$= \exp(\delta t)\varepsilon f(a\exp(-\delta t)ca, -a\exp(-\delta t)\left(\frac{2\omega}{\alpha+1}sa + \delta ca\right). \quad (4.206)$$

Equations (4.203) and (4.206) represent the transformed version of the differential equation (4.2) into a system of two first order differential equations with the new variables A and θ. After some modification and using the equality (see Rosenberg 1963) it is

$$\dot{a} = -\frac{1}{\omega}[\varepsilon f(a\exp(-\delta t)ca, -a\exp(-\delta t)(\frac{2\omega}{\alpha+1}sa + \delta ca)] \quad (4.207)$$
$$sa\exp(\delta t),$$
$$a\dot{\theta} = -\frac{\alpha+1}{2\omega}[\varepsilon f(a\exp(-\delta t)ca, -a\exp(-\delta t)(\frac{2\omega}{\alpha+1}sa + \delta ca)] \quad (4.208)$$
$$ca\exp(\delta t).$$

To solve the differential equations (4.207) and (4.208) is not an easy task. An asymptotic solving procedure has to be introduced. Unfortunately, the period of vibration is time dependent and it increases or decreases dependently on the order of nonlinearity, as it was discussed previously. It requires to average the differential equations (4.207) and (4.208) over the period $\bar{T}$ (3.130). Averaging the equations for the independent variable ψ over the period of the Ateb function we obtain

$$\dot{a} = -\frac{\varepsilon}{\omega}\left\langle\left[f(a\exp(-\delta t)ca, -a\exp(-\delta t)\left(\frac{2\omega}{\alpha+1}sa + \delta ca\right)\right]sa\right\rangle\exp(\delta t),$$
$$a\dot{\theta} = -\varepsilon\frac{\alpha+1}{2\omega}\left\langle\left[f(a\exp(-\delta t)ca, -a\exp(-\delta t)\left(\frac{2\omega}{\alpha+1}sa + \delta ca\right)\right]ca\right\rangle\exp(\delta t),$$

where $\langle ... \rangle = \frac{1}{\Pi}\int_0^{\Pi} ...d\psi$. (The averaging of the Ateb functions is given in the Appendix C).

As the special case, let us consider the pure nonlinear oscillator with linear viscous damping and an additional small nonlinearity of Van der Pol type.

4.4.1 Van der Pol Oscillator

Van der Pol oscillator, which is connected with the pure nonlinear oscillator which has a viscous damping, represents a mechanism which increases the energy, i.e. it is the source of a negative damping. Mathematical model of such an oscillator is

$$\ddot{x} + \kappa \dot{x} + c_\alpha^2 x \, |x|^{\alpha-1} = \varepsilon \dot{x}(1 - |x|^2), \tag{4.209}$$

where κ is the damping coefficient and ε is the coefficient of the Van der Pol nonlinearity. It is of interest to analyze the interaction between the positive κ and negative ε damping and also of the order of nonlinearity α on the motion of the oscillator.

According to the procedure given in the previous section (4.209) is transformed into a system of two first order coupled differential equations

$$\dot{a} = \frac{\varepsilon}{\omega} a \left(\frac{2\omega}{\alpha+1} sa^2 + \delta saca \right) (1 - a^2 \exp(-2\delta t) ca^2), \tag{4.210}$$

$$a\dot{\theta} = \frac{\alpha+1}{2\omega} \varepsilon a \left(\frac{2\omega}{\alpha+1} saca + \delta ca^2 \right) (1 - a^2 \exp(-2\delta t) ca^2). \tag{4.211}$$

The averaged differential equations are

$$\dot{a} = \frac{2\varepsilon a}{\alpha+1} (a_1 - a^2 a_2 \exp(-2\delta t)), \tag{4.212}$$

$$\dot{\theta} = \varepsilon \delta \frac{\alpha+1}{2\omega} a_3 - \varepsilon a^2 \delta \frac{\alpha+1}{2\omega} a_4 \exp(-(2\delta t), \tag{4.213}$$

where

$$a_1 = \langle sa^2 \rangle, \quad a_2 = \langle sa^2 ca^2 \rangle, \quad a_3 = \langle ca^2 \rangle, \quad a_4 = \langle ca^4 \rangle. \tag{4.214}$$

In the Appendix C, the averaged values of the Ateb functions (4.214) are calculated.

Introducing the new variable $u = 1/A^2$ into (4.212), the well known Bernoulli linear differential equation is obtained

$$\dot{u} + \frac{4\varepsilon a_1}{\alpha+1} u = \frac{4\varepsilon a_2}{\alpha+1} \exp(-2\delta t)). \tag{4.215}$$

In general the solution is

$$u \equiv \frac{1}{A^2} = \frac{2\varepsilon a_2}{2\varepsilon a_1 - \delta(\alpha+1)} \exp(-2\delta t) + C_1 \exp\left(-\frac{4\varepsilon a_1}{\alpha+1} t \right), \tag{4.216}$$

where C_1 is the constant of integration. For $a(0) = A$ the constant is $C_1 = \frac{1}{A^2} - \frac{2\varepsilon a_2}{2\varepsilon a_1 - \delta(\alpha+1)}$. Using (4.198) and (4.216), the amplitude–time function follows as

$$a\exp(-\delta t) = \left[\frac{2\varepsilon a_1 - \delta(\alpha+1)}{2\varepsilon a_2 + [2\varepsilon a_1 - \delta(\alpha+1)]C\exp\left(-\frac{2\varepsilon a_1-\delta(\alpha+1)}{\alpha+1}(\beta-1)t\right)}\right]^{1/2}. \tag{4.217}$$

Analyzing the relation (4.217), we have:

1. For $2\varepsilon a_1 - \delta(\alpha+1) \leq 0$, i.e., $\kappa \geq \varepsilon$, the amplitude of vibration of the damped pure nonlinear-Van der Pol oscillator tends to zero.
2. For $2\varepsilon a_1 - \delta(\alpha+1) > 0$, i.e., $\kappa < \varepsilon$, and when t tends to infinity, the amplitude of vibration tends to a constant steady state value

$$A_S = \left[\frac{2\varepsilon a_1 - \delta(\alpha+1)}{2\varepsilon a_2}\right]^{1/2}. \tag{4.218}$$

The motion is the limit cycle one with amplitude (4.218).

Substituting (3.121) and the relation for $a_1 = (\alpha+1)/(\alpha+3)$ shown in Appendix C, the amplitude of limit cycle motion is rewritten into

$$A_S = \sqrt{\frac{a_1}{a_2}}\sqrt{1-\frac{\kappa}{\varepsilon}}. \tag{4.219}$$

The amplitude of the limit cycle motion (4.219) depends on the ratio of the damping parameters κ/ε and on the order of the nonlinearity α expressed with ratio a_1/a_2. The steady state amplitude is independent on the initial amplitude A. Varying the ratio of damping coefficients it is obvious that the amplitude is higher for higher value of ε and smaller value of κ. For the certain damping coefficient ratio κ/ε,the steady state amplitude is higher for smaller values of α than for higher ones (see Figs. C.1 and C.2 in Appendix C).

For $\kappa = 0$ or when ε has extremely high value, i.e., $(\kappa/\varepsilon) = 0$, the boundary value of the amplitude of the limit cycle is $A_{Sb} = \sqrt{a_1/a_2}$. Thus, for the linear oscillator ($\alpha = 1$ with $a_1 = 1/2$ and $a_2 = 1/8$) without positive damping ($\kappa = 0$) the relation (4.219) simplifies into a constant $A_{Sb} = 2$. This result was previously obtained by Nayfeh and Mook (1979), Mickens (2010). For $\alpha > 1$, the boundary amplitude of the limit cycle is smaller than 2 ($A_{Sb} < 2$), and for $\alpha < 1$, it is higher than 2 ($A_{Sb} > 2$) (Fig. 4.12).

For $(\kappa/\varepsilon) = 1$ independently on the order of nonlinearity α, the motion of the system tends to zero.

Examples

1. Let us analyze the pure nonlinear oscillator with nonlinearity order $\alpha = 1/3$, with a linear damping whose coefficient is $\kappa = 0.1$ and with a Van der Pol type damping whose parameter ε is assumed to have various values. The mathematical model of the system is

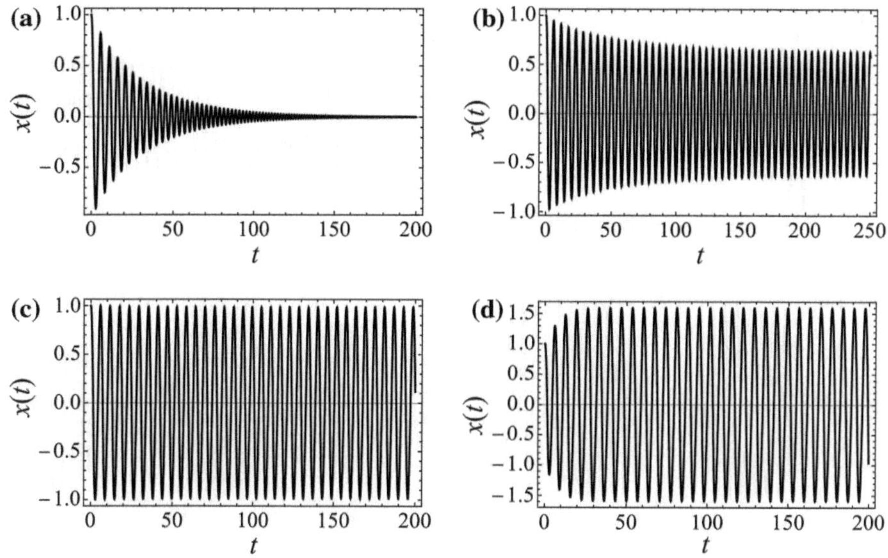

Fig. 4.12 $x - t$ diagrams for $\alpha = 1/3$ and: **a** $\varepsilon = 0.08$, **b** $\varepsilon = 0.13$, **c** $\varepsilon = 0.16$, **d** $\varepsilon = 0.25$

Table 4.5 Approximate and numerical amplitudes of limit cycles for various parameters of the Van der Pol oscillator and for nonlinearity order $\alpha = 1/3$

ε	0.08	0.13000	0.16000	0.2500
A_S	0	0.98708	1.25830	1.5916
A_{Sn}	0	1.00000	1.25000	1.5800

$$\ddot{x} + 0.1\dot{x} + x\,|x|^{-2/3} = \varepsilon\dot{x}(1 - x^2). \tag{4.220}$$

In Fig. 4.10 the $x - t$ curves for $\varepsilon = 0.08$, 0.13, 0.16 and 0.25 are plotted.

It is obvious, that the motion tends to zero for $\varepsilon = 0.08$ (see Fig. 4.10a) and is with limit cycle for other values of ε (see Fig. 4.10b–d). The value of the steady state amplitude is calculated according to (4.219)

$$A_S = 2.054\,8\sqrt{1 - \frac{0.1}{\varepsilon}}, \tag{4.221}$$

where $a_1 = \langle sa^2 \rangle = 0.4$ and $a_2 = \langle sa^2 ca^2 \rangle = 0.094737$ (see Appendix C). In Table 4.5 the approximate A_S limit values (4.221) and the numerically obtained A_{Sn} values by solving (4.220), are shown. It can be seen that the difference between analytical and numerical values is negligible.

Analyzing the relation (4.221) we find that for $\varepsilon = 0.131\,03$ the steady state amplitude is $A_S = 1$. For $\varepsilon \in (0.1, 0.13103)$ the steady state amplitude is $A_S \in (0, 1)$ and for $\varepsilon \in (0.13103, \infty)$, the amplitude of the steady state limit motion is in the interval $A_S \in (1, 2.054\,8)$.

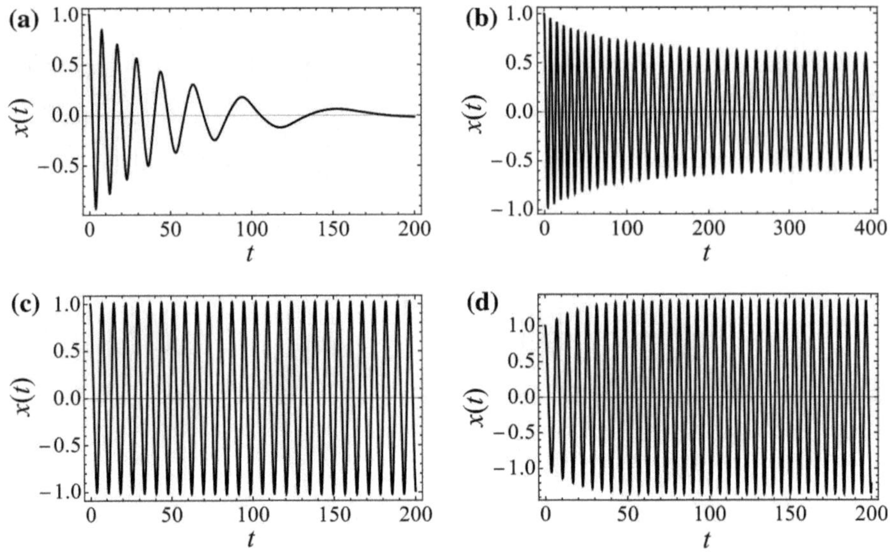

Fig. 4.13 $x - t$ diagrams for $\alpha = 3$ and: **a** $\varepsilon = 0.06$, **b** $\varepsilon = 0.11$, **c** $\varepsilon = 0.14$, **d** $\varepsilon = 0.15$

Table 4.6 Approximate A_S and numerical amplitudes A_{Sn} of limit cycles for various parameters ε of the Van der Pol oscillator and for nonlinearity order $\alpha = 3$

ε	0.06	0.11000	0.1400	0.1500
A_S	0	0.91746	1.0209	1.1026
A_{Sn}	0	0.60000	1.0000	1.1000

2. The mathematical model for the damped Duffing–Van der Pol oscillator is

$$\ddot{x} + 0.1\dot{x} + x^3 = \varepsilon\dot{x}(1 - x^2). \tag{4.222}$$

In Fig. 4.11, we plotted the $x - t$ diagrams for various values of parameter ε (0.06; 0.11; 0.14; 0.15). For $\varepsilon > 0.1$ the limit cycle motion is with amplitude (Fig. 4.13)

$$A_S = 1.9098\sqrt{1 - \frac{0.1}{\varepsilon}}, \tag{4.223}$$

where $a_1 = 0.66666$ and $a_2 = 0.182779$.

In Table 4.6 the steady state amplitudes obtained analytically A_S (4.223) and numerically A_{Sn} by solving the differential equation (4.222) are determined. For $\varepsilon \geq 0.1$ the motion is with a limit cycle, while for $\varepsilon < 0.1$ the positive damping properties are dominant and the motion tends to zero.

For $\varepsilon = 0.137\,77$ the steady state amplitude is $A_S = 1$, while for $\varepsilon \in (0.1, 0.13777)$ the steady state amplitude is $A_S \in (0, 1)$.

For $\varepsilon \in (0.13777, \infty)$, the steady state amplitude is in the interval $A_S \in (1, 1.9098)$.

4.4.2 Conclusion

Due to the previous consideration it is concluded:

1. For the coefficient of the viscous damping equal or higher than the coefficient of the Van der Pol oscillator, the amplitude of vibration decreases and tends to zero.
2. For the coefficient of viscous damping smaller than the coefficient of the Van der Pol oscillator, the motion of the pure nonlinear and damped-Van der Pol oscillator is with limit cycle. The steady state amplitude of vibration depends on the ratio of the damping coefficients and on the order of nonlinearity and is independent on the initial amplitude. The higher the coefficient of negative damping, the higher the steady state amplitude. The higher the order of nonlinearity, the smaller the amplitude of the limit cycle.
3. If the viscous damping is omitted or the coefficient of the Van der Pol oscillator is extremely high (tends to infinity), the steady state amplitude reaches the boundary value: for the linear oscillator it is 2, for the order of nonlinearity smaller than 1 it is the value higher than 2 and for the nonlinearity higher than 1, the steady state amplitude is smaller than 2.
4. Comparing the maximal amplitudes for the limit cycles without viscous damping it is concluded that the higher the order of nonlinearity, the smaller the maximal boundary value.
5. Analytical solving procedure based on the application of the Ateb function with time variable amplitude, frequency and phase gives the accurate results in comparison with numerically obtained ones.

4.5 Oscillators with Odd and Even Quadratic Nonlinearity

In this section two types of oscillators with quadratic nonlinearity are investigated: one, the so-called symmetric, with odd order of nonlinearity, and the other, asymmetric oscillator, with even order of nonlinearity. The mathematical model of the asymmetric oscillator is assumed to be

$$\ddot{x} + c_1 x + c_2 x^2 = 0, \tag{4.224}$$

and the mathematical model of the corresponding symmetric oscillator is then assumed to be

$$\ddot{x} + c_1 x + c_2 x\,|x| = 0, \tag{4.225}$$

where c_1 and c_2 are positive parameters of the linear and of the quadratic term, respectively. Meaning of 'symmetry' and 'asymmetry' in the oscillator is linked with the form of the equation of motion and also solution of the equation of motion (Cveticanin et al. 2016). For the symmetric quadratic oscillator (4.225) the function $F(x) = c_1x + c_2x|x|$ is odd, and that is $F(-x) = -F(x)$. The same relation is satisfied for the harmonic linear oscillator which is named 'symmetric oscillator', too. In symmetric oscillator the amplitude of vibration in positive direction is equal to the value in the negative direction of x. In (4.224) the quadratic term is non-negative independently on the sign of x and the condition of symmetry is not satisfied, that is $F(-x) \neq -F(x)$. As the result the amplitude in positive and negative direction of x-axis differs. The aim of this section is to investigate the resemblance between these two oscillators.

In general, the problem of symmetric and asymmetric oscillator is not a new one. Duffing, the famous German engineer, was dealing with the problem of the dynamics of a mechanism with periodic motion. Due to complex geometric properties of the mechanism, the model which Duffing formed was strongly nonlinear. For computational reasons he introduced some simplification to the restoring force and wrote it in a polynomial form and obtained the 'mixed-parity' type oscillator (Duffing 1918)

$$\ddot{x} + c_1x + c_2x^2 + c_3x^3 = f(x, \dot{x}, t), \tag{4.226}$$

where c_1, c_2 and c_3 are positive constants and $f(x, \dot{x}, t)$ is the function of position, velocity and time. Since that time, a significant number of publications appear dealing with primary resonance, interaction between parametric resonance and subharmonic resonances, internal resonances, 1/2 subharmonic resonances, secondary resonances, principal parametric resonances etc. Recently, He and Daqaq (2014) used this model for investigation of a vibratory energy harvester (VEH) with the aim to improve the broadband transduction capabilities.

Helmholtz was the first to introduce the model of an asymmetric oscillator (4.224) to explain vibrations of a prestressed membrane. The model is appropriate for description of many phenomena in physical and engineering sciences, general relativity, non-linear vibrations, solid-state physics, etc. Micro electromechanical systems (Bhushan et al. 2013) are also modelled with (4.224). These oscillators have numerous applications as ultra-sensitive sensors, switches, and signal-processing elements. The adopted Helmholtz oscillator is applied in optomechanical systems, too (see Isaar and Scheid 2002), where the extension to the theory of quantum mechanics was necessary. Pham et al. (2011) applied the model (4.224) for describing the vibrations of cable-stayed-bridges where nonlinear terms exist due to coupling with modes of the deck or of the pile of the bridge. In the paper of Lai and Chow (2012) the asymmetric quadratic oscillator (4.224) represents the model of a water pump with oscillating water column under the adiabatic compression, which has the practical important in the field of renewable energy. Unfortunately, experimental investigation shows that the obtained results for some of the systems differ from analytically obtained ones due to the fact that the nonlinearity is symmetric. It requires the modification of the model by introducing of the symmetric quadratic nonlinearity. For

$c_2 x\,|x|$, the model is given with (4.225). This model is appropriate to simulate the flight of the bird (Ramananarivo et al. 2011), but also for the wave energy converter (Zurkinden et al. 2014) and many electro- or optomechanical devices which are incorporated in the systems for cooling, squeezing, and quantum-state controlling of massive represent quadratic nonlinear oscillators (Zhang and Kong 2014). Sharma et al. (2012) signify that commercial semiconductor devices like photodiodes, laser diodes, light emitting diodes, single photon counting photodiode etc. are oscillators with symmetric quadratic nonlinearity. The theoretical explanation of the model and the properties of the oscillator are not discussed yet.

The aim of this section is to compare the oscillators with symmetric and asymmetric quadratic nonlinearity. First, the qualitative analysis for the both oscillators is given. Conditions for existence of periodic solutions are determined and calculated. Numerical simulation of vibration is shown. The theoretical result is applied for an optomechanical system.

4.5.1 Qualitative Analysis

Asymmetric Oscillator

Asymmetric oscillator (4.224) is a conservative system with total energy

$$E = \frac{1}{2}\dot{x}^2 + \frac{1}{2}c_1 x^2 + \frac{1}{3}c_2 x^3 = const. \tag{4.227}$$

For initial conditions

$$x(0) = x_0, \quad \dot{x}(0) = 0, \tag{4.228}$$

the energy constant is

$$E = \frac{1}{2}c_1 x_0^2 + \frac{1}{3}c_2 x_0^3 = const. \tag{4.229}$$

Transforming the relation (4.227), the velocity-position expression follows as

$$\dot{x} = \pm\sqrt{2E - x^2\left(c_1 + \frac{2}{3}c_2 x\right)}, \tag{4.230}$$

i.e., using (4.229) it is

$$\dot{x} = \pm\sqrt{c_1(x_0^2 - x^2) + \frac{2}{3}c_2(x_0^3 - x^3)}. \tag{4.231}$$

Potential energy of the system is a cubic position function

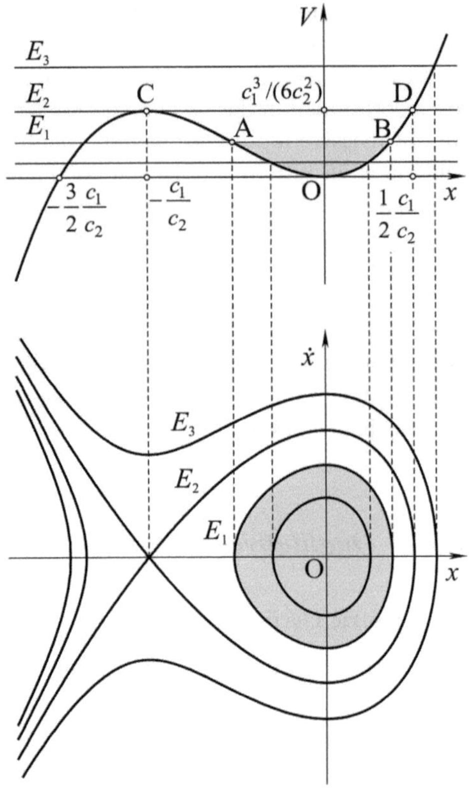

Fig. 4.14 $V-x$ and $\dot{x}-x$ diagrams for various values of the total energy of the asymmetric oscillator

$$V = \frac{1}{2}c_1x^2 + \frac{1}{3}c_2x^3, \tag{4.232}$$

whose first derivative is

$$\frac{dV}{dx} = c_1x + c_2x^2. \tag{4.233}$$

Let us analyze the potential energy function (4.232) and its derivative (4.233). Potential energy is zero for $x = -3c_1/2c_2$ or $x = 0$, while its first derivative is zero for $x = -c_1/c_2$ or $x = 0$. For $x = -c_1/c_2$ the potential energy has the local maximum $V_{\max} = c_1^3/6c_2^2$ and for $x = 0$ the local minimum $V_{\min} = 0$. Using the relation (4.232) it can be checked that the potential energy has the value $V = c_1^3/6c_2^2$ also for $x = c_1/2c_2$.

According to (4.230) and (4.232), the V versus x and the $\dot{x}$ versus x curves for various energy values E are plotted (see Fig. 4.14).

From the figure it can be observed that for the energy values $0 < E < V_{\max}$, the motion around the center $(0, 0)$ is periodical and the curves in the phase plane are of closed form. Thereby, for the initial conditions $\dot{x}(0) = 0$ and

$-3c_1/2c_2 < x_0 < c_1/2c_2$ the motion is periodical. Orbits are asymmetric and surround the center $(0, 0)$. For $E = E_2 = V_{\max}$, the separatrix curve exists which separates the periodic and aperiodic motions.

Symmetric Oscillator

The symmetric oscillator is a conservative system with constant total energy

$$E = \frac{1}{2}\dot{x}^2 + \frac{1}{2}c_1 x^2 + \frac{1}{3}c_2 \left|x\right| x^2 = const. \tag{4.234}$$

where for initial conditions (4.228) we have

$$E = \frac{1}{2}c_1 x_0^2 + \frac{1}{3}c_2 \left|x_0\right| x_0^2 = const. \tag{4.235}$$

Potential energy of the oscillator is a non-negative even function

$$V = x^2 \left(\frac{1}{2}c_1 + \frac{1}{3}c_2 \left|x\right|\right), \tag{4.236}$$

which satisfies the relation $V(-x) = V(x)$. Thereby, the potential energy-position diagram is a symmetric one (see Fig. 4.15).

The V versus x curve is without maximum, but with a minimum for $x = 0$. Thereby, the motion exists for all energy values not small than the potential energy, i.e., $E \geq V$.

Using (4.234) the velocity-position relation is obtained

$$\dot{x} = \pm\sqrt{2E - \left(c_1 + \frac{2}{3}c_2 \left|x\right|\right) x^2}. \tag{4.237}$$

For the energy constant (4.235), it is transformed into

$$\dot{x} = \pm\sqrt{c_1(x_0^2 - x^2) + \frac{2}{3}c_2(\left|x_0\right| x_0^2 - \left|x\right| x^2)}. \tag{4.238}$$

In Fig. 4.15, the V versus x diagram and the $\dot{x}$ versus x velocity-position curves for various values of initial conditions are plotted.

It is seen that curves are closed and the motion is periodical. All curves are central and surround the point (0, 0).

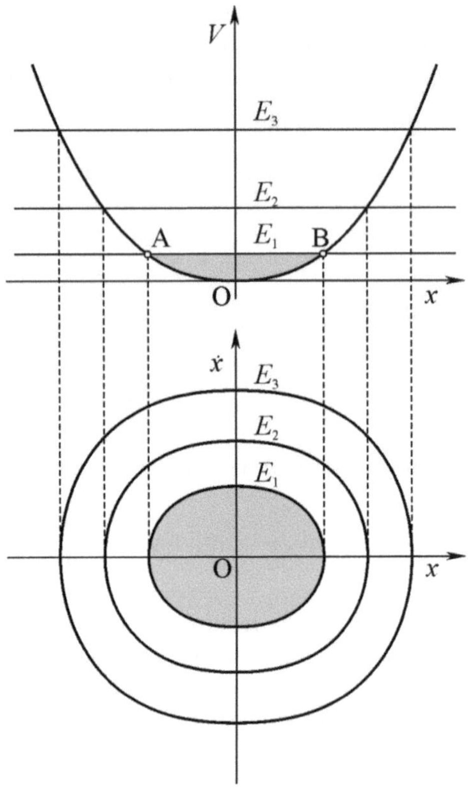

Fig. 4.15 $V - x$ and $\dot{x} - x$ diagrams for various values of the total energy of the symmetric oscillator

4.5.2 Exact Solution for the Asymmetric Oscillator

Chen et al. (1998) express the exact solution of the differential equation (4.224) with initial conditions (4.228) in the form

$$x = B + Acn^2(\omega t, k^2), \tag{4.239}$$

where $cn(\omega t, k^2)$ is the cosine Jacobian elliptic function (Byrd 1954), A, ω and k are amplitude, angular frequency and modulus of the elliptic function, respectively, and B is the so-called bias. The values of A, B, ω and k have to be determined. Using the relations (D.2) and (D.3), the time derivatives of (4.239) are

$$\begin{aligned}\dot{x} &= -2A\omega \,\mathrm{cn}\,\mathrm{sn}\,\mathrm{dn},\\ \ddot{x} &= -2A\omega^2(-\mathrm{sn}^2\,\mathrm{dn}^2 + \mathrm{cn}^2\,\,\mathrm{dn}^2 - k^2\mathrm{sn}^2\,\mathrm{cn}^2)\\ &= -2A\omega^2(-1 + k^2 + 2\mathrm{cn}^2 - 4k^2\,\mathrm{cn}^2 + 3k^2\mathrm{cn}^{\,4}).\end{aligned} \tag{4.240}$$

Substituting (4.239) and (4.240) into (4.224) and separating the terms with the same order of elliptic function cn, three algebraic equations are obtained

$$\begin{aligned} \mathrm{cn}^0 &: \quad 2A\omega^2(k^2-1)+c_1B+c_2B^2=0, \\ \mathrm{cn}^2 &: \quad -4A\omega^2(1-2k^2)+c_1A+2c_2AB=0, \\ \mathrm{cn}^4 &: \quad -6Ak^2\omega^2+c_2B^2=0. \end{aligned} \tag{4.241}$$

Substituting the initial condition (4.228), that is $x(0)=x_0$, into the solution (4.239), it is

$$B+A\mathrm{cn}^2(0,k^2)=x_0.$$

Using the value of the cosine Jacobian elliptic function $cn(0,k^2)=1$, the algebraic equation is obtained

$$B+A=x_0. \tag{4.242}$$

Solving Eqs. (4.241) and (4.242), we have

$$\omega=\sqrt{\frac{c_1+2c_2x_0}{4(k^2+1)}},\quad A=3x_0+3\frac{c_1k^2-2c_2x_0}{2c_2(k^2+1)},\quad B=-2x_0-3\frac{c_1k^2-2c_2x_0}{2c_2(k^2+1)}, \tag{4.243}$$

and

$$k^2=\frac{1}{2}\frac{(c_1+2c_2x_0)^2+2c_1^2}{(c_1+2c_2x_0)^2-c_1^2}\pm\frac{1}{2}\sqrt{(\frac{(c_1+2c_2x_0)^2+2c_1^2}{(c_1+2c_2x_0)^2-c_1^2})^2-4}. \tag{4.244}$$

The coefficients have to be real numbers. This condition is satisfied if the modulus of the Jacobi elliptic function is a real number, i.e.,

$$\left(\frac{(c_1-2c_2x_0)(3c_1+2c_2x_0)}{4c_2x_0(c_1+c_2x_0)}\right)\left(\frac{3(c_1+2c_2x_0)^2}{4c_2x_0(c_1+c_2x_0)}\right)\geq 0, \tag{4.245}$$

and

$$(c_1-2c_2x_0)(3c_1+2c_2x_0)\geq 0. \tag{4.246}$$

Thus, if $c_1=0$ the inequality (4.246) is not satisfied and the modulus is not a real number. It means, that for $c_1=0$ the periodical motion does not exist. It means, that in our calculation the constant c_1 cannot be omitted.

Transforming the inequality (4.246) into the form

$$c_1-2c_2x_0>0,\quad 3c_1+2c_2x_0>0.$$

the conditions for periodic solution are obtained. For $c_1 > 0$, $c_2 > 0$ and $x_0 > 0$ the solution is

$$-\frac{3c_1}{2c_2} < x_0 < \frac{c_1}{2c_2}, \tag{4.247}$$

which is already obtained due to the qualitative analysis of the oscillator. For the initial conditions, when the initial position satisfies the relation (4.247) and the initial velocity is zero, $\dot{x}_0 = 0$, the motion of the asymmetric oscillator is a periodical function

$$x = -\left(2x_0 + 3\frac{c_1k^2 - 2c_2x_0}{2c_2(k^2+1)}\right) + \left(3x_0 + 3\frac{c_1k^2 - 2c_2x_0}{2c_2(k^2+1)}\right)cn^2\left(t\sqrt{\frac{c_1 + 2c_2x_0}{4(k^2+1)}}, k^2\right). \tag{4.248}$$

As to the period of the Jacobi Elliptic Function cn^2 is $2K(k^2)$, where $K(k^2)$ is the total Jacobi elliptic integral of the first kind (Byrd 1954), the period of vibration is calculated as

$$T = \frac{2K(k^2)}{\omega} = 4K(k^2)\sqrt{\frac{k^2+1}{c_1 + 2c_2x_0}}. \tag{4.249}$$

Then, the frequency of vibration of the oscillator is

$$\Omega = \frac{2\pi}{T} = \frac{\pi}{2K(k^2)}\sqrt{\frac{c_1 + 2c_2x_0}{k^2+1}}. \tag{4.250}$$

4.5.3 *Solution for the Symmetric Oscillator*

If the linear term in (4.225) is neglected, i.e., $c_1 = 0$, the equation simplifies into

$$\ddot{x} + c_2\,|x|\,x = 0. \tag{4.251}$$

The exact solution of (4.251) has the form (3.35) (see Sect. 3.2). Due to complexity of *ca* Ateb function the approximate solution for (4.225), based on the Jacobi elliptic function, is developed. The limits for application of the suggested solving procedure are determined.

Approximate Solution of the Pure Quadratic Oscillator

The approximate solution of (4.251) and its time derivative are assumed in the form of the Jacobi elliptic functions (Cveticanin 2014), i.e.,

$$x = A\mathrm{cn}(\omega t, k^2), \tag{4.252}$$

$$\dot{x} = -A\omega\mathrm{sn}(\omega t, k^2)\mathrm{dn}(\omega t, k^2), \tag{4.253}$$

and

$$\ddot{x} = -A\omega^2 \text{cn}(1 - 2k^2 + 2k^2 cn^2), \tag{4.254}$$

where sn, cn and dn are Jacobi elliptic functions (Byrd and Friedman 1954), A is the amplitude of vibration, ω and k are the unknown frequency and modulus parameters of Jacobi functions. The assumed solution (4.252) and its derivatives have approximately to satisfy the Eq. (4.251), i.e.,

$$-A\omega^2 \text{cn}(1 - 2k^2 + 2k^2 \text{cn}^2) + c_2 \left|A \text{ cn}\right| A\text{cn} \approx 0. \tag{4.255}$$

To fulfill this requirement the two unknown parameters ω and k have to be determined based on the exact period of oscillator vibrations and on the exact value of the maximal velocity of oscillation. The period and the maximal velocity of vibration of the oscillator (4.251) will be determined exactly from the energy function and using properties of symmetric vibrations.

Exact Period of Vibration of the Pure Quadratic Oscillator

As it is shown in the previous section, the period of vibration for the asymmetric quadratic oscillator (4.224) is (4.249). Unfortunately, for the symmetric quadratic oscillator (4.225) the exact analytical solution is not known. However, for the Eq. (4.251) the exact period of vibration is given with (3.19)

$$T = \frac{6.8693}{\sqrt{c_2 \left|A\right|}}. \tag{4.256}$$

Frequency of vibration for (4.251) is

$$\Omega = \frac{2\pi}{T} = 0.914\,68\sqrt{c_2 \left|A\right|}. \tag{4.257}$$

Based on (4.256) and using the fact that the period of cosine Jacobi elliptic function ca is $4K(k^2)$, the frequency parameter of the Jacobi elliptic function is obtained

$$\omega = \frac{4K(k^2)}{T} = 0.5823K(k^2)\sqrt{c_2 \left|A\right|}, \tag{4.258}$$

where $K(k^2)$ is the complete elliptic integral of first kind (see Appendix D). The frequency parameter of the Jacobi function (4.258) depends on the amplitude A, coefficient of nonlinearity c_2 and the unknown modulus of the Jacobi elliptic function k^2. The modulus k^2 will be calculated from the condition of the maximal vibration velocity of the oscillator.

Exact Value of the Maximal Velocity of Vibration of the Pure Quadratic Oscillator

Analyzing the energy expression it can be verified that the velocity is maximal if the position is zero, i.e., $x = 0$. Then, the value of the maximal velocity of oscillation of the symmetrical quadratic oscillator is

$$\dot{x}_{\max} = \sqrt{\frac{2c_2}{3}\,|A|^3}. \tag{4.259}$$

For $x = 0$, the solution (4.252) transforms into an expression

$$A\mathrm{cn}(\omega t_V, k^2) = 0, \tag{4.260}$$

which is satisfied for

$$t_V = \frac{(2n-1)K(k^2)}{\omega}, \tag{4.261}$$

where t_V are values of time when the velocity of oscillation has its maximal value for $n = 1, 2, 3, \ldots$ The first time derivative of the approximate solution (4.252) is

$$\dot{x} = -x_0\omega\mathrm{sn}(\omega t, k^2)\mathrm{dn}(\omega t, k^2). \tag{4.262}$$

Substituting (4.261) into (4.262) it is

$$\dot{x}(t_V) = -x_0\omega\mathrm{sn}((2n-1)K(k^2), k^2)\mathrm{dn}((2n-1)K(k^2), k^2). \tag{4.263}$$

Values of these specified Jacobi elliptic functions are (see Appendix D)

$$\mathrm{sn}((2n-1)K(k^2), k^2) = \pm 1, \qquad \mathrm{dn}((2n-1)K(k^2), k^2) = k', \tag{4.264}$$

i.e.,

$$\mathrm{sn}((2n-1)K(k^2), k^2)\mathrm{dn}((2n-1)K(k^2), k^2) = \pm k', \tag{4.265}$$

and $k' = \sqrt{1-k^2}$ is the complementary modulus of the Jacobi elliptic function. Due to (4.263) and (4.265), the maximal value of the first time derivative is for (4.258)

$$\dot{x}_{\max} \equiv \dot{x}(t_V) = x_0\omega k' = 0.5823K(k^2)k'\sqrt{c_2\,|A|^3}. \tag{4.266}$$

Equating the relations (4.259) and (4.266) and using the complementary modulus of the Jacobi elliptic function, we obtain

$$\sqrt{1-k^2}K(k^2) = 1.4022. \tag{4.267}$$

Equation (4.267) is solved numerically. The solution of (4.267) gives the value of the modulus of the Jacobi elliptic function $k^2 = 0.35723$, i.e., $k = 0.597687$, and $K(k^2) = 1.74897$. The modulus is independent on the initial amplitude of vibration and the coefficient of nonlinearity c_2.

Trial Solution

Based on the generating solution of (4.251), an approximate solution procedure for (4.225) is developed. The suggested method is the extended version of Krylov and Bogolubov. The basic idea is to rewrite the second order differential equations of motion into a system of two first order differential equations which are suitable for solving.

The trial solution of (4.225) and its first time derivative are assumed to have the same form as the generating solution (4.252) and its first time derivative (4.253), but the amplitude A and phase ψ are supposed to be time dependent:

$$x = A(t)\text{cn}(\psi(t), k^2), \tag{4.268}$$

$$\dot{x} = -A(t)\omega(t)\text{sn}(\psi(t), k^2)\text{dn}(\psi(t), k^2), \tag{4.269}$$

where

$$\dot{\psi}(t) = \dot{\psi} = \omega(t) + \dot{\theta}(t), \tag{4.270}$$

and

$$\omega(t) = \omega^* \sqrt{c_2} A(t)^{1/2}, \tag{4.271}$$

where $\omega^* = 0.5823K(k^2) = 1.01842$. The main task is to rewrite the Eq. (4.225) into two new variables A and ψ. Let us calculate the time derivative of (4.268)

$$\dot{x} = \dot{A}(t)\text{cn}(\psi(t), k^2) - A(t)(\omega(t) + \dot{\theta}(t))\text{sn}(\psi(t), k^2)\text{dn}(\psi(t), k^2). \tag{4.272}$$

Comparing (4.272) with the assumed relation (4.269) the difference is verified which is as follows

$$\dot{A}cn - A\dot{\theta}\text{sn dn} = 0, \tag{4.273}$$

where

$$\psi = \psi(t), \qquad \omega = \omega(t), \qquad A(t) = A, \qquad \theta(t) = \theta, \tag{4.274}$$

and $\text{cn} = \text{cn}(\psi, k^2)$, $\text{sn} = \text{sn}(\psi, k^2)$ and $\text{dn} = \text{dn}(\psi, k^2)$. Relation (4.273) is a constraint described with a first order differential equation with variables A and θ, i.e., ψ.

Time derivative of (4.269) is

$$\begin{aligned} \ddot{x} = & -\dot{A}\omega\text{sndn} - A\dot{\omega}\text{sn dn} - A\omega^2\text{cn}(1 - 2k^2 + 2k^2\text{cn}^2) \\ & -A\omega\dot{\theta}\text{cn}(1 - 2k^2 + 2k^2\text{cn}^2). \end{aligned} \tag{4.275}$$

According to (4.271)

$$\dot{\omega} = \frac{1}{2}\frac{\omega^*}{A^{1/2}}\sqrt{c_2}\dot{A} = \frac{1}{2}\frac{\omega}{A}\dot{A}.$$

and (4.275) modifies into

$$\ddot{x} = -\frac{3}{2}\dot{A}\omega \mathrm{sndn} - A\omega^2 \mathrm{cn}(1 - 2k^2 + 2k^2\mathrm{cn}^2)$$
$$-A\omega\dot{\theta}\mathrm{cn}(1 - 2k^2 + 2k^2\mathrm{cn}^2). \tag{4.276}$$

Substituting (4.268) and (4.276) into (4.225) we have

$$-\frac{3}{2}\dot{A}\omega \mathrm{sndn} - A\omega(\omega + \dot{\theta})\, \mathrm{cn}(1 - 2k^2 + 2k^2\mathrm{cn}^2)$$
$$+c_2 \left|A\mathrm{cn}\right| A\mathrm{cn} = -\varepsilon c_1 A\mathrm{cn},$$

and applying the assumption (4.255) it is

$$-A\dot{\theta}\omega\mathrm{cn}(1 - 2k^2 + 2k^2\mathrm{cn}^2) - \frac{3}{2}\dot{A}\omega\ \mathrm{sn}\ \mathrm{dn} = -\varepsilon c_1 A\mathrm{cn}. \tag{4.277}$$

Equations (4.273) and (4.277) are the two first order differential equations with two new variables A and ψ, which correspond to the second order differential equation (4.225).

Let us solve Eqs. (4.273) and (4.277) for variables $\dot{A}$ and $\dot{\theta}$. After some modification we obtain

$$\dot{\theta}\omega\left(\frac{3}{2}\mathrm{sn}^2\mathrm{dn}^2 + \mathrm{cn}^2(1 - 2k^2 + 2k^2\mathrm{cn}^2)\right) = \varepsilon c_1 \mathrm{cn}^2, \tag{4.278}$$

$$\dot{A}\omega\left(\frac{3}{2}\mathrm{sn}^2\ \mathrm{dn}^2 + \mathrm{cn}^2(1 - 2k^2 + 2k^2\mathrm{cn}^2)\right) = \varepsilon c_1 A\mathrm{cn}\ \mathrm{sn}\ \mathrm{dn}. \tag{4.279}$$

These equations are coupled and complicated to be analytically solved. This is the reason that the averaging over the period of the Jacobi elliptic function $4K(k^2)$ is suggested. The averaged differential equations (4.278) and (4.279) are

$$\dot{\theta}\omega = \frac{\varepsilon c_1 A}{r}\int_0^{4K(k^2)} \mathrm{cn}^2 d\psi = \frac{\varepsilon c_1}{r} C_2, \tag{4.280}$$

$$\dot{A} = \frac{\varepsilon c_1 A}{r}\int_0^{4K(k^2)} \mathrm{cn}\ \mathrm{sn}\ \mathrm{dn}\ \mathrm{d}\psi = 0, \tag{4.281}$$

where

$$r = \frac{3}{2}(1 - k^2 + C_2(2k^2 - 1) - k^2 C_4) + C_2(1 - 2k^2) + 2k^2 C_4, \tag{4.282}$$

and

$$C_2 = \int_0^{4K(k^2)} \mathrm{cn}^2 d\psi = \frac{1}{k^2}\left(\frac{E}{K} - 1 + k^2\right) \approx \frac{1}{2} + \frac{k^2}{16} + \ldots, \tag{4.283}$$

$$C_4 = \int_0^{4K(k^2)} \mathrm{cn}^4 d\psi = \frac{1}{3k^2}\left[2(2k^2 - 1)C_2 + (1 - k^2)\right] \approx \frac{3}{8} + \frac{k^2}{32} + \ldots. \tag{4.284}$$

$E \equiv E(k^2)$ is the complete elliptic integral of the second kind (Appendix D). Solving the Eq. (4.281) and using the initial conditions (4.228), the constant amplitude of vibration is obtained

$$A = x_0 = const. \tag{4.285}$$

Substituting (4.285) and (4.271) into (4.280) and integrating, the averaged correction of the frequency of the Jacobi elliptic function follows as

$$\dot{\theta} = \frac{\varepsilon c_1}{1.01842\sqrt{c_2 |x_0|}} \frac{C_2}{r} = const., \tag{4.286}$$

which gives the corrected frequency (4.270) of the Jacobi elliptic function

$$\omega = 1.01842\sqrt{c_2 |x_0|} + \frac{\varepsilon c_1}{1.01842\sqrt{c_2 |x_0|}} \frac{C_2}{r}. \tag{4.287}$$

Based on (4.285) and (4.287) the approximate trial solution (??) is

$$x = x_0 \mathrm{cn}\left((1.01842\sqrt{c_2 |x_0|} + \frac{\varepsilon c_1}{1.01842\sqrt{c_2 |x_0|}} \frac{C_2}{r})t, k^2\right). \tag{4.288}$$

Remark Using the results of the qualitative analysis and of the energy function (4.235) for the symmetric quadratic oscillator it is concluded that the value of the first and of the second term depend not only on the coefficients of the linear and quadratic term, but also on the initial position. If the value of the first term for the given initial conditions is smaller than the second term, we conclude that the oscillator is a strong nonlinear one and the solving procedure suggested in this Section has to be applied. If the numerical value of the first term in (4.235) is much higher than the second one, we suggest to analyze the oscillator as a linear one. Namely, according to the energy value (4.235) we calculate an 'energetically equivalent' coefficient

$$c_1^* = \frac{2E}{x_0^2}, \tag{4.289}$$

which is used to 'transform' the nonlinear to a 'linear' oscillator

$$\ddot{x} + c_1^* x = 0. \tag{4.290}$$

The solution is assumed in the form of a cosine function

$$x = x_0 \cos(t\sqrt{c_1^*}), \tag{4.291}$$

where $\sqrt{c_1^*}$ is the approximate frequency of vibration.

4.5.4 Oscillations in an Optomechanical System

Recently, many optomechanical systems, which are incorporated into photonic crystals (Eichenfield et al. 2009), vibrating suspended nanomechanical beams (Peano and Thorwart 2004), optically levitated nanosphere (Chang et al. 2010), phonon induced EIT (Borkje et al. 2013), optomechanical crystals (Eichenfield et al. 2009), ultracold atoms (Purdy et al. 2010), membrane in optomechanical cavity (Sankey et al. 2010), quantum information storage (Safavi-Naeini et al. 2011; Bagheri et al. 2011), cavity Bose-Einstein condensate (Brennecke et al. 2008), ultrafast driven systems (Xiong et al. 2013), cavity optomechanical cooling (Liu et al. 2013), GaAs optomechanical disk (Xiong et al. 2014), optomechanical system with a coherent mechanical pump (Suzuki et al. 2015), semiconductor quantum well (Wang et al. 2016), are investigated. The light interacts with matter and leads to mechanical effects which are suitable to be applied in many practical devices. Systems work on cavity and realize low-energy-consuming information processing. Very often the optomechanical systems are modeled as an oscillator coupled to an optical cavity where the connection is nonlinear. Such model is also applied by Ludwig et al. (2008) for the optomechanical system consisting of a driven optical cavity and a movable mirror attached to a cantilever. Motion of a mechanical oscillator is much slower than that of the cavity mode and the cavity mode can be adiabatically eliminated. Intensity of cavity mode will closely follow the position of the mechanical oscillator. Experimental investigation show that for linear and quadratic mechanic and optic coupling the control of behavior of the system is the most appropriate (Sankey et al. 2010). In that case, when both linear ω_1 and quadratic ω_2 connections are present, the adiabatic dynamic equation reads (Zhang and Kong 2014)

$$\ddot{x} + (1 + 4\omega_2 I)x = -2\omega_1 I, \tag{4.292}$$

where the intracavity field $I(t)$ is given as

$$I(t) \approx 2\left(\frac{\eta}{(\kappa/2)^2 + \delta_c^2}\right)^2 \delta_c(\omega_1 x - \omega_2 x^2), \tag{4.293}$$

where η is the amplitude related to the input optic power, κ is the damping rate and $\delta_c = \omega_0 - \omega_p$ is the cavity-pump detuning which depends on resonant frequency ω_0 and frequency of energy pumping ω_p. Substituting (4.293) into (4.292) and using the assumption that the connection is asymmetric, the following equation is obtained

$$\ddot{x}+\left(1-4\delta_c\omega_1^2\left(\frac{\eta}{(\kappa/2)^2+\delta_c^2}\right)^2\right)x+4\delta_c\omega_1\omega_2\left(\frac{\eta}{(\kappa/2)^2+\delta_c^2}\right)^2x^2=0. \quad (4.294)$$

If the quadratic connection is symmetric, the equation of motion is

$$\ddot{x}+\left(1-4\delta_c\omega_1^2\left(\frac{\eta}{(\kappa/2)^2+\delta_c^2}\right)^2\right)x+4\delta_c\omega_1\omega_2\left(\frac{\eta}{(\kappa/2)^2+\delta_c^2}\right)^2x\,|x|=0. \quad (4.295)$$

Equations (4.294) and (4.295) correspond to (4.224) and (4.225).

1. For numerical values $\omega_1 = 0.03$, $\omega_2 = 0.03333$, $\delta_c = 1$, $\varkappa = 2$, $\eta^2 = 1000$, which are obtained experimentally (see Ludwig et al. 2008), equations are

$$\ddot{x} + 0.1x + x^2 = 0, \quad (4.296)$$

and

$$\ddot{x} + 0.1x + x\,|x| = 0. \quad (4.297)$$

Let us solve them for the initial conditions

$$x(0) = 0.01, \quad \dot{x}(0) = 0. \quad (4.298)$$

Asymmetric Oscillator

As the initial conditions satisfy the relation (4.247), the exact periodic solution of (4.296) is

$$x = -0.010718 + 0.020718\,\mathrm{cn}^2\,(0.16293t, 0.13007)\,. \quad (4.299)$$

The period and frequency of vibration are

$$T = 19.959, \quad \Omega = 0.3148. \quad (4.300)$$

Symmetric Oscillator

For initial values (4.298) the total energy of the system (4.235) has two terms: one, for the linear connection ($E_1 = 5 * 10^{-6}$) and the second, for the quadratic connection ($E_2 = \frac{1}{3} * 10^{-6}$). Using the Remark for the case when the energy of the linear term is higher than of the quadratic term, the 'energetically equivalent' coefficient which corresponds to the linear oscillator is calculated $c_1^* = 0.10667$. The approximate solution of (4.297) is

$$x = 0.01\cos(0.3266t). \quad (4.301)$$

of the asymmetric oscillator are around a position (so-called 'bias') which is not zero. The bias depends on the initial deflection of the system. Besides, the amplitude of vibration also depends on the initial deflection and is different for different values of bias. The higher the value of bias, the smaller the amplitude of vibration.

3. Frequency of vibration of the asymmetric oscillator depends on the coefficient of nonlinearity: the increase of quadratic coefficient affects the decrease of the frequency of the oscillator.

4. Period of vibration of the symmetric oscillator depends on the coefficient of nonlinearity: the increase of quadratic coefficient affects the decrease of the period of vibration of the oscillator.

5. If the symmetric oscillator has a linear and a quadratic term, its vibration properties may be close to those of the linear or of the pure quadratic oscillator. To conclude which solving procedure to apply for solving of the symmetric quadratic oscillator, the so-called 'nonlinearity rate', which is defined as the product of coefficient of quadratic term and initial position, is introduced. If the nonlinearity rate is smaller than the coefficient of the linear term, the motion of the quadratic oscillator is closer to the linear one. Then, the approximation of the solution has to be determined using the linear oscillator with small nonlinearity. If the nonlinearity rate is higher than the coefficient of the linear term, the oscillator is closer to the pure quadratic vibrations, and the calculation of the periodical properties have to be done based on the solution of the pure quadratic oscillator.

6. The approximate solving procedure based on the Jacobi elliptic function which is developed in the paper gives good results. The analytically obtained results are compared with numerically obtained ones and show a good agreement.

7. Comparing the experimental results for the self-excited vibrations of the cantilever in the optomechanical system with those obtained for the symmetric and asymmetric oscillator it is concluded that the appropriate model is that of the symmetric oscillator as it describes the real system more realistic than the asymmetric one: the decrease of the cantilever frequency is by decreasing of the quadratic coefficient

References

Askey, R. A., & Roy, R. (2010). Gamma function. In F. W. J. Olver, D. W. Lozier, R. F. Boisvert, & C. W. Clark (Eds.), *NIST handbook of mathematical functions* (pp. 135–147). New York: Cambridge University Press.

Bagheri, M., Poot, M., Li, M., Pernice, W. P. H., & Tang, H. X. (2011). Dynamic manipulation of nanomechanical resonators in the high-amplitude regime and non-volatile mechanical memory operation. *Nature Nanotechnology*, *6*(11), 726–732.

Belendez, A., Pascual, C., Gallego, S., Ortufio, M., & Neipp, C. (2007). Application of a modified He's homotopy perturbation method to obtain higher-order approximations of an $x^{1/3}$ force nonlinear oscillator. *Physica Letters A*, *371*, 421–426.

Belhaq, M., & Lakrad, F. (2000a). On the elliptic harmonic balance method for mixed parity nonlinear oscillators. *Journal of Sound and Vibration*, *233*, 935–937.

Belhaq, M., & Lakrad, F. (2000b). The elliptic multiple scales method for a class of autonomous strongly non-linear oscillators. *Journal of Sound and Vibration, 234*, 547–553.

Belhaq, M., & Lakrad, F. (2000c). Prediction of homoclinic bifurcation: The elliptic averaging method. *Chaos, Solitons and Fractals, 11*, 2251–2258.

Bogolubov, N. N., & Mitropolski, J. A. (1974). *Asimptoticheskie metodi v teorii nelinejnih kolebanij*. Moscow: Nauka.

Bhushan, A., Inamdar, M. M., & Pawaskar, D. N. (2013). Dynamic analysis of a double-sided actuated MEMS oscillator using second-order averaging. *Lecture Notes in Engineering and Computer Science*, 3 LNECS (pp. 1640–1645).

Borkje, K., Nunnenkamp, A., Teufel, J. D., & Girvin, S. M. (2013). Signatures of nonlinear cavity optomechanics in the weak coupling regime. *Physical Review Letters, 111*(5), 053603.

Brennecke, F., Ritter, S., Donner, T., & Esslinger, T. (2008). Cavity optomechanics with a Bose–Einstein condensate. *Science, 322*, 235–238.

Byrd, P. F. (1954). *Handbook of elliptic integrals for engineers and physicists*. Berlin: Springer.

Byrd, P. F., & Friedman, D. (1954). *Handbook of elliptic integrals for engineers and physicists*. Berlin: Springer.

Chang, D. E., Regal, C. A., Papp, S. B., Wilsonb, D. J., Ye, J., Painter, O., et al. (2010). Cavoty opto-mechanics using an optically levitated nanosphere. *Proceedings of the National Academy of Sciences of the United States of America, 107*(3), 1005–1010.

Chen, C. H., Yang, X. M., & Cheung, Y. K. (1998). Periodic solutions of strongly quadratic nonlinear oscillators by the elliptic perturbation method. *Journal of Sound and Vibration, 212*, 771–780.

Cheung, Y. K., Chen, S. H., & Lau, S. L. (1991). A modified Lindstedt–Poincaré method for certain strongly non-linear oscillators. *International Journal of Non-Linear Mechanics, 26*, 367–378.

Coppola, V. T., & Rand, R. H. (1990). Averaging using elliptic functions: approximation of limit cycles. *Acta Mechanica, 81*, 125–142.

Cveticanin, L. (2000). Vibrations in a parametrically excited system. *Journal of Sound and Vibration, 229*, 245–271.

Cveticanin, L. (2006). Homotopy-perturbation method for pure non-linear differential equation. *Chaos, Solitons and Fractals, 30*, 1221–1230.

Cveticanin, L. (2009). Oscillator with fraction order restoring force. *Journal of Sound and Vibration, 320*, 1064–1077.

Cveticanin, L. (2014). *Strongly nonlinear oscillators—Analytical solutions*. Berlin: Springer.

Cveticanin, L. (2014). An analytical method for truly nonlinear oscillators. In A. B. Gumel (Ed.), *Mathematics of continuous and discrete dynamics systems*, Series Contemporary Mathematics, AMS 618 (pp. 229–245).

Cveticanin, L., & Pogany, T. (2012). Oscillator with a sum of non-integer order non-linearities. *Journal of Applied Mathematics, 2012*(649050), 1–20.

Cveticanin, l., Kalmi-Yazdi, M., Saadatnia, Z., & Askari, H. (2010). Application of Hamiltonian approach to the generalized nonlinear oscillator with fractional power. *International Journal of Nonlinear Sciences and Numerical Simulation, 11*(12), 997–1002.

Cveticanin, L., Zukovic, M., Mester, Gy, Biro, I., & Sarosi, J. (2016). Oscillators with symmetric and asymmetric quadratic nonlinearity. *Acta Mechanica, 227*(6), 1727–1742.

Drogomirecka, H. T. (1997). Integrating a special Ateb-function. *Visnik Lvivskogo Universitetu. Serija mehaniko-matematichna, 46*, 108–110. (in Ukrainian).

Duffing, G. (1918). *Erzwungene Schwingungen bei veranderlicher Eigenfrequenz und ihre technische Bedeutung*. Braunschweig: Vieweg & Sohn.

Eichenfield, M., Chan, J., Camacho, R. M., Vahal, K. J., & Painter, O. (2009). Optomechanical crystals. *Nature, 462*, 78–82.

He, J. H. (1999). Homotopy perturbation technique. *Computer Methods in Applied Mechanics and Engineering, 178*, 257–262.

He, J. H. (2000). A new perturbation technique which is also valid for large parameters. *Journal of Sound and Vibration, 229*, 1257–1263.

He, J. H. (2001a). Modified Lindstedt–Poincaré methods for some strongly non-linear oscillations Part III: Double series expansion. *International Journal of Nonlinear Sciences and Numerical Simulation*, *2*, 317–320.
He, J. H. (2001b). Bookkeeping parameter in perturbation methods. *International Journal of Nonlinear Sciences and Numerical Simulation*, *2*, 257–264.
He, J. H. (2002a). Modified Lindstedt–Poincaré methods for some strongly non-linear oscillations Part I: Expansion of a constant. *International Journal of Non-Linear Mechanics*, *37*, 309–314.
He, J. H. (2002b). Modified Lindstedt–Poincaré methods for some strongly non-linear oscillations Part II: A new transformation. *International Journal of Non-Linear Mechanics*, *37*, 315–320.
He, J. H. (2003a). Determination of limit cycles for strongly nonlinear oscillators. *Physical Review Letters, 90,* Art. No. 174301.
He, J. H. (2003b). Homotopy perturbation method: A new nonlinear analytical technique. *Applied Mathematics and Computation*, *135*, 73–79.
He, J. H. (2004). Asymptology by homotopy perturbation method. *Applied Mathematics and Computation*, *156*, 591–596.
He, J. H. (2010). Haniltonian approach to nonlinear oscillators. *Physics Letters A*, *374*(23), 2312–2314.
He, Q., & Daqaq, M. F. (2014). Influence of potential function asymmetries on the performance of nonlinear energy harvesters under white noise. *Journal of Sound and Vibration*, *333*, 3479–3489.
Hu, H. (2004). A classical perturbation technique which is valid for large parameters. *Journal of Sound and Vibration*, *269*, 409–412.
Isaar, A., & Scheid, W. (2002). Deformed quantum harmonic oscillator with diffusion and dissipation. *Physica A*, *310*, 364–376.
Kamke, A. H. (1959). *Differentialgleichungen—Losungsmethoden und Losungen.* Leipzig: Akademische Verlagsgesellschaft.
Krylov, N., & Bogolubov, N. (1943). *Introduction to Nonlinear Mechanics*. Princeton, NJ: Princeton University Press.
Lai, S. K., & Chow, K. W. (2012). Exact solutions for oscillators with quadratic damping and mixed-parity nonlinearity. *Physica Scripta*, *85*(045006), 1–10.
Lakrad, F., & Belhaq, M. (2002). Periodic solutions of strongly non-linear oscillators by the multiple scales method. *Journal of Sound and Vibration*, *258*, 677–700.
Liao, S. J. (1995). An approximate solution technique not depending on small parameters: A special example. *International Journal of Non-Linear Mechanics*, *30*, 371–380.
Liao, S. J., & Tan, Y. (2007). A general approach to obtain series solutions of nonlinear differential equations. *Studies in Applied Mathematics*, *119*, 297–355.
Lim, C. W., & Wu, B. S. (2002). A modified Mickens procedure for certain non-linear oscillators. *Journal of Sound and Vibration*, *257*, 202–206.
Liu, H. M. (2005). Approximate period of nonlinear oscillators with discontinuities by modified Linstedt–Poincaré method. *Chaos, Solitons & Fractals*, *23*, 577–579.
Liu, Y.-C., Hu, Y.-W., Wong, C.-W., & Xiao, Y.-F. (2013). Review of cavity optomechanical cooling. *Chinese Physics B*, *22*(11), 114213.
Ludwig, M., Kubala, B., & Marquardt, F. (2008). The optomechanical instability in the quantum regime. *New Journal of Physics*, *10*(095013), 1–12.
Margallo, G. J., & Bejarano, J. D. (1987). A generalization of the method of harmonic balance. *Journal of Sound and Vibration*, *116*, 591–595.
Mickens, R. E. (2001). Oscillations in an $x^{4/3}$ potential. *Journal of Sound and Vibration*, *246*, 375–378.
Mickens, R. E. (2002). Analysis of non-linear oscillators having non-polynomial elastic terms. *Journal of Sound and Vibration*, *255*, 789–792.
Mickens, R. E. (2010). *Truly nonlinear oscillations*. Singapore: World Scientific.
Navarro, H. A., & Cveticanin, L. (2016). Amplitude–frequency relationship obtained using Hamiltonian approach for oscillators with sum of non-integer order nonlinearities. *Applied Mathematics and Computation*, *291*, 162–171.

Nayfeh, A. H., & Mook, D. T. (1979). *Nonlinear oscillations*. New York: Wiley.

Peano, V., & Thorwart, M. (2004). Macroscopic quantum effects in a strongly driven nanomechanical resonator. *Physical Review B: Condensed Matter and Materials Physics*, *70*(23), 1–5.

Pham, T. T., Lamarque, C. H., & Pernot, S. (2011). Passive control of one degree of freedom nonlinear quadratic oscillator under combination resonance. *Communications in Nonlinear Science and Numerical Simulations*, *16*, 2279–2288.

Purdy, T. P., Brooks, D. W. C., Botter, T., Brahms, N., Ma, Z.-Y., & Stamper-Kurn, D. M. (2010). Turnable cavity optomechanics with ultracold atoms. *Physical Review Letters*, *105*(13), 133602.

Qaisi, M. I. (1996). A power series approach for the study of periodic motion. *Journal of Sound and Vibration*, *196*, 401–406.

Ramananarivo, S., Godoy-Diana, R., & Thiria, B. (2011). Rather than resonance, flapping wing flyers may play on aerodynamics to improve performance. *Proceedings of the National Academy of Sciences of the United States of America*, *108*(15), 5964–5969.

Rosenberg, R. M. (1963). The Ateb(h)-functions and their properties. *Quarterly of Applied Mathematics*, *21*, 37–47.

Rudowski, J., & Szemplinska-Stupnicka, W. (1987). On an approximate criterion for chaotic motion in a model of a buckled beam. *Ingenieur-Archiv*, *57*, 243–255.

Safavi-Naeini, A. H., Alegre, T. P. M., Chan, J., Chang, D. E., & Painter, O. (2011). Electromagnetically induced transparency and slowlight with optomechanics. *Nature*, *472*(7341), 69–73.

Sankey, J. C., Yang, C., Zwickl, B. M., Jayich, A. M., & Harris, J. G. E. (2010). Strong and tunable nonlineaer optomechanical coupling in a low-loss system. *Nature Physics*, *6*(9), 707–712.

Sarma, M. S., & Rao, B. N. (1997). A rational harmonic balance approximation for the Duffing equation of mixed parity. *Journal of Sound and Vibration*, *207*, 597–599.

Sensoy, S., & Huseyin, K. (1998). On the application of IHB technique to the analysis of non-linear oscillations and bifurcations. *Journal of Sound and Vibration*, *215*, 35–46.

Sharma, A. K., Patidar, R. K., Raghuramaiah, M., Joshi, A. S., Naik, P. A., & Gupta, P. D. (2012). A study on wavelength dependence and dynamic range of the quadratic response of commercial grade light emitting diodes. *Optics Communications*, *285*, 3300–3305.

Suzuki, H., Brown, E., & Sterling, R. (2015). Nonlinear dynamics of an optomechanical system with a coherent mechanical pump: Second order sideband generation. *Physical Review A—Atomic, Molecular and Optical Physics*, *92*(3), 033823.

Wang, G., Fang, Z., & Wu, F. (2016). Control of two-dimensional electron population in semiconductor quantum well. *Physica E: Low-Dimensional System and Nanostructures*, *75*(12136), 241–245.

Wu, B. S., & Lim, C. W. (2004). Large amplitude non-linear oscillations of a general conservative system. *International Journal of Non-Linear Mechanics*, *39*, 859–870.

Xiong, H., Si, L.-G., Lu, X.-Y., Yang, X., & Wu, Y. (2013). Carrier-envelope phase-dependent effect of high-order sideband generation in ultrafast driven optomechanical system in sideband generation. *Optics Letters*, *38*(3), 353–355.

Xiong, H., Si, L.-G., & Lu, X.-Y. (2014). Nanosecond-pulse-controlled higher order side band comb in a GaAs optomechanical disk resonator in the perturbative regime. *Annals of Physics*, *349*, 43–53.

Yuste, S. B., & Bejarano, J. D. (1990). Improvement of a Krylov–Bogoliubov method that use Jacobi elliptic function. *Journal of Sound and Vibration*, *139*, 151–163.

Zhang, L., & Kong, H. Y. (2014). Self-sustained oscillation and harmonic generation in optomechanical systems with quadratic couplings. *Physical Review A*, *89*(023847), 1–12.

Zurkinden, A. S., Ferri, F., Beatty, S., Kofoed, J. P., & Kramer, M. M. (2014). Non-linear numerical modeling and experimental testing of a point absorber wave energy converter. *Ocean Engineering*, *78*, 11–21.

Chapter 5
Oscillators with the Time Variable Parameters

In this chapter the motion of the oscillator with time variable parameters is considered. The parameter variation may be a periodic time function or a continually varying slow time function. The motion of the oscillator caused by the first group of parameters is named parametrically excited vibration, while by the latter type of parameters cause nonstationary vibrations. In this chapter the both types of oscillators will be considered.

The mathematical model of the one-degree-of-freedom oscillator with slow time variable parameter is

$$\ddot{x} + \omega^2(\tau)x\,|x|^{\alpha-1} = \varepsilon f(\tau, x, \dot{x}), \tag{5.1}$$

where α is the order of nonlinearity (a rational number - an integer or non-integer), $\omega^2(\tau)$ is the time variable parameter, $\tau = \varepsilon t$ is the slow time, $\varepsilon \ll 1$ is a small parameter and εf is the additional small nonlinear function. This type of oscillators is widely applied in textile, carpet and cable industry and also in various machines like sieves, separators, centrifuges, etc. (see Cveticanin 1998). In this chapter the averaging solution procedure based on various types of periodical functions with time variable parameters is presented. The method is applied for solving certain vibration problems. As a special case the Van der Pol oscillator with time variable parameter is considered. Some of the presented procedures are already published in Cveticanin (2012, 2013).

In this chapter, the parametrically excited vibration of the pure nonlinear oscillator is also considered. An approximate solution procedure is developed for solving of the second order differential equation with time periodic parameters and negative cubic nonlinearity. The results of investigation on parametric vibrations, given in the paper of Cveticanin and Kovacic (2007), are presented.

L. Cveticanin, *Strong Nonlinear Oscillators*, Mathematical Engineering,
DOI 10.1007/978-3-319-58826-1_5

5.1 Oscillators with Slow Time Variable Parameters

In this section the pure nonlinear oscillator with slow time variable parameters and additional nonlinear terms (5.1) is considered. Equation (5.1) is assumed to represent the perturbed version of the generating Eq. (3.2). Thus, the trial solution of (5.1) has to be close to the generating solution of (3.2). The trial solution is assumed in the same form as the generation solution but the amplitude, frequency and phase angle are the slow time variable functions. The solution is assumed in the form of the Jacobi elliptic function, periodical Ateb function or a trigonometric function. The solution procedure requires averaging over the period of corresponding function. The so obtained approximate averaged solution is applied for the nonlinear oscillator with linear damping and time variable mass where the reactive force is of Levi-Civita type (1928). Besides, the linear oscillator with time variable parameter is considered and the obtained solution is compared with already published one (see Meshcherskij 1952; Leach 1983; Abdalla 1986a, b; Crespo et al. 1990; Xie et al. 1995; Sanchez-Ortiz and Salas-Brito 1995; Cveticanin 1998; Flores et al. 2003; Irschik and Holl 2004) where the Bogolubov and Mitropolski (1963), Krylov and Bogolubov (1943) and its modification (see Cveticanin 1992, 1995) or the multiple scales method (Nayfeh and Mook 1979) is applied.

5.2 Solution in the Form of the Ateb Function

Let us assume the solution of (5.1) and its first time derivative in the forms (4.36) and (4.37), respectively, but with time variable amplitude, $a(t)$, and time variable phase, $\psi(t)$, i.e.,

$$x = a(t)ca(\alpha, 1, \psi(t)), \quad \dot{\psi}(t) = \frac{\sqrt{\alpha+1}}{\sqrt{2}}\omega(\tau)a(t)^{\frac{\alpha-1}{2}} + \dot{\theta}(t) \tag{5.2}$$

and

$$\dot{x} = -\sqrt{\frac{2}{\alpha+1}}\,|\omega(\tau)|\,a(t)^{\frac{\alpha+1}{2}}sa\,(1, \alpha, \psi(t)) \tag{5.3}$$

The time derivative of the approximate solution (5.2)

$$\begin{aligned}\dot{x} = \dot{a}ca\,(\alpha, 1, \psi(t)) - \sqrt{\frac{2}{\alpha+1}}\,|\omega(\tau)|\,a(t)^{\frac{\alpha+1}{2}}sa\,(1, \alpha, \psi(t)) \\ -\frac{2}{\alpha+1}a(t)\dot{\theta}(t)sa\,(1, \alpha, \psi(t)),\end{aligned} \tag{5.4}$$

is compared with (5.3). The relation (5.4) is equal to (5.3), if the following relation is satisfied:

$$\dot{a}ca - \frac{2}{\alpha+1}a\dot{\theta}sa = 0 \tag{5.5}$$

where $a \equiv a(t)$, $\theta \equiv \theta(t)$, $\omega \equiv \omega(\tau)$, $\psi \equiv \psi(t)$, $ca \equiv ca(\alpha, 1, \psi)$ and $sa \equiv sa(1, \alpha, \psi)$. Substituting (5.2), (5.3) and the time derivative of (5.3) into (5.1) it follows

$$\dot{a}sa + \frac{2}{\alpha+1}a\dot{\theta}ca^{\alpha} = -\frac{\varepsilon\omega'}{\omega}\frac{2}{\alpha+1}asa - \sqrt{\frac{2}{\alpha+1}}\frac{1}{\omega a^{\frac{\alpha-1}{2}}}\varepsilon f, \tag{5.6}$$

where $\omega' = d\omega/d\tau$ and $\varepsilon f \equiv \varepsilon f\left(\tau, aca, -\omega\sqrt{\frac{2}{\alpha+1}}a^{\frac{\alpha+1}{2}}sa\right)$.

As we see the second order differential equation (5.1) is replaced with two first order differential equations (5.5) and (5.6). After some modification (5.5) and (5.6) are rewritten as

$$\dot{a} = -\frac{\varepsilon\omega'}{\omega}\frac{2}{\alpha+1}asa^2 - \sqrt{\frac{2}{\alpha+1}}\frac{\varepsilon f}{\omega a^{\frac{\alpha-1}{2}}}sa, \tag{5.7}$$

$$a\dot{\theta} = -\frac{\varepsilon\omega'}{\omega}\frac{2}{\alpha+1}asaca - \sqrt{\frac{\alpha+1}{2}}\frac{\varepsilon f}{\omega a^{\frac{\alpha-1}{2}}}ca. \tag{5.8}$$

Usually, to find the exact solution of the Eqs. (5.7) and (5.8) is very complicated. For simplification, the averaging of Ateb functions over the period $2\Pi_\alpha$ is done. Introducing $\Pi_\alpha = B\left(\frac{1}{\alpha+1}, \frac{1}{2}\right)$ where B is the Beta function, and using (5.7) and (5.8) we obtain

$$\dot{a} = -\frac{\varepsilon a\omega'}{\omega}\left(\frac{2}{\alpha+1}\right)\frac{1}{2\Pi_\alpha}\int_0^{2\Pi_\alpha} sa^2 d\psi - \sqrt{\frac{2}{\alpha+1}}\frac{1}{\omega a^{\frac{\alpha-1}{2}}}\frac{1}{2\Pi_\alpha}\int_0^{2\Pi_\alpha}\varepsilon f sa d\psi \tag{5.9}$$

$$\dot{\theta} = -\frac{\varepsilon\omega'}{\omega}\left(\frac{2}{\alpha+1}\right)\frac{1}{2\Pi_\alpha}\int_0^{2\Pi_\alpha} saca d\psi - \sqrt{\frac{\alpha+1}{2}}\frac{1}{\omega a^{\frac{\alpha-3}{2}}}\frac{1}{2\Pi_\alpha}\int_0^{2\Pi_\alpha}\varepsilon f ca d\psi \tag{5.10}$$

i.e., according to (5.2)

$$\dot{\psi} = \sqrt{\frac{\alpha+1}{2}}\omega a^{\frac{\alpha-1}{2}} - \frac{1}{2\Pi_\alpha}\frac{\varepsilon\omega'}{\omega}\int_0^{2\Pi_\alpha} saca d\psi - \sqrt{\frac{\alpha+1}{2}}\frac{a^{-\frac{\alpha+1}{2}}}{\omega}\frac{1}{2\Pi_\alpha}\int_0^{2\Pi_\alpha}\varepsilon f ca d\psi \tag{5.11}$$

Using the results given by Drogomirecka (1997)

$$\int_0^{2\Pi_\alpha} saca d\psi = 0, \quad \int_0^{2\Pi_\alpha} sa^2 d\psi = 2B\left(\frac{3}{\alpha+1}, \frac{1}{2}\right),$$

we have the averaged differential equations of motion

$$\dot{a} = -\frac{\omega'}{\omega}\left(\frac{2\varepsilon a}{\alpha+1}\right)\frac{1}{\Pi_\alpha}B\left(\frac{3}{\alpha+1},\frac{1}{2}\right) - \sqrt{\frac{2}{\alpha+1}}\frac{1}{\omega a^{\frac{\alpha-1}{2}}}\frac{1}{2\Pi_\alpha}\int_0^{2\Pi_\alpha}\varepsilon f s a d\psi \tag{5.12}$$

$$\dot{\psi} = \sqrt{\frac{\alpha+1}{2}}\omega a^{\frac{\alpha-1}{2}} - \sqrt{\frac{\alpha+1}{2}}\frac{a^{-\frac{\alpha+1}{2}}}{\omega}\frac{1}{2\Pi_\alpha}\int_0^{2\Pi_\alpha}\varepsilon f c a d\psi \tag{5.13}$$

As (5.12) is an uncoupled first order differential equation, it has to be solved first. Substituting the obtained $a(t)$ function into (5.2) and (5.3) and using the initial conditions (3.3) the approximate solution (5.2) of the differential equation (5.1) is determined.

Using the suggested solution procedure some special types of oscillators will be analyzed.

Pure Nonlinear Oscillator with Time Variable Parameter

Let us consider the pure nonlinear oscillator, when the function f in the differential equation (5.1) is zero,

$$\ddot{x} + \omega^2(\tau)x\,|x|^{\alpha-1} = 0 \tag{5.14}$$

If $f = 0$, the amplitude and phase expressions are according to (5.12) and (5.13)

$$a = A\left(\frac{\omega_0}{\omega}\right)^{\beta} \tag{5.15}$$

$$\psi = \psi_0 + \sqrt{\frac{\alpha+1}{2}}\omega_0 A^{\frac{\alpha-1}{2}}\int\left(\frac{\omega}{\omega_0}\right)^{1-\gamma}dt \tag{5.16}$$

where A and ψ_0 are arbitrary constants, $\omega_0 \equiv \omega(0)$ is the constant parameter for $\tau = 0$,

$$\gamma = \beta\left(\frac{\alpha-1}{2}\right) \tag{5.17}$$

and

$$\beta = \frac{1}{\Pi_\alpha}\left(\frac{2}{\alpha+1}\right)B\left(\frac{3}{\alpha+1},\frac{1}{2}\right) \tag{5.18}$$

It can be concluded that the amplitude and phase of vibration depend, not only on ω which is time variable, but also on the order of nonlinearity α. Besides:

1. Analyzing the relation (5.18) it can be seen that the parameter β is positive for all values of α and has a tendency to decrease by increasing of the parameter α: for

$\alpha = 0$ we have $\beta = 16/15$, and for extremely high values of α, β tends to zero. Thereby, the amplitude of vibration (5.15) decreases in time for the case when ω is a monotone increasing time function and increases in time for ω monotone decreasing time function independently on the order of nonlinearity α.

2. The parameter γ, given in (5.17), increases from $(-8/15)$ for $\alpha = 0$ to $\lim_{\alpha\to\infty} \gamma = 1/3$. For $\alpha < 1$ the parameter γ is negative, for $\alpha > 1$ is positive and for $\alpha = 1$ is zero. The influence of the nonlinearity order on the phase properties could not be in general discussed, but only for certain problems.

5.2.1 Oscillator with Linear Time Variable Parameter

Let us consider the special case of the pure nonlinear oscillator (5.14) where ω is a linear time function

$$\omega = \omega_0(1 \pm \varepsilon t) \tag{5.19}$$

The amplitude–time expression is already given as (5.15), but the phase–time relation (5.16) transforms into

$$\psi = \psi_0 \pm \sqrt{\frac{\alpha+1}{2}} A^{\frac{\alpha-1}{2}} \omega_0 \frac{(\omega/\omega_0)^{2-\gamma} - 1}{\varepsilon(2-\gamma)} \tag{5.20}$$

where "+" sign in (5.20) corresponds to the "+" sign in (5.19) and the "−" sign in (5.20) to the "−" sign in (5.19).

For the initial amplitude and phase $A = 1$ and $\psi_0 = 0$ and parameter values $\omega_0 = 1$ and $\varepsilon = 0.01$, the amplitude–time (5.15) and phase–time (5.20) diagrams for various orders of nonlinearity α are plotted (see Figs. 5.1 and 5.2).

Analyzing the curves in Figs. 5.1 and 5.2, it can be concluded:

1. For the case when ω increases the amplitude of vibration decreases in time proportional to α as given in Fig. 5.1a. The phase angle increases in time: the higher α, the increase is faster (Fig. 5.1b).

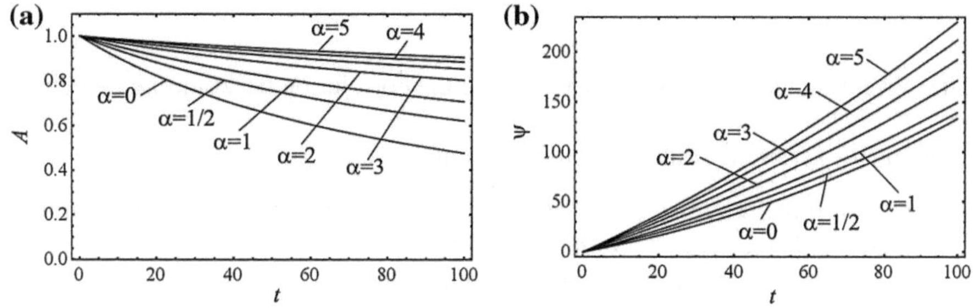

Fig. 5.1 **a** Amplitude–time and **b** phase–time diagrams for various values of α and increasing parameter

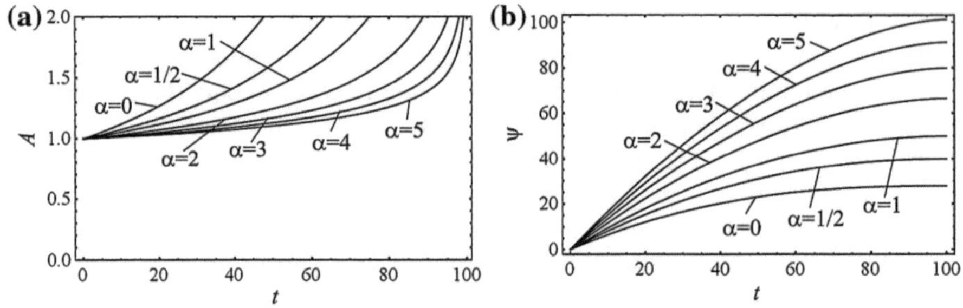

Fig. 5.2 **a** Amplitude–time and **b** phase–time diagrams for various values of α and decreasing parameter

2. If ω decreases with time the corresponding amplitude of vibration increases: the smaller the α, the amplitude increase is faster (Fig. 5.2a). The phase angle increases for the case when ω is a linear decreasing function of time: the increase of phase angle is faster for higher order of nonlinearity α (Fig. 5.2b).

Let us compare the pure nonlinear oscillator ($f = 0$) with time variable parameter (5.14) with the corresponding oscillator with constant parameter (3.2). Comparing the amplitude and phase properties (5.15) and (5.20) for the oscillator with time variable parameter with the relations for the system with constant parameters (3.8) and (3.9), the amplitude and phase angle variations are obtained

$$a = Ak_A \tag{5.21}$$

$$\psi = \psi_0 + k_\psi A^{\frac{\alpha-1}{2}} \omega_0 \sqrt{\frac{\alpha+1}{2}} t \tag{5.22}$$

where the correction terms for the amplitude and phase are

$$k_A = \left(\frac{1}{1 \pm \varepsilon t}\right)^\beta \tag{5.23}$$

$$k_\psi = \frac{(1 \pm \varepsilon t)^{2-\gamma} - 1}{(\pm)\varepsilon t(2-\gamma)} \tag{5.24}$$

The correction terms (5.23) and (5.24) give the correlation between the decrement of parameter variation εt and the order of nonlinearity α.

1. For $\varepsilon t \to 0$ the relations (5.23) and (5.24) are independent on the order of nonlinearity α. The both correction parameters, k_A and $k\psi$, tend to 1 and the motion of the oscillator is near that with constant parameters.

2. According to (5.23), for $1+\varepsilon t \to \infty$ the amplitude correction parameter k_A is zero while, it is one when $\varepsilon t \to 0$. If $1-\varepsilon t \to 0$ the amplitude correction parameter tends to infinity independently on the order of nonlinearity.
3. The phase correction term $k\psi$ depends on the value of (εt). For high value of (εt) the limes of (5.24) is calculated as

$$\lim_{\varepsilon t\to\infty} k_\psi \approx \lim_{\varepsilon t\to\infty} (1 \pm \varepsilon t)^{1-\gamma} \tag{5.25}$$

4. Using the result 2 from the previous section, i.e., $\gamma \in [-8/15, 1/3)$ for $\alpha \in [0, \infty)$, and analyzing the relation (5.25) it can be concluded that for $1+\varepsilon t \to \infty$ the phase correction term k_ψ is independently on the value of α and tends to infinity. For $1-\varepsilon t \to 0$ the phase correction term tends to zero for all values of parameter α: the smaller the parameter α the faster the decrease of k_ψ to zero.

The method suggested in this section is based on the Ateb function (see Appendix A and Cveticanin and Kovacic 2007).

5.3 Solution in the Form of a Trigonometric Function

To avoid the complication connected to Ateb functions, let us suggest the solution of (5.1) in the form of a trigonometric function which is a more appropriate for calculation. The trial solution of (5.1) and its time derivative are assumed in the form of trigonometric functions (3.104) with time variable amplitude, frequency and phase

$$x = a(t)\cos\psi(t) \tag{5.26}$$

$$\dot{x} = -a(t)\Omega(\tau, a(t))\sin\psi(t) \tag{5.27}$$

with

$$\dot{\psi} = \dot{\theta} + \Omega(\tau, a(t)) \tag{5.28}$$

where $a(t)$, $\theta(t)$ and $\psi(t)$ are time variable functions and according to (3.106) and (3.88)

$$\Omega(\tau, a(t)) = \Omega_\alpha \omega(\tau)(a(t))^{\frac{\alpha-1}{2}} \tag{5.29}$$

Calculating the first time derivative of (5.26) and equating it with expression (5.27), it follows

$$\dot{a}(t)\cos\psi(t) - a(t)\dot{\theta}(t)\sin\psi(t) = 0 \tag{5.30}$$

Substituting (5.26), (5.27) and the time derivative of (5.27) into (5.1) and using the relation (5.29), we obtain

$$-\dot{a}\Omega\sin\psi - a\dot{a}\frac{\partial\Omega}{\partial a}\sin\psi - \varepsilon a\frac{\partial\Omega}{\partial\omega}\frac{\partial\omega}{\partial\tau}\sin\psi - a\Omega\dot{\theta}\cos\psi = \varepsilon f(\tau, a\cos\psi, -a\Omega\sin\psi) \tag{5.31}$$

where $a \equiv a(t), \theta \equiv \theta(t), \psi \equiv \psi(t)$. Hence, two first order differential equations (5.30) and (5.31) replace the second order differential equation (5.1). Solving (5.30) and (5.31) with respect to $\dot{a}$ and $\dot{\theta}$, we have

$$\dot{a}(\Omega - a\frac{\partial\Omega}{\partial a}\sin^2\psi) = -\varepsilon a\frac{\partial\Omega}{\partial\omega}\frac{\partial\omega}{\partial\tau}\sin^2\psi - \varepsilon f(\tau, a\cos\psi, -a\Omega\sin\psi)\sin\psi \tag{5.32}$$

$$a\dot{\theta}(\Omega - a\frac{\partial\Omega}{\partial a}\sin^2\psi) = -\varepsilon a\frac{\partial\Omega}{\partial\omega}\frac{\partial\omega}{\partial\tau}\sin\psi\cos\psi - \varepsilon f(\tau, a\cos\psi, -a\Omega\sin\psi)\cos\psi \tag{5.33}$$

Averaging the differential equations in the period 2π we obtain the following equations

$$\dot{a} = -\frac{2\varepsilon a}{(5-\alpha)\omega}\frac{d\omega}{d\tau} - \frac{4\varepsilon a^{\frac{1-\alpha}{2}}}{(5-\alpha)\omega\Omega_\alpha}\frac{1}{2\pi}\int_0^{2\pi} f(\tau, a\cos\psi, -a\Omega\sin\psi)\sin\psi d\psi \tag{5.34}$$

$$a\dot{\theta} = -\frac{2\varepsilon a^{\frac{1-\alpha}{2}}}{(5-\alpha)\omega\Omega_\alpha}\frac{1}{2\pi}\int_0^{2\pi} f(\tau, a\cos\psi, -a\Omega\sin\psi)\cos\psi d\psi \tag{5.35}$$

and from (5.28) and (5.29)

$$\dot{\psi} = \Omega_\alpha\omega a^{\frac{\alpha-1}{2}} - \frac{2\varepsilon a^{\frac{1-\alpha}{2}}}{(5-\alpha)\omega\Omega_\alpha}\frac{1}{2\pi}\int_0^{2\pi} f(\tau, a\cos\psi, -a\Omega\sin\psi)\cos\psi d\psi \tag{5.36}$$

Solving the averaged differential equation (5.34) and substituting the obtained solution for A into (5.36) the approximate function ψ is obtained which gives the solution (5.26).

Remark 3 The averaging procedure is valid for $\alpha \neq 5$, because the expressions (5.34)–(5.36) are singular for $\alpha = 5$.

For $f = 0$ and initial conditions A and ψ_0, the solution of (5.34) is

$$a = A\left(\frac{\omega_0}{\omega}\right)^{\frac{2}{5-\alpha}} \tag{5.37}$$

which gives

$$\psi = \psi_0 + \Omega_\alpha A^{\frac{\alpha-1}{2}}(\omega_0)^{\frac{\alpha-1}{5-\alpha}}\int \omega^{1-\frac{\alpha-1}{5-\alpha}}dt \tag{5.38}$$

The relation (5.37) represents qualitatively the same result as (5.15) obtained by the previous method.

5.3.1 Linear Oscillator with Time Variable Parameters

For the linear oscillator, when $\alpha = 1$, the averaged differential equations (5.34) and (5.36) are

$$\dot{a} = -\frac{1}{2}\frac{\varepsilon a}{\omega}\frac{d\omega}{d\tau} - \frac{\varepsilon}{2\pi\omega}\int_0^{2\pi} f(\tau, a\cos\psi, -a\Omega\sin\psi)\sin\psi d\psi \tag{5.39}$$

$$\dot{\psi}(t) = \omega(\tau) - \frac{\varepsilon}{2\pi\omega a}\int_0^{2\pi} f(\tau, a\cos\psi, -a\Omega\sin\psi)\cos\psi d\psi \tag{5.40}$$

This result is previously published by Bessonov (1967). For $f = 0$ and initial conditions $a(0) = A$ and $\psi(0) = \psi_0$, the solution in the first approximation is

$$x = A\left(\frac{\omega_0}{\omega}\right)^{1/2} \cos\left(\psi_0 + \int \omega dt\right) \tag{5.41}$$

Considering the relations (5.15) and (5.16) for the linear case, when $\alpha = 1$, the approximate solution (5.2) is obtained as

$$x = A\left(\frac{\omega_0}{\omega}\right)^{1/2} ca\left(1, 1, \psi_0 + \int \omega dt\right) \tag{5.42}$$

Due to the fact that the cosine Ateb function for $\alpha = 1$ corresponds to the trigonometric cosine function (see Gricik and Nazarkevich 2007), the relation (5.42) transforms into (5.41). The amplitude of vibration is not constant, but depends on the variation of the parameter ω. The same is valid for the phase variation.

5.3.2 Non-integer Order Nonlinear Oscillator

Let us consider the oscillator

$$\ddot{x} + (1 + 0.01t)^2 x^{4/3} = 0, \tag{5.43}$$

with initial conditions $x(0) = A = 0.1$ and $\dot{x}(0) = 0$.

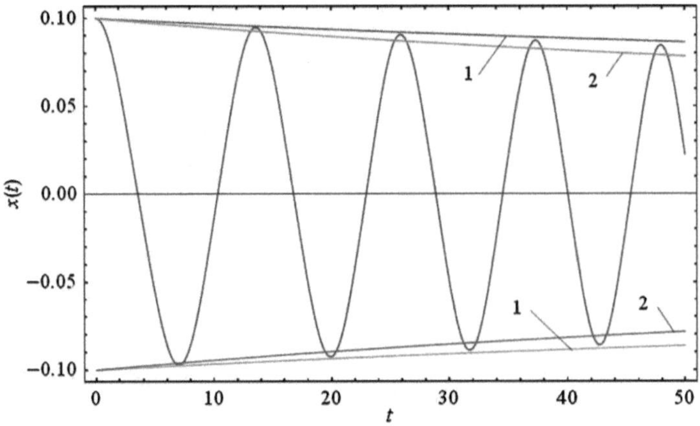

Fig. 5.3 The x–t diagram obtained numerically and a–t diagrams: 1—for (5.44) and 2—for (5.45)

According to the first procedure given in Sect. 5.2.1 (based on the exact solution of the oscillator with constant parameters) the amplitude–time variation (5.15) is

$$a = A\left(\frac{1}{1+0.01t}\right)^{0.42085}. \tag{5.44}$$

and according to (5.37) it is

$$a = A\left(\frac{1}{1+0.01t}\right)^{6/11}. \tag{5.45}$$

In Fig. 5.3 the numerical solution of (5.43) and the amplitude–time diagrams (5.44) and (5.45) are plotted. It can be seen that the first analytical procedure gives a very accurate result for the long time period, but the approximate solution (5.45) obtained by using the second method is not so accurate.

5.3.3 Levi-Civita Oscillator with a Small Damping

One of the mechanical systems with time variable parameters is the mass variable oscillator. If the absolute velocity of mass adding or separating is zero, the mathematical model of the oscillator is in general

$$m(\tau)\ddot{x} + k_\alpha x\,|x|^{\alpha-1} = -\varepsilon\frac{dm(\tau)}{d\tau}\dot{x} - \varepsilon b\dot{x} \tag{5.46}$$

where the first term on the right side of (5.46) represents the reactive force which exists according to mass variation (Levi-Civita 1928) and is the damping force. As the mass variation is slow and the damping coefficient is small, the reactive and damping force are small in comparison to the elastic force.

Comparing the Eqs. (5.46) with (5.1) we have

$$\omega = \sqrt{\frac{k_\alpha}{m}} \tag{5.47}$$

and

$$f = -\left(\frac{b}{m} + \frac{1}{m}\frac{dm}{d\tau}\right)\dot{x} \tag{5.48}$$

Substituting (5.47) and (5.48) into (5.34) and (5.36) the differential equation (5.46) transforms into a system of two averaged first order differential equations

$$\frac{\dot{a}}{a} = -\frac{\varepsilon}{(5-\alpha)m}\left(\frac{dm}{d\tau}\right) - \frac{2\varepsilon b}{(5-\alpha)\,m} \tag{5.49}$$

$$\dot{\psi}(t) = \Omega_\alpha a^{\frac{\alpha-1}{2}}\sqrt{\frac{k_\alpha}{m}} \tag{5.50}$$

In general, the averaged amplitude variation is the solution of (5.49)

$$a = A\left(\frac{m_0}{m}\right)^{\frac{1}{5-\alpha}} \exp\left(-\frac{2\varepsilon b}{5-\alpha}\int\frac{dt}{m}\right) \tag{5.51}$$

which gives the phase angle function

$$\psi = \psi_0 + \Omega_\alpha\sqrt{k_\alpha}\left(Am_0^{\frac{1}{5-\alpha}}\right)^{\frac{\alpha-1}{2}}\int m^{-\frac{\alpha-1}{2(5-\alpha)}-\frac{1}{2}}\left(\exp\left(-\frac{2\varepsilon b}{5-\alpha}\int\frac{dt}{m}\right)\right)^{\frac{\alpha-1}{2}} dt \tag{5.52}$$

The amplitude and the phase of vibration vary in time due to damping, but also due to mass variation. The order of nonlinearity has a significant influence on the velocity of amplitude and phase increase or decrease.

5.3.3.1 Linear Mass Variation

Let us consider the case when the mass variation is linear, as it is suggested by Yuste (1991)

$$m = m_0 + m_1\tau = m_0 + \varepsilon m_1 t \tag{5.53}$$

where m_1 is a constant and ε is a small parameter. According to (5.49), we obtain the differential equation for the amplitude variation

$$\frac{\dot{a}}{a}=-\frac{\varepsilon(2b+m_1)}{(5-\alpha)m} \tag{5.54}$$

(a) For the special parameter values, when $m_1/b=-2$ the amplitude of vibration is constant i.e.,

$$a=A=const. \tag{5.55}$$

and the relation (5.52) transforms into

$$\psi=\psi_0+\frac{2}{\varepsilon m_1}\Omega_\alpha A^{\frac{\alpha-1}{2}}\sqrt{k_\alpha}\left(m^{1/2}-m_0^{1/2}\right) \tag{5.56}$$

For this special case in spite of the action of the linear damping the amplitude of vibration is constant due to the fact that the linear mass separation makes the compensation to the effect of damping. Using the series expansion of the function ψ we have

$$\psi=\psi_0+2\Omega_\alpha A^{\frac{\alpha-1}{2}}\sqrt{\frac{k_\alpha}{m_0}}t \tag{5.57}$$

and the approximate period of vibration is independent on the mass variation and damping, as follows

$$T=\frac{2\pi}{2\Omega_\alpha A^{\frac{\alpha-1}{2}}\sqrt{\frac{k_\alpha}{m_0}}} \tag{5.58}$$

The approximate period value depends only on the order of nonlinearity.

(b) For $m_1/b\neq-2$ the amplitude–time and phase–time functions are

$$a=A\left(\frac{m}{m_0}\right)^{-\frac{1}{5-\alpha}\left(1+\frac{2b}{m_1}\right)} \tag{5.59}$$

and

$$\psi=\frac{\sqrt{m_0k_\alpha}}{\varepsilon m_1}\frac{A^{\frac{\alpha-1}{2}}\Omega_\alpha}{\frac{1}{2}-\frac{\alpha-1}{2(5-\alpha)}\left(1+\frac{2b}{m_1}\right)}\left(\left(\frac{m}{m_0}\right)^{\frac{1}{2}-\frac{\alpha-1}{2(5-\alpha)}\left(1+\frac{2b}{m_1}\right)}-1\right)+\psi_0 \tag{5.60}$$

which give the approximate solution (5.26)

$$x=A\left(\frac{m}{m_0}\right)^{-\frac{m_1+2b}{(5-\alpha)m_1}}\cos\left(\frac{\sqrt{m_0k_\alpha}}{\varepsilon m_1}\frac{A^{\frac{\alpha-1}{2}}\Omega_\alpha\left(\left(\frac{m}{m_0}\right)^{\frac{1}{2}-\frac{(\alpha-1)(m_1+2b)}{2(5-\alpha)m_1}}-1\right)}{\frac{1}{2}-\frac{(\alpha-1)(m_1+2b)}{2(5-\alpha)m_1}}+\psi_0\right) \tag{5.61}$$

The amplitude and phase variation depend on the relation m_1/b, parameter m_1 and order of nonlinearity α.

Let us analyze the influence of the relation m_1/b on the motion of the oscillator. A numerical example is considered. If the order of nonlinearity is $\alpha = 4/3$, the rigidity $k_{4/3} = 1$ and the mass decrease is $m = 1 - 0.01t$, where $m_0 = 1$, $m_1 = 1$ and $\varepsilon = 0.01$, the differential equation of motion for damping coefficient b is

$$\ddot{x} + \frac{x\,|x|^{1/3}}{1 - 0.01t} = 0.01\,(1 - b)\,\dot{x} \tag{5.62}$$

For the initial conditions $x(0) = A = 0.1$ and $\dot{x}(0) = 0$ the analytical solution (5.61) transforms into

$$x = \frac{0.1}{(1 - 0.01t)^{0.27273(1-2b)}} \cos\left(\frac{132.056\left(1 - (1 - 0.01t)^{0.45455+0.0909b}\right)}{0.9091 + 0.1818b}\right) \tag{5.63}$$

In Fig. 5.4 the approximate solution (5.63) and the numerical solution of (5.62), obtained by using of the Runge–Kutta procedure, are plotted. The x–t diagrams for various values of the damping parameter b are shown.

It can be concluded that for $b = 1/2$ the amplitude of vibration is constant as it is previously stated (see (5.52)). For the case when the damping is neglected ($b = 0$),

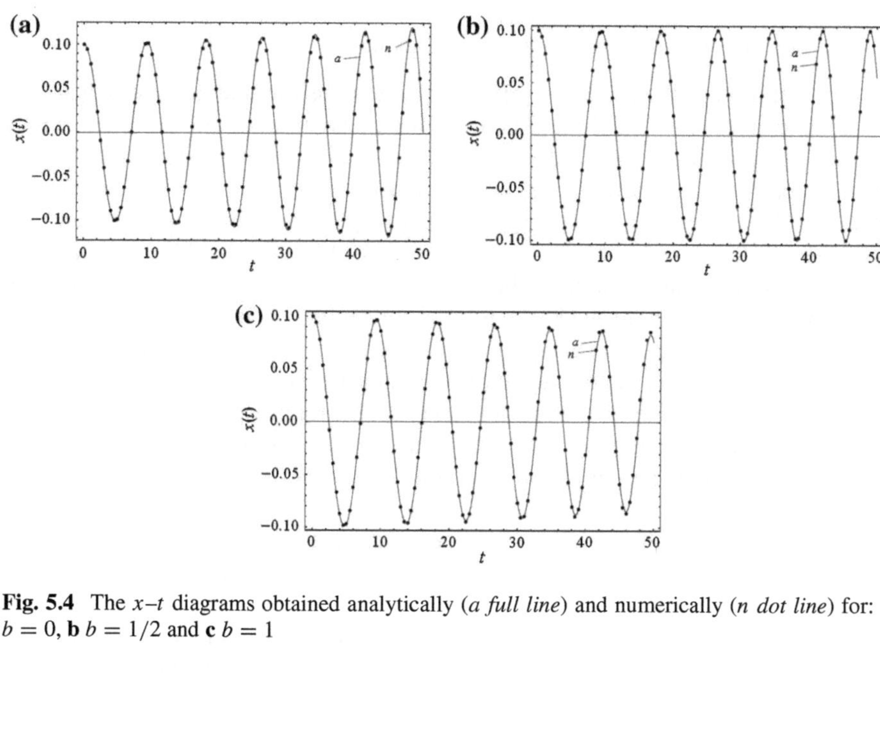

Fig. 5.4 The x–t diagrams obtained analytically (*a full line*) and numerically (*n dot line*) for: **a** $b = 0$, **b** $b = 1/2$ and **c** $b = 1$

due to mass decrease and existence of the reactive force, the amplitude of vibration increases. For certain damping ($b = 1$) which is higher than the limit value ($b = 1/2$) the amplitude of vibration decreases. The analytical solution is in a very good relation to the numeric one in spite of the long time interval of consideration.

5.3.3.2 Linear Oscillator

For the linear oscillator when $\alpha = 1$

$$a = A\left(\frac{m_0}{m}\right)^{\frac{1}{4}(1+\frac{2b}{m_1})} \tag{5.64}$$

and

$$\psi = \frac{2\sqrt{k_1}}{\varepsilon m_1}\left(\sqrt{m} - \sqrt{m_0}\right) + \psi_0 \tag{5.65}$$

If the damping parameter is zero the amplitude variation is $a = A\left(\frac{m_0}{m}\right)^{1/4}$, as it was previously published by Krylov and Bogolubov (1943). Using the series expansion of the functions in (5.65) the approximate frequency of vibration is $\sqrt{k_1/m_0}$ which corresponds to the systems with constant mass and without damping.

5.3.4 Conclusion

Due to previous consideration it can be concluded:

1. The vibration of the oscillator with monotone time variable parameter has time variable amplitude and phase. The free vibrations for all of the oscillators with a strong nonlinearity of any order and with the certain monotone slow time variable parameters are qualitatively the same independently on the order of the nonlinearity. The order of nonlinearity quantitatively changes the amplitude and the phase of vibrations but has no influence on the character of vibrations. Namely, for certain parameter variation the higher the order of nonlinearity, the faster or slower is the amplitude and phase increase or decrease. The tendency of increase or decrease of amplitude and phase i.e., frequency of vibration variation is not directed by the order of nonlinearity but with the type of time parameter variation.
2. It is evident that in the oscillator with variable mass for the special relation between the coefficient of damping and parameter of mass variation (which affects the reactive force) the amplitude of vibration is constant, but the phase angle varies independently on the order of nonlinearity.
3. The approximate solution of the nonlinear differential equation with strong nonlinearity of any order (integer or non-integer) and time variable parameter can be obtained analytically.
4. The approximate analytic method for solving the differential based on the exact solution of the corresponding differential equation with constant parameters and

strong nonlinearity of any order (integer or non-integer) gives very accurate results in comparison to the numerical one.

5. The solving method based on the approximate solution with exact period of vibration of the corresponding oscillator with constant parameter gives very convenient results for the oscillator with time variable parameters. For technical purpose the solution is accurate enough and appropriate for practical use. This solution has the form of trigonometric function and satisfies the requirements for simplicity and usefulness for application in techniques.

5.4 Solution in the Form of a Jacobi Elliptic Function

In this section an approximate solution procedure for (5.1) which is based on the solution with exact amplitude, period and velocity of vibration of a pure nonlinear constant parameter oscillator is presented (Cveticanin 2013). The solution is assumed in the form of a Jacobi elliptic function. The trial solution and the first time derivative of the solution for (5.1) are assumed in the form of the Jacobi elliptic functions as for (3.2), but with time variable parameters, i.e.,

$$x = acn(\psi, k^2), \tag{5.66}$$

and

$$\dot{x} = -a\Omega_1 sndn, \tag{5.67}$$

where $a = a(t)$ is the unknown time variable amplitude,

$$\psi = \psi(t) = \int_0^t \Omega_1(t)dt + \theta(t), \tag{5.68}$$

$\theta = \theta(t)$ is the unknown time variable phase and using (5.47) the time variable frequency is

$$\Omega_1 = \frac{2K(k^2)}{\pi}\Omega_\alpha a^{(\alpha-1)/2}\sqrt{\frac{k_\alpha}{m(\tau)}}. \tag{5.69}$$

Let us determine the first time derivative of (5.66)

$$\dot{x} = -a\Omega_1 sndn + \dot{a}cn - a\dot{\theta}sndn. \tag{5.70}$$

Comparing the time derivatives (5.67) and (5.70) it is evident that they are equal for

$$\dot{a}cn - a\dot{\theta}sndn = 0. \tag{5.71}$$

Introducing (5.66), (5.67) and the time derivative of (5.67) into (5.1), and using the relation

$$-a\Omega_1^2 cn(1-2k^2+2k^2cn^2)+\omega^2 acn\approx 0, \tag{5.72}$$

we obtain a first order differential equation

$$-\dot{a}\Omega_1 sndn - a\dot{\Omega}_1 sndn - a\Omega_1\dot{\theta}cn(1-2k^2+2k^2cn^2)=\varepsilon\frac{f}{m}+\frac{\dot{m}}{m}a\Omega_1 sndn. \tag{5.73}$$

The relations (5.71) and (5.73) represent the rewritten version of the second order nonlinear differential equation (5.1) as a system of two coupled first order nonlinear differential equations. After some modification, we have

$$\dot{a}\Omega_1[sn^2dn^2+cn^2(1-2k^2+2k^2cn^2)]=-\varepsilon\frac{f}{m}sndn-\frac{\dot{m}}{m}a\omega sn^2dn^2-a\dot{\Omega}_1 sn^2dn^2, \tag{5.74}$$

$$a\dot{\theta}\Omega_1[sn^2dn^2+cn^2(1-2k^2+2k^2cn^2)]=-\varepsilon\frac{f}{m}cn-\frac{\dot{m}}{m}a\Omega_1 cnsndn-a\dot{\Omega}_1 cnsndn. \tag{5.75}$$

Solving (5.74) and (5.75), two unknown functions a and θ are calculated. For simplification the averaging of the differential equations is done. The differential equations (5.74) and (5.75) are averaged over the period of the Jacobi elliptic function $4K(k^2)$. The averaged differential equations follow as

$$\dot{a}\left\langle sn^2dn^2+cn^2(1-2k^2+2k^2cn^2)\right\rangle=-\frac{\varepsilon}{m\Omega_1}\langle fsndn\rangle-a\left(\frac{\dot{m}}{m}+\frac{\dot{\Omega}_1}{\Omega_1}\right)\left\langle sn^2dn^2\right\rangle, \tag{5.76}$$

$$a\dot{\theta}\left\langle sn^2dn^2+cn^2(1-2k^2+2k^2cn^2)\right\rangle=-\frac{\varepsilon}{m\Omega_1}\langle fcn\rangle, \tag{5.77}$$

where $\langle\cdots\rangle=\frac{1}{4K(k^2)}\int_0^{4K(k^2)}(\cdots)\,d\psi$. After integration, the Eqs. (5.76) and (5.77) transform into

$$\begin{aligned}\dot{a}\left[(1-k^2)+k^2C_4\right]&=-\frac{\varepsilon}{m\Omega_1}\langle fsndn\rangle\\&\quad-a\left(\frac{\dot{m}}{m}+\frac{\dot{\Omega}_1}{\Omega_1}\right)\left[(1-k^2)-(1-2k^2)\,C_2-k^2C_4\right],\\a\dot{\theta}\left[(1-k^2)+k^2C_4\right]&=-\frac{\varepsilon}{m\Omega_1}\langle fcn\rangle,\end{aligned} \tag{5.78}$$

where

$$C_2=\frac{1}{k^2}\left(\frac{E}{K}-1+k^2\right)\approx\frac{1}{2}+\frac{k^2}{16}+\cdots \tag{5.79}$$

$$C_4 = \frac{1}{3k^2}\left[2(2k^2-1)C_2 + (1-k^2)\right] \approx \frac{3}{8} + \frac{k^2}{32} + \cdots \tag{5.80}$$

and $E \equiv E(k^2)$ is the complete elliptic integral of the second kind (Byrd and Friedman 1954).

Integrating the differential equations (5.78) and substituting the obtained relations into (5.66), the approximate solution of the differential equation (5.1) is obtained.

The suggested solving procedure is applied for the Van der Pol oscillator with time variable mass.

5.4.1 Van der Pol Oscillator with Time Variable Mass

As a specific example, let us consider the vibrations of a Van der Pol oscillator with time variable mass

$$m\ddot{x} + k_\alpha x\,|x|^{\alpha-1} = \varepsilon(b - cx^2)\dot{x} - \dot{m}\dot{x}, \tag{5.81}$$

with

$$x(0) = A, \qquad \dot{x}(0) = 0, \tag{5.82}$$

where εb and εc are constants. If the mass variation is the function of the slow time $\tau = \varepsilon t$, where $\varepsilon \ll 1$ is the small parameter, the right hand side terms of (5.81) are small. Using the previously mentioned procedure, the differential equation is rewritten into two first order averaged differential equations

$$\dot{a}\left[(1-k^2) + k^2C_4\right] = -\left(\frac{\varepsilon c a^3 \Omega_1 sn^2 cn^2 dn^2}{m\Omega_1} + a\left(\frac{\dot{m}}{m} + \frac{\dot{\Omega}_1}{\Omega_1} - \frac{\varepsilon b}{m}\right)\right)$$

$$((1-k^2) - (1-2k^2)\,C_2 - k^2C_4), \tag{5.83}$$

$$\dot{\theta} = 0, \tag{5.84}$$

where

$$C_6 = \frac{4(2k^2-1)C_4 + 3(1-k^2)C_2}{5k^2} \approx \frac{5}{16} + \frac{5}{256}k^2 + \cdots \tag{5.85}$$

Integrating (5.84) it is obvious that the phase angle is constant, i.e. $\theta = const$. Using the relation (5.69) and its time derivative

$$\frac{\dot{\Omega}_1}{\Omega_1} = -\frac{1}{2}\frac{\dot{m}}{m} + \frac{\alpha-1}{2}\frac{\dot{a}}{a}, \tag{5.86}$$

the differential equation (5.83) transforms into

$$\dot{a}a_1 + a\left(\frac{1}{2}\dot{m} - \varepsilon b\right)\frac{a_2}{m} + \varepsilon c a^3 \frac{a_3}{m} = 0, \tag{5.87}$$

where

$$a_1 = \left(1 - k^2\right) + k^2 C_4 + \frac{\alpha - 1}{2} a_2, \tag{5.88}$$

$$a_2 = \left(1 - k^2\right) - \left(1 - 2k^2\right) C_2 - k^2 C_4, \tag{5.89}$$

$$a_3 = \left(1 - k^2\right) C_2 - \left(1 - 2k^2\right) C_4 - k^2 C_6. \tag{5.90}$$

The differential equation (5.87) is of the Bernoulli type (Kamke 1971). Introducing the new variable

$$p = a^{-2}, \tag{5.91}$$

the nonlinear differential equation (5.87) is transformed into the linear one

$$p' - \frac{\dot{m} - 2\varepsilon b}{m}\frac{a_2}{a_1} p = \frac{2a_3 \varepsilon c}{a_1 m}. \tag{5.92}$$

In general, the solution of (5.92) with (5.91) is

$$\frac{1}{a^2} \equiv p = \exp(-F)\left(C + \frac{2a_3 \varepsilon c}{a_1} \int_0^t \exp(F)\frac{dt}{m}\right), \tag{5.93}$$

where

$$F = \frac{2\varepsilon b a_2}{a_1} \int_0^t \frac{dt}{m} - \ln\left(\frac{m}{m_0}\right)^{\frac{a_2}{a_1}}, \tag{5.94}$$

and C is the arbitrary constant dependent on the initial conditions.

For the linear mass variation

$$m(\tau) = m_0(1 + \tau), \tag{5.95}$$

with $\dot{m} = \varepsilon m_0 = \text{const.}$, and

$$F = \ln\left(\frac{m}{m_0}\right)^{\frac{(2b - m_0)a_2}{m_0 a_1}}, \tag{5.96}$$

the solution of (5.93) is in general

$$\frac{1}{a^2} = \left(\frac{m_0}{m}\right)^q \left(C + \frac{2a_3c}{a_1q}\left(\left(\frac{m}{m_0}\right)^q - 1\right)\right), \tag{5.97}$$

where

$$q = \frac{(2b - m_0)a_2}{m_0a_1}, \tag{5.98}$$

and m_0 is the initial mass.Finally, using the initial conditions (5.82), we have

$$\frac{m^q}{a^2} - \frac{m_0^q}{A^2} = \frac{2ca_3}{m_0a_1q}(m^q - m_0^q). \tag{5.99}$$

5.4.1.1 Discussion of the Results

1. Substituting the mass variation (5.95), the Eq. (5.99) is rewritten in the form

$$a^2 = \frac{(1+\varepsilon t)^q A^2 m_0a_1q}{m_0a_1q + 2ca_3A^2((1+\varepsilon t)^q - 1)}. \tag{5.100}$$

The amplitude of vibration depends on q (5.98), which gives the relation between the negative damping coefficient b and the initial mass m_0. Namely, this relation takes into the consideration the effect of reactive force on the vibrations of the Van der Pol oscillator. The following three cases are evident:

(a) For $2b = m_0$, when $q = 0$, the amplitude of vibration tends to the initial amplitude A independently on the order of nonlinearity of the oscillator.

(b) According to (5.98) and (5.100), the amplitude–time curve is for $2b > m_0$ given as

$$a = \sqrt{\frac{(1+\varepsilon t)^{\frac{(2b-m_0)a_2}{m_0a_1}} A^2(2b - m_0)a_2}{(2b - m_0)a_2 + 2ca_3A^2((1+\varepsilon t)^{\frac{(2b-m_0)a_2}{m_0a_1}} - 1)}}. \tag{5.101}$$

The amplitude of vibration is time variable and tends to a steady state value

$$A_S = \sqrt{\frac{m_0a_1q}{2ca_3}} = \sqrt{\frac{(2b - m_0)a_2}{2ca_3}}, \tag{5.102}$$

which does not depend on the initial displacement A.

(c) For the case when $2b < m_0$, i.e., $q < 0$, the amplitude–time relation is according to (5.98) and (5.100)

$$a = \sqrt{\frac{(1+\varepsilon t)^{-\frac{(m_0-2b)a_2}{m_0a_1}} A^2(m_0 - 2b)a_2}{(m_0 - 2b)a_2 + 2ca_3A^2(1 - (1+\varepsilon t)^{\frac{(m_0-2b)a_2}{m_0a_1}})}}.$$

The amplitude of vibration is time variable and tends to zero

$$A_S = 0. \tag{5.103}$$

Namely, independently on the initial conditions, properties of the Van der Pol oscillator and mass variation, after some time the vibration disappear.

Remarks

1. For the case of the linear mass variation the reactive force acts as the positive damping force.
2. For the case when the reactive force is zero, the Eq. (5.99) is

$$\frac{m^{q_1}}{a^2} - \frac{m_0^{q_1}}{A^2} = \frac{2ca_3}{m_0 a_1 q_1}(m^{q_1} - m_0^{q_1}), \tag{5.104}$$

where

$$q_1 = \frac{2ba_2}{m_0 a_1}. \tag{5.105}$$

The amplitude varies in time according to the relation

$$a = \sqrt{\frac{2b(1+\varepsilon t)^{\frac{2ba_2}{m_0 a_1}} A^2 a_2}{2ba_2 + 2ca_3 A^2((1+\varepsilon t)^{\frac{2ba_2}{m_0 a_1}} - 1)}}, \tag{5.106}$$

and tends to the steady state one

$$A_S = \sqrt{\frac{ba_2}{ca_3}}, \tag{5.107}$$

which is independent on the value of the initial mass and mass variation. For all values of the initial displacements and mass variation, only one steady state motion exists. The steady state amplitude depends on the properties of the oscillator and is independent on mass variation properties.
3. For the oscillator with reactive force and with linear negative damping, i.e., (5.81) for $c = 0$,

$$m\ddot{x} + k_\alpha x\,|x|^{\alpha-1} = \varepsilon b\dot{x} - \dot{m}\dot{x}, \tag{5.108}$$

the amplitude–time relation is according to (5.100)

$$a = A\left(\frac{m}{m_0}\right)^{\frac{(2b-m_0)a_2}{2m_0 a_1}}.$$

For the linear mass variation (5.95), when $(m/m_0) \geqslant 1$, the amplitude of vibration increases if $(2b - m_0) > 0$ and decreases for $(2b - m_0) < 0$. For $(2b - m_0) = 0$,

in spite of mass variation, the amplitude has the constant value equal to the initial amplitude.
If in the oscillator with time variable mass the reactive force is zero, due to the fact that the relative velocity of the mass which is added of separated is zero, but the negative linear damping acts, the amplitude of vibration increases for all values of the initial mass and mass variation parameters and yields,

$$a = A\left(\frac{m}{m_0}\right)^{\frac{2ba_2}{m_0 a_1}}. \tag{5.109}$$

4. If in the mass variable oscillator the reactive force and only the nonlinear (positive) damping acts ($b = 0$ and $c \neq 0$), we have the mathematical model

$$m\ddot{x} + k_\alpha x\,|x|^{\alpha-1} = -(\varepsilon c x^2 - \dot{m})\dot{x}. \tag{5.110}$$

According to (5.100), the amplitude–time relation is

$$a = \sqrt{\frac{A^2 m_0 a_2}{(m_0 a_2 + 2ca_3 A^2)(1+\varepsilon t)^{\frac{a_2}{a_1}} - 2ca_3 A^2}}. \tag{5.111}$$

Analyzing the relation (5.111), it is evident that the amplitude of vibration decreases in time and tends to zero.
For the case of mass variable oscillator without reactive force, but with negative damping, the amplitude–time relation is according to (5.93)

$$a = \sqrt{\frac{A^2 a_1 m_0}{a_1 m_0 + 2ca_3 A^2 \ln|1+\varepsilon t|}}, \tag{5.112}$$

and the vibration decreases.
5. The obtained values can be compared with those obtained for the Van der Pol oscillator with constant mass

$$m\ddot{x} + x\,|x|^{2/3} = \varepsilon(b - cx^2)\dot{x}. \tag{5.113}$$

Thus, the relation (5.92) gives the amplitude–time variation

$$a^2 = \frac{ba_2 A^2}{ca_3 A^2 + (ba_2 - ca_3 A^2)\exp(-\frac{2\varepsilon b}{m}\frac{a_2}{a_1}t)}, \tag{5.114}$$

and the steady state amplitude

$$A_S = \sqrt{\frac{ba_2}{ca_3}}, \tag{5.115}$$

which is independent on the initial displacement. For parameter values $b = 1$ and $c = 1$, the steady state amplitude (5.115) for the linear oscillator ($\alpha = 1$) is $A_S = 2\sqrt{3}$, as it is previously published by Nayfeh and Mook (1979), while for the pure cubic nonlinear one ($\alpha = 3$) is $A_S = 1.7076$. The value, given by Mickens (2010), is $A_S = 2$. Comparing the solution of (5.113) obtained numerically, applying the Runge Kutta procedure, with the approximate amplitude (5.115) it can be concluded that the suggested solution in the form of the Jacobi elliptic function gives the more accurate result than the approximate trigonometric solution.

5.4.1.2 Oscillator with Linear Mass Variation

To illustrate the obtained results, let us consider an example where the mass variation is linear

$$m = m_0(1 + 0.01t), \tag{5.116}$$

and the coefficients of the Van der Pol oscillator are $\varepsilon b = 0.01$ and $\varepsilon c = 0.01$.

Substituting the suggested values into (5.81) it follows

$$m_0(1 + 0.01t)\ddot{x} + x\,|x|^{2/3} = -0.01 m_0 \dot{x} + 0.01(1 - x^2)\dot{x}. \tag{5.117}$$

Equation (5.117) is solved numerically by applying of the Runge–Kutta solving procedure and the solutions are compared with the approximate analytical results for the amplitude–time relation (5.99).

In Figs. 5.5, 5.6, 5.7, 5.8 the numerically calculated x–t and analytically obtained a–t curves are plotted for various values of the initial displacement A and initial mass m_0.

In Fig. 5.5 the amplitude–time a–t curves obtained by solving of the analytical relation (5.100) for the values of the initial mass $m_0 = 0.5$, 1 and 1.5, respectively, and various values of the initial amplitude are plotted. As $m_0 < 2$ the steady-state amplitude of vibration satisfies the relation (5.102) and have the values $A_S = 1.691$, $A_S = 1.381$ and $A_S = 0.976$, respectively. Besides, for $A < A_S$, the amplitude of vibration increases to the limit value A_S, and for $A > A_S$ it decreases to the limit value. For $A = A_S$ the motion is with constant amplitude.

In Fig. 5.6 the x–t curve obtained numerically by solving (5.117) and amplitude–time a–t curves (5.102) for $m_0 = 1$ and the initial displacements: (a) $A = 2.5$; (b) $A = 1.381$; (c) $A = 0.5$ are plotted. It is evident that the analytically obtained results are on the top of the numerical ones. The difference is negligible.

In Fig. 5.7, the a–t curves for $m_0 = 3$ and initial displacements $A = 0.5$, 1.5 and 2.5 are plotted. For this case, where $m_0 > 2$, the amplitude of vibration decreases to zero, independently on the initial amplitude.

In Fig. 5.8, beside the a–t curves also the x–t curves for $m_0 = 3$ and initial displacements: (a) $A = 0.5$ and (b) $A = 2.5$ are plotted.

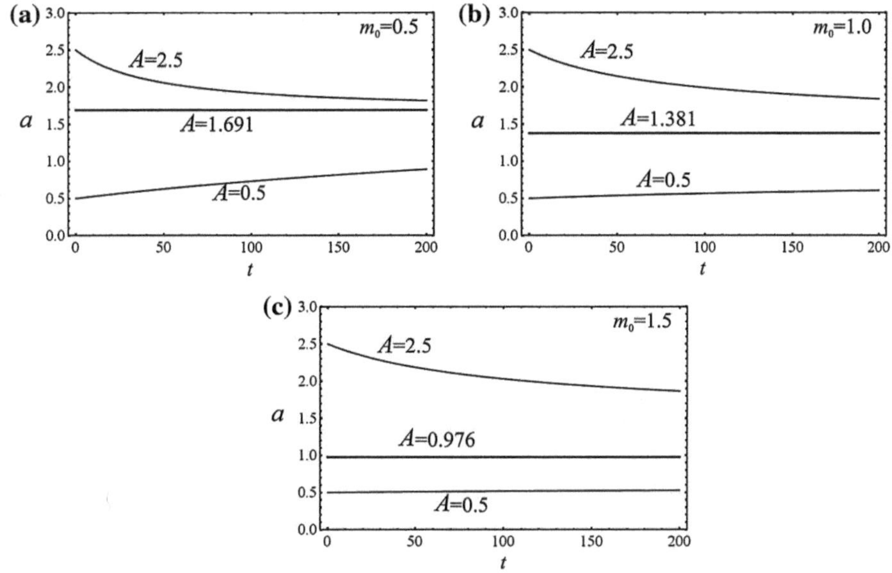

Fig. 5.5 a–t curves for various initial displacements A and various initial masses: **a** $m_0 = 0.5$; **b** $m_0 = 1$ and **c** $m_0 = 1.5$

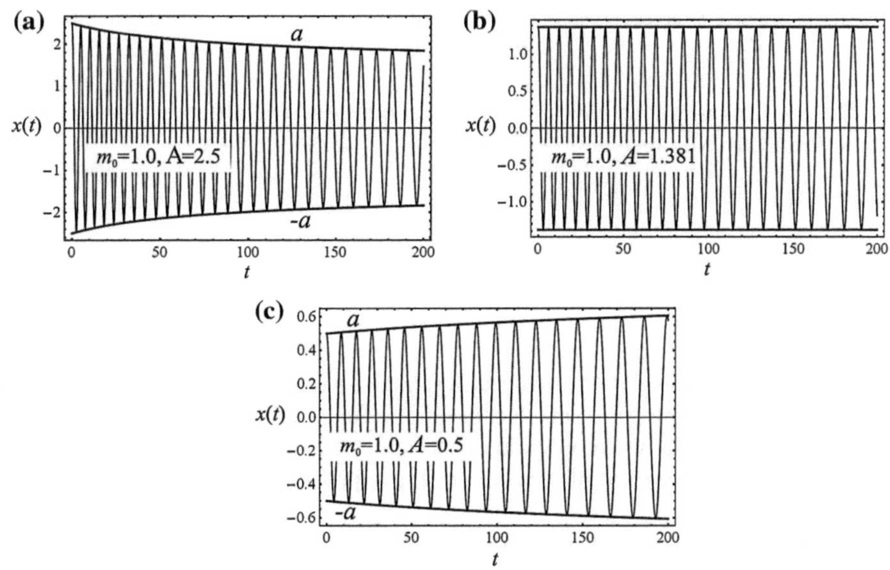

Fig. 5.6 x–t and a–t curves for $m_0 = 1$ and initial displacements: **a** $A = 2.5$; **b** $A = 1.381$; **c** $A = 0.5$

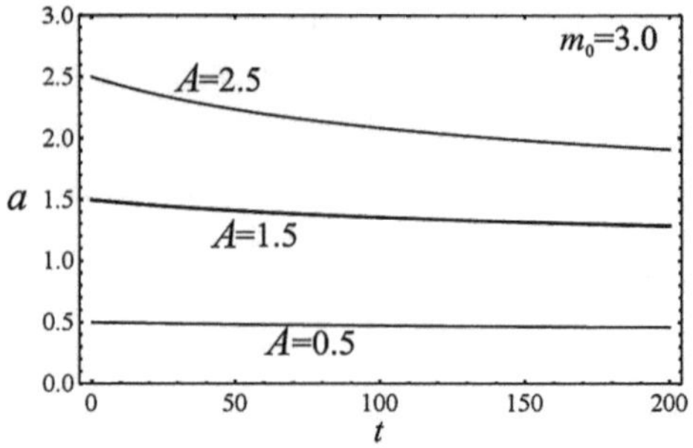

Fig. 5.7 a–t curves for $m_0 = 3$ and initial displacements $A = 0.5$, 1.5 and 2.5

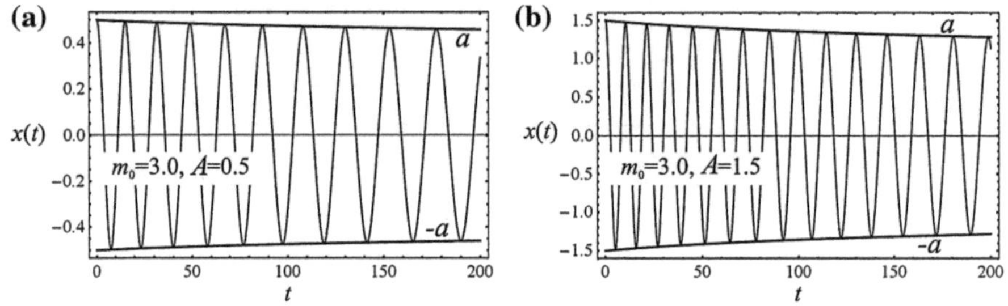

Fig. 5.8 x–t and a–t curves for $m_0 = 3$ and initial displacements: **a** $A = 2.5$ and **b** $A = 0.5$

The a–t curves are the envelopes of the x–t curves. The difference between the numeric and analytic results is negligible.

5.4.2 Conclusion

It can be concluded:

1. The approximate solution of the differential equation with slow time variable parameters in the form of the Jacobi elliptic function is close to the exact one, due to the fact that the amplitude of vibration, the amplitude of the velocity of vibration and the period of vibration of the approximate solution is close to the exact one.
2. For the case of the linear mass variation the reactive force acts as the positive damping force.
3. The motion of the Van der Pol oscillator with time variable parameters deeply depends on the properties of mass variation and specially on the initial mass. A boundary value of the initial mass for which the constant limit value exists is determined. If the initial mass is larger than the limit value, the amplitude of vibration of the Van der Pol oscillator with variable mass tends to zero indepen-

dently on the initial displacements, while, for the case when the initial mass is smaller than the limit value, the limit cycle motion with the steady state amplitude occurs.

a. If the initial mass is smaller than the limit one and:
   - for the initial displacement smaller than the limit value, the amplitude of vibration increases and the motion tends to the limit cycle motion,
   - if the initial displacement is higher than the limit value, the amplitude of vibration decreases and the motion tends to the limit cycle motion,
   - if the initial displacement is equal to the limit amplitude, the limit cycle motion is with this constant amplitude.
b. If the initial mass is higher than the limit one, for all values of initial displacement the amplitude of vibration decreases and tends to zero. Then, there is no limit cycle motion.

4. The value of the limit amplitude is independent on the initial displacement as it is the case for the Van der Pol oscillator with constant mass.
5. For the case when the reactive force is zero only one limit cycle motion exists with the amplitude which is independent on the mass variation and initial displacement, but depends on the mechanical properties of the oscillator.

5.5 Parametrically Excited Strong Nonlinear Oscillator

Parametrically excited systems are widely spread in many branches of physics and engineering. In mechanical and elastic systems, parametrically excited vibrations occur due to time varying loads, especially periodic ones. These vibrations appear in columns made of nonlinear elastic material (Mond and Cederbaum 1993), beams with a harmonically variable length (Esmailzadeh et al. 1997), beams with harmonic motion of their support (Esmailzadeh and Jalili 1998), floating offshore structures (Esmailzadeh and Goodarzi 2001), parametrically excited pendulums (El-Dib 2001), cables being towed by a submarine (Ng and Rand 2002a; Ng and Rand 2002b), etc. Parametric excitations occur in electrostatically driven micro-electro-mechanical oscillators (Rhoads et al. 2006), which are produced by fluctuating voltages applied across comb drives. In practical engineering situations the properties of parametric oscillations are widely used, for example, in the radio, the computer and laser engineering, in vibromachines with special design (Belovodsky et al. 2002), Paul trap mass spectrometers (Abraham and Chatterjee 2003) and a simulator for proving the equivalence of inertia and passive gravitational mass (Wiegand et al. 2004). Parametric resonance has been well established in many areas of science, including the stability of ships, the forced motion of a swing and Faraday surface wave patterns on water. The highly sensitive mass sensor is studied as an in-plane parametrically resonant oscillator (Zhang et al. 2002).

The simplest mathematical model of the system with a parametric periodic load is usually a linear Mathieu differential equation. Due to the nonlinear properties of a real system, nonlinear terms are added to the equation (Younesian et al. 2005). Usually, they are of a cubic type and the differential equation is transformed to the Mathieu–Duffing equation (Cveticanin 2000). To determine the combined effect of nonlinearity and parametric excitations, numerous analytical techniques have been developed. Two classes of these techniques are dominant. One class is based on the integral of energy and numerical integration (Jazar 2004). The method is suitable for obtaining the boundaries between bounded and unbounded solutions of the equation. The advantage of the method is that it gives accurate stability charts, but the procedure is time consuming. If the energy integral represents the Lyapunov function, then Lyapunov stability theory is also applicable. With this approach, it is possible to determine qualitatively the general stability of the system, but one cannot determine qualitatively the system response. The second technique, which is much more developed, consists of the perturbation methods that are based on the assumption that the variable-coefficient terms are small in some sense. The most widely applied is the method of multiple scales (Nayfeh and Mook 1979). The method is used to obtain solutions that are valid in neighborhoods close to the transient curves. The method of multiple scales is extended for solving the stochastic Mathieu–Duffing equation, too. The almost sure-stability criterion and instability criterion are determined (Li et al. 2005). Ng and Rand (2002a, b) investigated the Mathieu–Duffing equation using another perturbation method. They showed that the averaging method is suitable for solving the deterministic Mathieu oscillator.

The method of strained parameters (Nayfeh and Mook 1979) is also an asymptotic analytical method, which is well suited for the determination of the transient curves between stable and unstable solutions. This method yields a solution which is valid right on the transient curve and does not yield a solution that is valid in the neighborhood close to the transient curve. Following this method and based on Floquet theory, if one assumes that the solutions have periods of π and 2π, then, the values of the parameters for which this assumption is true can be determined (Nayfeh and Mook 1979).

All the previous techniques have been applied to solve differential equations with a small parametric excitation and small nonlinearity. Zounes and Rand (2002) considered the Mathieu–Duffing oscillator, assuming that the parametric perturbation is small but the coefficient of the nonlinear term is positive and not necessary small.

In this section the Mathieu–Duffing equation with a small parametric excitation and strong negative nonlinearity is investigated. The mathematical model corresponds to the parametrically excited oscillator with a softening spring. The aim is to obtain the transient values of the parameters of the system which lead the periodic solutions. The two-dimensional Lindstedt–Poincaré perturbation technique is adopted for solving the differential equation

$$\ddot{x} + (\delta + 2\varepsilon \cos 2t)x - \varphi x^3 = 0, \tag{5.118}$$

with the initial conditions

$$x(0) = X_0, \qquad \dot{x}(0) = \dot{X}_0, \tag{5.119}$$

where φ is the parameter of nonlinearity and $\varepsilon \ll 1$ is a small parameter. In contrast to the previous perturbation treatments, the unperturbed system is nonlinear and the use of the elliptic functions, instead of the trigonometric functions, which have been applied for the linear systems, is introduced. The regions of stability and instability, i.e., the bounded and unbounded solutions of the Mathieu–Duffing equation are discussed.

5.5.1 Solution Procedure

The perturbation expansions of the function x and parameter δ are introduced

$$x(t, \varepsilon) = x_0(t) + \varepsilon x_1(t) + \cdots , \tag{5.120}$$

$$\delta = \delta_0 + \varepsilon \delta_1 + \cdots \tag{5.121}$$

Substituting (5.120) and (5.121) into (5.118) and (5.119) and equating coefficients of similar powers of ε, we obtain

$$\begin{aligned} \ddot{x}_0 + \delta_0 x_0 - \varphi x_0^3 &= 0, \\ \ddot{x}_1 + \delta_0 x_1 - 3\varphi x_0^2 x_1 &= -\delta_1 x_0 - 2x_0 \cos 2t, \\ &\dots \end{aligned} \tag{5.122}$$

and

$$\begin{aligned} x_0(0) = X_0, &\quad \dot{x}(0) = \dot{X}_0, \\ x_1(0) = 0, &\quad \dot{x}_1(0) = 0, \\ &\dots \end{aligned} \tag{5.123}$$

Equation $(5.123)_1$ is a differential equation with strong cubic nonlinearity, which has a periodic solution of the form

$$x_0 = A_0 sn(\omega_0 t + \alpha_0, k_0), \tag{5.124}$$

where sn is the elliptic function (Appendix D and Byrd and Friedman 1954) with the frequency ω_0 and the modulus k_0 given respectively by

$$\omega_0^2 = \delta_0 - \frac{\varphi A_0^2}{2}, \quad k_0^2 = \frac{\varphi A_0^2}{2(\delta_0 - \frac{\varphi A_0^2}{2})}, \tag{5.125}$$

and A_0 and α_0 are arbitrary constants. The motion is periodic for $A_0 < \sqrt{2\delta_0/\varphi}$.

The elliptic function sn has the period $4nK(k)$, where $K(k)$ is the complete elliptic integral of the first kind and $n = 0, 1, 2, \ldots$.

(1) For $n = 0$ we obtain $\omega_0 = 0$. If the initial velocity is zero ($\dot{X}_0 = 0$) Eqs. (5.124) and (5.125) results in

$$x_0 = X_0 = const., \tag{5.126}$$

and

$$\delta_0 = \varphi X_0^2. \tag{5.127}$$

Because x_0 is constant, Eq. (5.122) becomes an ordinary non-homogenous differential equations, of the form

$$\ddot{x}_1 - 2\varphi X_0^2 x_1 = -\delta_1 X_0 - 2X_0 \cos 2t. \tag{5.128}$$

Solving the Eq. (5.128) with the initial conditions given in $(5.123)_2$, we obtain

$$x_1 = \frac{\delta_1}{2\varphi X_0} - \left(\frac{\delta_1}{2\varphi X_0} + \frac{X_0}{2+\varphi X_0^2})ch(X_0 t\sqrt{2\varphi}\right) + \frac{X_0}{2+\varphi X_0^2}\cos 2t. \tag{5.129}$$

The periodic solution

$$x_1 = -\frac{X_0}{2+\varphi X_0^2}(1-\cos 2t), \tag{5.130}$$

exists for

$$\delta_1 = -\frac{2\varphi X_0^2}{2+\varphi X_0^2}. \tag{5.131}$$

Using Eqs. (5.121), (5.127) and (5.131), the transient curve in the first approximation is given by

$$\delta = \varphi X_0^2\left(1 - \frac{2\varepsilon}{2+\varphi X_0^2}\right), \tag{5.132}$$

and along this curve the solution is

$$x = X_0 - \frac{\varepsilon X_0}{2+\varphi X_0^2} + \frac{\varepsilon X_0}{2+\varphi X_0^2}\cos 2t. \tag{5.133}$$

To confirm the correctness of the analytical procedure, the analytical result (5.133) is compared with the numerical solution obtained by solving (5.118) with (5.132) for the initial conditions $x_0(0) = X_0 = 0.1,\ \ \dot{x}(0) = 0$.

In Fig. 5.9 the time histories $x(t)$ are plotted for $X_0 = 0.1$, $\varphi = 2$ and various values of the small parameter: $\varepsilon = 0.01$ (Fig. 5.9a) and $\varepsilon = 0.1$ (Fig. 5.9b). It is evident that during the initial time period the difference between the analytical and numerical solutions is negligible.

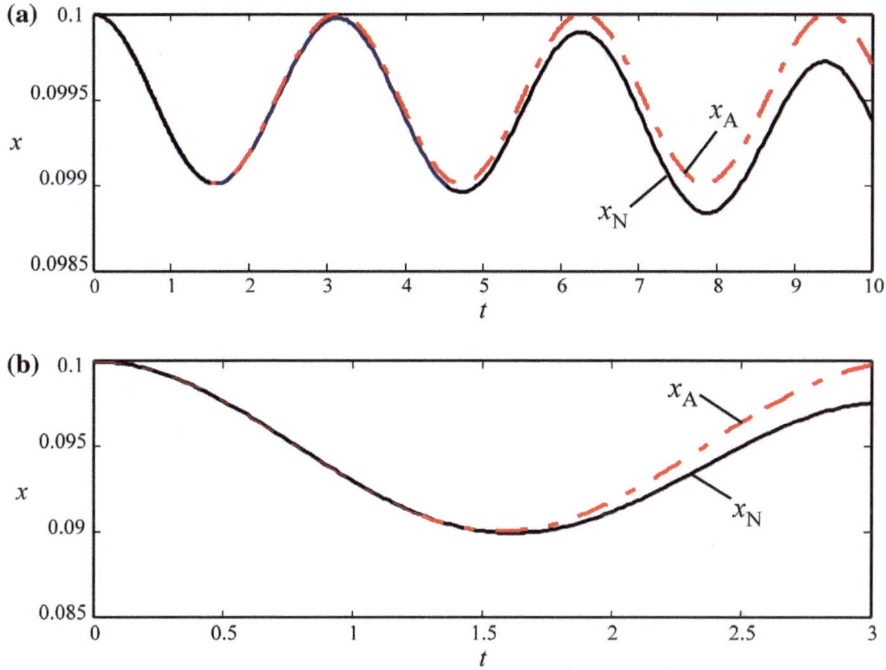

Fig. 5.9 History diagrams x-t obtained analytically (x_A) and numerically (x_N) for: $\varphi = 2, x(0) = 0.1, \dot{x}(0) = 0$ and: **a** $\varepsilon = 0.01$, **b** $\varepsilon = 0.1$

(2) For the period $4K(k)$ the frequency is $\omega_0 = 1$ and

$$\delta_0 = 1 + \frac{\varphi A_0^2}{2}, \qquad k_0^2 = \frac{\varphi A_0^2}{2}, \tag{5.134}$$

where the arbitrary amplitude is (see $(5.123)_1$ and (5.124))

$$A_0 = \sqrt{X_0^2 + \frac{\dot{X}_0^2}{1 - \frac{\varphi}{2} X_0^2}}. \tag{5.135}$$

The initial phase α_0 satisfies the relation

$$sn(\alpha_0, \frac{\varphi A_0^2}{2}) = \frac{X_0}{A_0}. \tag{5.136}$$

Using the transformation given in Abramowitz and Stegun (1979), the relation (5.124) is transformed into

$$x_0 = A_0 \frac{sn_0 C_0 D_0 - cn_0 dn_0 S_0}{1 - k_0^2 sn_0^2 S_0^2}, \tag{5.137}$$

where $cn_0 \equiv cn(t, k_0)$, $sn_0 \equiv sn(t, k_0)$, $dn_0 \equiv dn(t, k_0)$, $S_0 = sn(\alpha_0, k_0)$, $C_0 = cn(\alpha_0, k_0)$, $D_0 = dn(\alpha_0, k_0)$.

Substituting (5.137) into $(5.122)_2$, yields

$$\ddot{x}_1 + (1 + \frac{\varphi A_0^2}{2})x_1 - 3\varphi A_0^2 \frac{(sn_0 C_0 D_0 - cn_0 dn_0 S_0)^2}{(1 - k_0^2 sn_0^2 S_0^2)^2} x_1$$
$$= -\frac{A_0(sn_0 C_0 D_0 - cn_0 dn_0 S_0)}{1 - k_0^2 sn_0^2 S_0^2}(\delta_1 + 2\cos 2t). \tag{5.138}$$

This equation is a linear parametrically excited differential equation. For the case when $k_0^2 \ll 1$, i.e., $\frac{\varphi A_0^2}{2} \ll 1$, the elliptic functions are transformed to the harmonic functions

$$sn_0 \approx \sin t - \frac{k^2}{4}\cos t(t - \sin t \cos t), \qquad cn_0 \approx \cos t - \frac{k_0^2}{4}\cos t(t - \sin t \cos t),$$
$$dn_0 \approx 1 - \frac{k^2}{2}\sin^2 t. \tag{5.139}$$

The simplification of the expression (5.137) and the differential equation (5.138) with (5.139) gives

$$x_0 = A_0\big(C_0 D_0 \sin\sqrt{\Delta}t - S_0 \cos\sqrt{\Delta}t\big),$$

and

$$\ddot{x}_1 + \Delta x_1 - \varepsilon_1 \frac{S_0^2 - C_0^2 D_0^2}{2} x_1 \cos 2t + \varepsilon_1 S_0 C_0 D_0 x_1 \sin 2t$$
$$= -A_0(C_0 D_0 \sin t - S_0 \cos t)(\delta_1 + 2\cos 2t), \tag{5.140}$$

where ε_1 is a new small parameter

$$\varepsilon_1 = 3\varphi A_0^2 \ll 1, \tag{5.141}$$

while

$$\Delta = 1 + \frac{\varepsilon_1}{6}(1 - 3S_0^2 - 3C_0^2 D_0^2). \tag{5.142}$$

At this point, the two dimensional Lindstedt–Poincaré expansion is introduced once more. The series expansion with the small parameter ε_1

$$\delta_1 = \delta_{10} + \varepsilon_1 \delta_{11} + \cdots, \quad x_1 = x_{10} + \varepsilon_1 x_{11} + \cdots, \tag{5.143}$$

is substituted into (5.140). After separating the terms with the same order of the small parameter ε_1, the following equations are obtained

$$\varepsilon_1^0 : \quad \ddot{x}_{10} + \Delta x_{10} = -\delta_{10} A_0 (C_0 D_0 \sin\sqrt{\Delta}t - S_0 \cos\sqrt{\Delta}t) \\ -2A_0 (C_0 D_0 \sin\sqrt{\Delta}t - S_0 \cos\sqrt{\Delta}t) \cos 2\sqrt{\Delta}t, \tag{5.144}$$

$$\varepsilon_1^1 : \quad \ddot{x}_{11} + \Delta x_{11} = -\delta_{11} A_0 (C_0 D_0 \sin\sqrt{\Delta}t - S_0 \cos\sqrt{\Delta}t) \\ + \frac{S_0^2 - C_0^2 D_0^2}{2} x_{10} \cos 2\sqrt{\Delta}t - S_0 C_0 D_0 x_{10} \sin 2\sqrt{\Delta}t, \tag{5.145}$$

where

$$\sqrt{\Delta} \approx 1 + \frac{\varepsilon_1}{12}(1 - 3S_0^2 - 3C_0^2 D_0^2). \tag{5.146}$$

To ensure that x_{10} is periodic, the terms with $\sin\sqrt{\Delta}t$ and $\cos\sqrt{\Delta}t$ in (5.144) which lead to secular terms must vanish. This is attainable either for

$$\delta_{10} = -1, \quad \text{and} \quad A_0 = X_0, \quad C_0 = 0, \tag{5.147}$$

or

$$\delta_{10} = 1, \quad \text{and} \quad A_0 = \dot{X}_0, \quad S_0 = 0. \tag{5.148}$$

If $\delta_{10} = -1$ and the amplitude and phase angle satisfy the relations $A_0 C_0 D_0 = 0$, $A_0 S_0 = X_0$, we obtain the solution of Eq. (5.144) with the secular terms eliminated

$$x_{10} = \frac{X_0}{8\Delta}(\cos\sqrt{\Delta}t - \cos 3\sqrt{\Delta}t), \tag{5.149}$$

since the initial conditions are

$$x_{10}(0) = 0, \quad \dot{x}_{10}(0) = 0. \tag{5.150}$$

For $\delta_{10} = 1$ and $A_0 C_0 D_0 = \dot{X}_0$, $A_0 S_0 = 0$, the solution of (5.144) yields

$$x_{10} = -\frac{\dot{X}_0}{8\Delta}(3\sin\sqrt{\Delta}t - \sin 3\sqrt{\Delta}t). \tag{5.151}$$

Substituting (5.149) and (5.151) into (5.145) and eliminating the secular terms, we obtain

$$\delta_{11} = 0, \quad \text{for} \quad A_0 = X_0 \quad \text{and} \quad C_0 = 0, \tag{5.152}$$

and

$$\delta_{11} = -\frac{1}{8\Delta}, \quad \text{for} \quad A_0 = \dot{X}_0 \quad \text{and} \quad S_0 = 0. \tag{5.153}$$

Using (5.121), (5.134), (5.143), (5.147) and (5.152) or (5.148) and (5.153), the transient curves in the first approximation are obtained

$$\delta = 1 - \varepsilon + \frac{\varphi X_0^2}{2}, \tag{5.154}$$

and

$$\delta = 1 + \varepsilon + \frac{\varphi \dot{X}_0^2}{2}\left(1 - \frac{3\varepsilon}{4\Delta}\right). \tag{5.155}$$

For the special case when the value of the initial deflection is equal to the initial velocity, i.e.,

$$\varepsilon_1 = 3\varphi A_0^2 = 3\varphi X_0^2 = 3\varphi \dot{X}_0^2, \tag{5.156}$$

the special forms of the transient curves are obtained

$$\delta = 1 + \frac{\varepsilon_1}{6} - \varepsilon, \tag{5.157}$$

and

$$\delta = 1 + \frac{\varepsilon_1}{6} + \varepsilon\left(1 - \frac{\varepsilon_1}{8(1 - \frac{\varepsilon_1}{3})}\right). \tag{5.158}$$

In Fig. 5.10 the transient surfaces in $\varepsilon_1 \delta \varepsilon$ space are plotted.

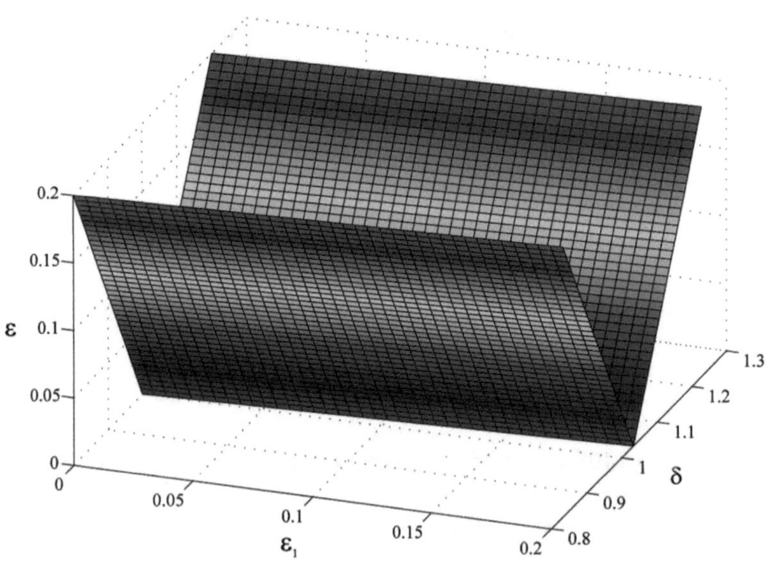

Fig. 5.10 Transient surfaces corresponding to the solution with the frequency $\omega_0 = 1$

The transient values depend strongly on the coefficient of nonlinearity and the initial conditions. This is a main difference in comparison to the linear systems described by the Mathieu equation, whose transient values are independent of the initial conditions. The dependence of the transient values on the initial conditions goes along with the general characteristic of nonlinear systems that their dynamic properties are affected by the initial conditions. For the combinations of parameter δ, ε_1 and ε defined by (5.157) and (5.158), i.e. for those lying on the surfaces, the motion is periodic with the period $4K(\sqrt{\frac{\varepsilon_1}{6}})$.

(3) For the period $2K(k)$ the frequency is $\omega_0 = 2$ and the solution of the differential equation $(5.122)_1$ is

$$x_0 = A_1 \frac{sn_1 C_1 D_1 - cn_1 dn_1 S_1}{1 - k_1^2 sn_1^2 S_1^2}, \tag{5.159}$$

where $cn_1 \equiv cn(2t, k_1)$, $sn_1 \equiv sn(2t, k_1)$, $dn_1 \equiv dn(2t, k_1)$, $S_1 = sn(\alpha_1, k_1)$, $C_1 = cn(\alpha_1, k_1)$, $D_1 = dn(\alpha_1, k_1)$ with the modulus of the elliptic function and the parameter value

$$k_1^2 = \frac{\varphi A_1^2}{8}, \qquad \delta_0^1 = 4 + \frac{\varphi A_1^2}{2}. \tag{5.160}$$

The initial amplitude and phase satisfy the relations

$$A_1 = \sqrt{X_0^2 + \frac{\dot{X}_0^2}{4 - \frac{\varphi}{2}X_0^2}}, \quad sn(\alpha_1, \frac{\varphi A_1^2}{8}) = \frac{X_0}{A_1}. \tag{5.161}$$

Assuming that $k_1^2 \ll 1$ and using the transformations of the elliptic into circular function (5.139), the Eq. $(5.122)_2$ is transformed into

$$\begin{aligned}
&\ddot{x}_1 + 4\Delta_1 x_1 + \varepsilon_1 x_1 S_1 C_1 D_1 \sin 4\sqrt{\Delta_1} t - \frac{\varepsilon_1}{2} x_1 (S_1^2 - C_1^2 D_1^2) \cos 4\sqrt{\Delta_1} t \\
&= -\delta_1 A_1 (C_1 D_1 \sin 2\sqrt{\Delta_1} t - S_1 \cos 2\sqrt{\Delta_1} t) - A_1 C_1 D_1 \sin 4\sqrt{\Delta_1} t \\
&\quad + A_1 S_1 (1 + \cos 4\sqrt{\Delta_1} t),
\end{aligned} \tag{5.162}$$

where

$$\Delta_1 = 1 + \frac{\varepsilon_1}{24}(1 - 3C_1^2 D_1^2 - 3S_1^2). \tag{5.163}$$

Using the series expansion in equation (5.143) with respect to the small parameter ε_1, the differential equation for ε_1^0 is obtained

$$\begin{aligned}
\ddot{x}_{10} + 4\Delta_1 x_{10} &= -\delta_{10} A_1 (C_1 D_1 \sin 2\sqrt{\Delta_1} t - S_1 \cos 2\sqrt{\Delta_1} t) \\
&\quad - A_1 C_1 D_1 \sin 4\sqrt{\Delta_1} t + A_1 S_1 (1 + \cos 4\sqrt{\Delta_1} t).
\end{aligned} \tag{5.164}$$

Eliminating the secular terms in (5.164) it is clear that

$$\delta_{10} = 0, \tag{5.165}$$

and the solution is

$$x_{10} = \frac{A_1 S_1}{\Delta_1}\left(1 - \frac{14}{15}\cos 2\sqrt{\Delta_1}t - \frac{1}{15}\cos 4\sqrt{\Delta_1}t\right) \\ - \frac{A_1 C_1 D_1}{15\Delta_1}(2\sin 2\sqrt{\Delta_1}t - \sin 4\sqrt{\Delta_1}t), \tag{5.166}$$

where

$$\sqrt{\Delta_1} \approx 1 + \frac{\varepsilon_1}{48}(1 - 3C_1^2 D_1^2 - 3S_1^2). \tag{5.167}$$

Substituting (5.166) into (5.162) and eliminating the secular terms, yields

$$\delta_{11} = -\frac{1}{24\Delta_1}, \quad \text{for} \quad A_1 = \frac{\dot{X}_0}{2} \quad \text{and} \quad S_1 = 0, \tag{5.168}$$

and

$$\delta_{11} = \frac{1}{24\Delta_1}, \quad \text{for} \quad A_1 = X_0 \quad \text{and} \quad C_1 = 0. \tag{5.169}$$

According to the previous considerations, the following two transient curves are obtained

$$\delta = 4 + \frac{\varphi \dot{X}_0^2}{8} - \varepsilon \frac{\varphi \dot{X}_0^2}{32 - 2\varphi \dot{X}_0^2}, \tag{5.170}$$

and

$$\delta = 4 + \frac{\varphi X_0^2}{2} + \varepsilon \frac{\varphi X_0^2}{8 - 2\varphi X_0^2}. \tag{5.171}$$

The special case when the value of the initial deflection is twice the initial velocity and

$$\varepsilon_1 = 3\varphi A_1^2 = 3\varphi X_0^2 = \frac{3}{4}\varphi \dot{X}_0^2, \tag{5.172}$$

gives the transient curves

$$\delta = 4 + \frac{\varepsilon_1}{6} - \frac{\varepsilon\varepsilon_1}{24 - 2\varepsilon_1}, \tag{5.173}$$

and

$$\delta = 4 + \frac{\varepsilon_1}{6} + \frac{\varepsilon\varepsilon_1}{24 - 2\varepsilon_1}. \tag{5.174}$$

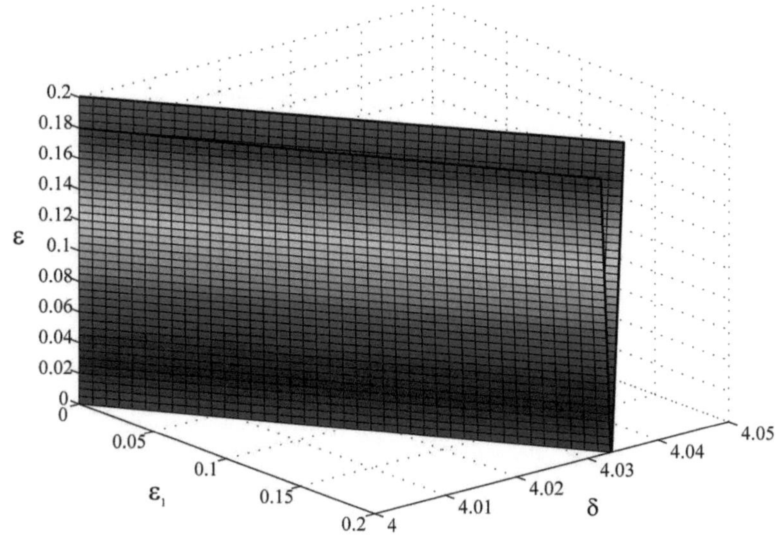

Fig. 5.11 Transient surfaces corresponding to the solution with the frequency $\omega_0 = 2$

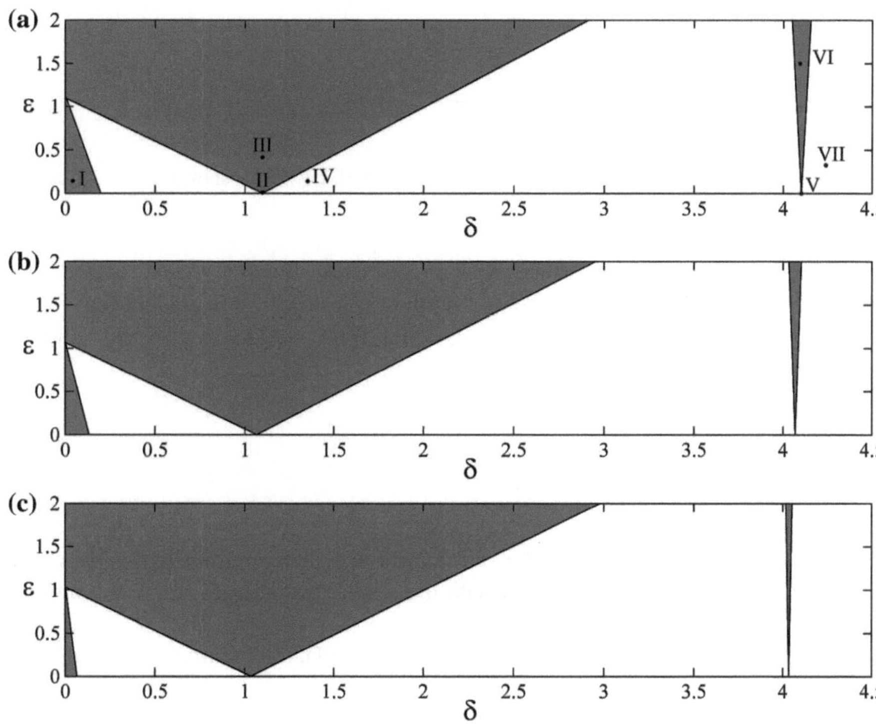

Fig. 5.12 Parameter $\delta\varepsilon$-plane for various values of ε_1: **a** $\varepsilon_1 = 0.6$, **b** $\varepsilon_1 = 0.4$, **c** $\varepsilon_1 = 0.2$

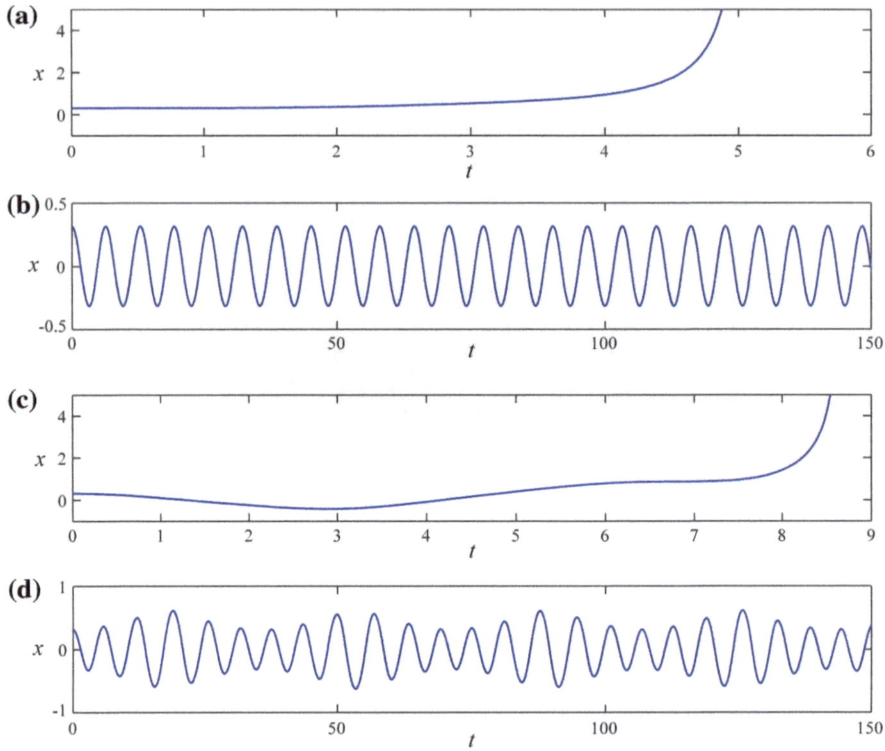

Fig. 5.13 Time-history diagrams x–t for the points: **a** I, **b** II, **c** III, **d** IV shown in Fig. 5.12a

In Fig. 5.11 the transient surfaces in $\varepsilon_1\delta\varepsilon$ space are shown.

The transient values depend strongly on the values of ε_1. It is evident that for the linear case when $\varepsilon_1 = 0$ we obtain $\delta = 4$, which is the well known critical value for a linear system.

5.5.2 *Numerical Simulation*

As a demonstration and check of the validity of the approximations in this paper, the theoretical predictions are compared with the results from direct numerical integration.

Using (5.132), (5.157), (5.158), (5.173) and (5.174), the parameter $\delta\varepsilon$-planes for various values of ε_1 are plotted in Fig. 5.12.

For the values in the shaded regions the solution implies unbounded motion. For the values in the non-shaded regions it is bounded. The higher the values of the parameter ε_1, the more the shaded regions are translated to the right. This is because

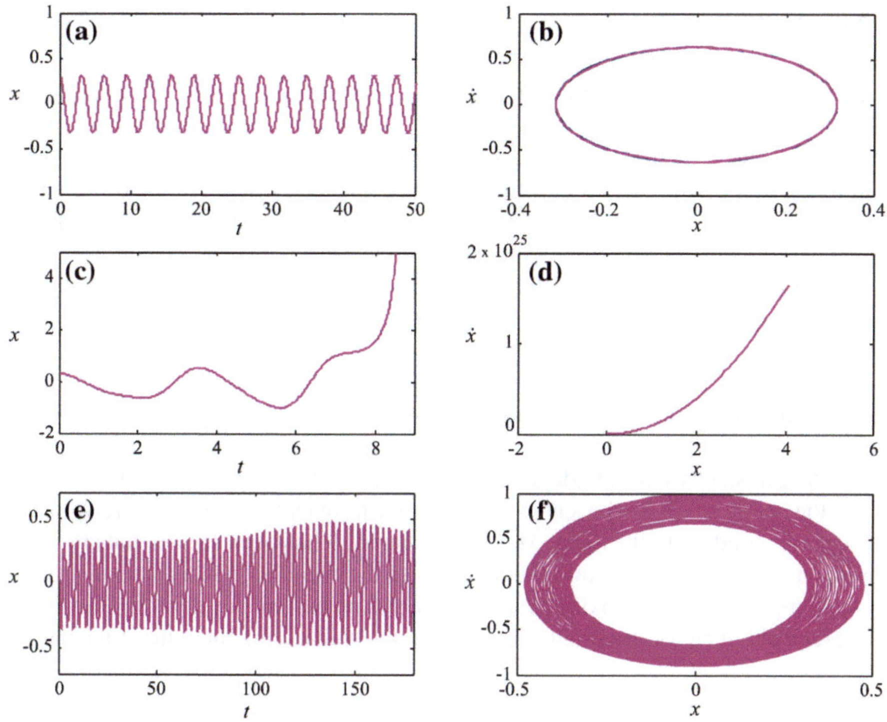

Fig. 5.14 x–t and x–$\dot{x}$ curves for the points shown in Fig. 5.12a: V (**a**, **b**), VI (**c**, **d**), and VII (**e**, **f**)

the parameter describing nonlinearity together with the initial conditions cause the critical values of δ to be higher than the corresponding value for the linear case.

Also, the regions corresponding to unbounded solutions are wider if ε_1 is larger. Furthermore, for small positive values of δ a region of unbounded ones exists. This is the main difference in comparison to the linear parametrically excited oscillator, for which there is a region of bounded motion in the small positive neighborhood of the origin.

Choosing some points from the stability chart (see Fig. 5.12a, points I–VII), the numerical procedure is carried out for $\varphi = 2$ and the initial conditions $x(0) = 0.3162$, $\dot{x}(0) = 0$ (Figs. 5.13 and 5.14). The parameters corresponding to these points are: point I: $\varepsilon = 0.1$, $\delta = 0.08$; point II: $\varepsilon = 0$, $\delta = 1.1$; point III: $\varepsilon = 0.4$, $\delta = 1.1$; point IV: $\varepsilon = 0.1$, $\delta = 1.25$ (Fig. 5.13); point V: $\varepsilon = 0$, $\delta = 4.1$; point VI: $\varepsilon = 1.5$, $\delta = 4.1$; point VII: $\varepsilon = 0.3$, $\delta = 4.2$ (Fig. 5.14). While Fig. 5.13 shows the time history diagrams, in Fig. 5.14 both time history diagrams and the corresponding phase planes are plotted. The numerical results convey the results of the analytical analysis.

5.5.3 Conclusion

A first order analytical solution of the Mathieu–Duffing differential equation has been obtained by using the two dimensional Linstedt–Poincaré perturbation method. The unperturbed equation includes strong negative cubic nonlinearity, whose exact solution is given with Jacobi elliptic functions. The transition curves along which the periodic solutions with period $2K$ and $4K$ exist, have been constructed. Based on the obtained results, the following can be concluded:

1. The analytical results show that the nonlinearity significantly changes the characteristics of parametric resonance and the presence of a negative cubic nonlinearity alters the dynamic behavior of the system.

2. The use of the elliptic functions instead of the trigonometric functions has a principle advantage since the subharmonic resonances of all orders are accounted for.

3. The transient values are dependent on the initial conditions. In the special case when the value of φ is chosen to be zero, the transient curves are equivalent to those corresponding to the linear Mathieu equation. It has been shown that any small deviations from those transition curves leads to the complete loss of periodicity.

4. Comparing the approximate analytical and numerical results it is shown that the results obtained for the small values of ε agree well during the initial time period.

References

Abdalla, M. S. (1986a). Canonical treatment of harmonic oscillator with variable mass. *Physical Review A*, *33*, 2870–2876.

Abdalla, M. S. (1986b). Time-dependent harmonic oscillator with variable mass under the action of a driving force. *Physical Review A*, *34*, 4598–4605.

Abraham, G. T., & Chatterjee, A. (2003). Approximate asymptotic solution for nonlinear Mathieu equation using harmonic balance based averaging. *Nonlinear Dynamics*, *31*, 347–365.

Abramowitz, M., & Stegun, I. A. (1979). *Handbook of mathematical functions with formulas, graphs and mathematical tables*. Moscow: Nauka.

Belovodsky, V. N., Tsyfansky, S. L., & Beresnevich, V. I. (2002). The dynamics of a vibromachine with parametric excitation. *Journal of Sound and Vibration*, *254*, 897–910.

Bessonov, A. P. (1967). *Osnovji dinamiki mehanizmov s peremennoj massoj zvenjev*. Moscow: Nauka.

Bogolubov, N., & Mitropolski, J. A. (1963). *Asymptotical methods in the theory of nonlinear oscillations*. Delhi: Hindustan Publishing Co.

Byrd, P. F., & Friedman, M. D. (1954). *Handbook of elliptic integrals for engineers and physicists*. Berlin: Springer.

Crespo, G., Proto, A. N., Plastino, A., & Otero, D. (1990). Information-theory approach to the variable-mass harmonic oscillator. *Physical Review A*, *42*, 3608–3617.

Cveticanin, L. (1992). The influence of the reactive force on a nonlinear oscillator with variable parameter. *Transaction of ASME Journal of Vibration and Acoustics*, *114*, 578–580.

Cveticanin, L. (1995). Approximate solution of a time-dependent differential equation. *Meccanica*, *30*, 665–671.

Cveticanin, L. (1998). *Dynamics of machines with variable mass*. London: Gordon and Breach.

Cveticanin, L. (2000). Vibrations in a parametrically excited system. *Journal of Sound and Vibration, 229*, 245–271.

Cveticanin, L. (2012). Oscillator with non-integer order nonlinearity and time variable parameters. *Acta Mechanica, 223*, 1417–1429.

Cveticanin, L. (2013). Van der Pol oscillator with time-variable parameters. *Acta Mechanica, 224*, 945–955.

Cveticanin, L., & Kovacic, I. (2007). Parametrically excited vibrations of the oscillator with strong cubic negative non-linearity. *Journal of Sound and Vibration, 304*, 201–212.

Drogomirecka, H. T. (1997). On integration of a special Ateb-function. *Visnik Lvivskogo Universitetu, Serija mehaniko-matematichna, 46*, 108–110. (in Ukranian).

El-Dib, Y. O. (2001). Nonlinear Mathieu equation and coupled resonance mechanism. *Chaos, Solitons and Fractals, 12*, 705–720.

Esmailzadeh, E., & Goodarzi, A. (2001). Stability analysis of a CALM floating offshore structure. *International Journal of Non-Linear Mechanics, 36*, 917–926.

Esmailzadeh, E., & Jalili, N. (1998). Parametric response of cantilever Timoshenko beams with tip mass under harmonic support motion. *International Journal of Non-Linear Mechanics, 33*, 765–781.

Esmailzadeh, E., Jazar, G. N., & Mehri, B. (1997). Existence of periodic solution for beams with harmonically variable length. *ASME Journal of Vibration and Acoustics, 119*, 485–488.

Flores, J., Solovey, G., & Gill, S. (2003). Variable mass oscillator. *American Journal of Physics, 71*, 721–725.

Gricik, V. V., & Nazarkevich, M. A. (2007). Mathematical models algorythms and computation of Ateb-functions. *Dopovidi NAN Ukraini Series A, 12*, 37–43. (in Ukranian).

Irschik, H., & Holl, H. J. (2004). Mechanics of variable-mass systems—Part 1: Balance of mass and linear momentum. *Applied Mechanics Reviews, 57*, 145–161.

Jazar, G. N. (2004). Stability chart of parametric vibrating systems using energy-rate method. *International Journal of Non-Linear Mechanics, 39*, 1319–1331.

Kamke, E. (1971). *Spravocnik po objiknovennjim differencialjnjim uravnenijam*. Moscow: Nauka.

Krylov, N., & Bogolubov, N. (1943). *Introduction to nonlinear mechanics*. New York: Princenton University Press.

Leach, P. G. L. (1983). Harmonic oscillator with variable mass. *Journal of Physics A: General Physics,16*, Art. No. 019, 3261–3269.

Levi-Civita, T. (1928). Sul moto di un corpo di massa variabile. *Rendiconti del Linci, 329–333*, 621–622.

Li, J., Xu, W., Ren, Z., & Lei, Y. (2005). Maximal Lyapunov exponent and almost-sure stability for stochastic Mathieu–Duffing systems. *Journal of Sound and Vibration, 286*, 395–402.

Meshcherskij, I. V. (1952). *Rabotji po mehanike tel peremennoj massji*. Moscow: Gos. Izd. tehniko-teoret.lit.

Mickens, R. E. (2010). *Truly nonlinear oscillations*. Singapore: World Scientific.

Mond, M., & Cederbaum, G. (1993). Stability analysis of the non-linear Mathieu equation. *Journal of Sound and Vibration, 167*, 77–89.

Nayfeh, A. H., & Mook, D. T. (1979). *Nonlinear oscillations*. New York: Wiley.

Ng, L., & Rand, R. (2002a). Bifurcations in a Mathieu equation with cubic nonlinearities. *Chaos, Solitons and Fractals, 14*, 173–181.

Ng, L., & Rand, R. (2002b). Bifurcations in a Mathieu equation with cubic nonlinearities (Part II). *Communications in Nonlinear Science and Numerical Simulation, 7*, 107–121.

Rhoads, J. F., Shaw, S. W., Turner, K. L., Moehlis, J., DeMartini, B. E., & Zhang, W. (2006). Generalized parametric resonance in electrostatically actuated microelectromechanical oscillators. *Journal of Sound and Vibration, 296*, 797–829.

Sanchez-Ortiz, G. I., & Salas-Brito, A. L. (1995). Chaos in a variable mass relaxation oscillator model for the leaky tap. *Physica D: Nonlinear Phenomena, 89*, 151–168.

Wiegand, M., Scheithauer, S., & Theil, S. (2004). Step proof mass dynamics. *Acta Astronautica, 54*, 631–638.

Xie, G.-Q., Qian, S.-W., & Gu, Z.-Y. (1995). Separation of variables treatment of the time-dependent damped harmonic oscillator with an arbitrary varying mass and with a force quadratic in the velocity under the action of an arbitrary time-varying force. *Physics Letters A*, *207*, 11–16.

Younesian, D., Esmailzadeh, E., & Sedaghati, R. (2005). Existence of periodic solutions for the generalized form of Mathieu Equation. *Nonlinear Dynamics*, *39*, 335–348.

Yuste, B. S. (1991). On Duffing oscillators with slowly varying parameters. *International Journal of Non-Linear Mechanics*, *26*, 671–677.

Zhang, W., Baskaran, R., & Turner, K. L. (2002). Effect of cubic nonlinearity on auto-parametrically amplified resonant MEMS mass sensor. *Sensors and Actuators A*, *102*, 139–150.

Zounes, R. S., & Rand, R. H. (2002). Subharmonic resonance in the non-linear Mathieu equation. *International Journal of Non-Linear Mechanics*, *37*, 43–73.

Chapter 6
Forced Vibrations

In this Chapter the excited vibrations of a pure nonlinear oscillator are analyzed. Two types of excitation forces are considered: the constant force and the periodical force.

For the pure nonlinear oscillator (without a linear term) excited with a constant force the mathematical model is a second order nonhomogeneous differential equation with strong nonlinear term. Using the first integral, the exact value of period of vibration is determined. The pure odd-integer-order oscillator with constant excitation is considered. The approximate solution of the differential equation is obtained in the form of a trigonometric function. The solution includes the exact value of the period of vibration. For the case when additional small perturbation of the pure oscillator acts, the averaging method is applied. Two special cases are considered: one, when the additional term is a function of deflection, and the second, when damping acts.

In this Chapter the vibration of the excited vibrations due to periodical force are also considered. The pure cubic oscillator excited with a harmonic force is analyzed. The method of harmonic balance, as one of the simplest classical solving techniques, is applied for solving of the differential equation of motion. Due to the feature of the method, the form of solution is specified in advance. It is troublesome because the system is strongly nonlinear and the principle of superposition does not work as in the linear system. Attention is paid not only to the so called 'resonant case' and bifurcation points, but also to the case when the frequency of excitation force Ω and the frequency of the excited motion ω differ. The new explanation of the physical properties of the system and also the extension and a generalization to the previous results are given based on the analytical approximate solution.

In this Chapter the special type of external excitation is designed to yield the closed-form solution in terms of the Ateb or Jacobi cn function. Frequency-response curves have been presented to illustrate the response in the frequency domain and the forced vibrations in the time domain.

The investigations reported in this Chapter are of special interest for those who are dealing with the problem of vibration reduction in the oscillator with constant excitation and pure nonlinear restoring force the examples of which can be found in various scientific and engineering systems.

L. Cveticanin, *Strong Nonlinear Oscillators*, Mathematical Engineering,
DOI 10.1007/978-3-319-58826-1_6

6.1 Oscillator with Constant Excitation Force

The mathematical model of the pure nonlinear oscillator with a constant excitation is

$$\ddot{x} + c_\alpha^2 x \, |x|^{\alpha-1} = F_0, \tag{6.1}$$

where F_0 is the additional term to the differential equation (3.2) which considered in Chap. 3. The forced vibrations are considered for the zeroth initial conditions

$$x(0) = 0, \qquad \dot{x}(0) = 0. \tag{6.2}$$

For (6.1) with (6.2) the first integral exists and has the following form

$$\frac{\dot{x}^2}{2} + |x| \, (c_\alpha^2 \frac{|x|^\alpha}{\alpha + 1} - F_0) = 0. \tag{6.3}$$

Assuming that the direction of excitation is constant and does not depend on motion direction, the curves in the $x - \dot{x}$ plane are determined

$$\frac{\dot{x}^2}{2} \pm |x| \left(c_\alpha^2 \frac{|x|^\alpha}{\alpha + 1} - F_0 \right) = 0, \tag{6.4}$$

where the upper sign is for $x \geq 0$ and the other for $x \leqslant 0$. Due to equality of the curves, the analyses is done only for one of them. The curves in $x - \dot{x}$ which correspond to (6.4), when $x \geq 0$ and $x \leqslant 0$, are closed ones and it gives the conclusion that the solution of (6.1) is periodic.

For calculating the period of vibration let us rewrite the relation (6.4) into

$$\frac{dx}{dt} = \sqrt{2F_0 \, |x| - 2c_\alpha^2 \frac{|x|^{\alpha+1}}{\alpha + 1}}. \tag{6.5}$$

i.e.,

$$\frac{dx}{dt} = \sqrt{2F_0 \, |x|} \sqrt{1 - \frac{c_\alpha^2}{F_0} \frac{|x|^\alpha}{\alpha + 1}}. \tag{6.6}$$

For mathematical expediency, we introduce a new variable

$$u = \frac{c_\alpha^2}{F_0} \frac{|x|^\alpha}{\alpha + 1}, \tag{6.7}$$

with time derivative

$$\frac{du}{dt} = \frac{c_\alpha^2}{F_0} \frac{\alpha}{\alpha + 1} |x|^{\alpha-1} \frac{dx}{dt}. \tag{6.8}$$

Substituting (6.7) and (6.8) into (6.5) and separating the variables the following expression is obtained

$$dt = \frac{1}{\alpha u\sqrt{2F_0(1-u)}}\left(\frac{F_0(\alpha+1)u}{c_\alpha^2}\right)^{1/2\alpha} du. \tag{6.9}$$

After suitable transformation and integration half the period of vibration is shown to be

$$\frac{T}{2} = \frac{1}{\alpha\sqrt{2F_0}}\left(\frac{F_0(\alpha+1)}{c_\alpha^2}\right)^{1/2\alpha}\int_0^1 u^{\frac{1-2\alpha}{2\alpha}}(1-u)^{-1/2} du. \tag{6.10}$$

According to the definition of the Beta function given in the Appendix E

$$\int_0^1 u^{\frac{1}{2\alpha}-1}(1-u)^{1/2-1} du = B\left(\frac{1}{2\alpha}, \frac{1}{2}\right), \tag{6.11}$$

the period of vibration is

$$T = \frac{2}{\alpha\sqrt{2F_0}}\left(\frac{F_0(\alpha+1)}{c_\alpha^2}\right)^{1/2\alpha} B\left(\frac{1}{2\alpha}, \frac{1}{2}\right). \tag{6.12}$$

The expression (6.12) is rewritten by using the Gamma function into

$$T = \frac{\sqrt{2}}{\alpha F_0^{(\alpha-1)/2\alpha}}\left(\frac{(\alpha+1)}{c_\alpha^2}\right)^{1/2\alpha}\frac{\Gamma(\frac{1}{2\alpha})\Gamma(\frac{1}{2})}{\Gamma(\frac{1+\alpha}{2\alpha})}. \tag{6.13}$$

Based on the exact period of vibration (6.13) the exact frequency of vibration of (6.1) is obtained

$$\Omega = \frac{2\pi}{T} = \sqrt{2}\pi\alpha F_0^{(\alpha-1)/2\alpha}\left(\frac{c_\alpha^2}{\alpha+1}\right)^{1/2\alpha}\frac{\Gamma(\frac{1+\alpha}{2\alpha})}{\Gamma(\frac{1}{2\alpha})\Gamma(\frac{1}{2})}. \tag{6.14}$$

Let us rewrite (6.13) as

$$T = \left(\frac{F_0^{(1-\alpha)}}{c_\alpha^2}\right)^{1/2\alpha} T^*, \tag{6.15}$$

where

$$T^* = \sqrt{2}\frac{(\alpha+1)^{1/2\alpha}}{\alpha}\frac{\Gamma(\frac{1}{2\alpha})\Gamma(\frac{1}{2})}{\Gamma(\frac{1+\alpha}{2\alpha})}. \tag{6.16}$$

In Table 6.1, the values of T^* for various orders of nonlinearity α are shown.

Table 6.1 The $T^* - \alpha$ values

α	1/3	2/3	1	4/3	5/3	3	5
T^*	10.2604	7.45642	6.28319	5.61931	5.18671	4.32736	3.83111

Analyzing the relations (6.13) i.e. (6.15) and (6.14), it can be concluded:

(1) The period of vibrations depend not only on the order of nonlinearity, but also on the amplitude of the excitation force and coefficient of nonlinearity. For higher values of parameter F_0, the period of vibration is shorter. If F_0 tends to infinity, the period tends to zero. If F_0 is zero the period of vibration is infinitely large and no oscillatory motion exists.

(2) For the linear oscillator ($\alpha = 1$) the period of vibration does not depend on the excitation frequency F_0 and has the well known value

$$T = \frac{2}{c_\alpha} B(\frac{1}{2}, \frac{1}{2}) = \frac{2\pi}{c_\alpha}, \tag{6.17}$$

as $\Gamma(1/2) = \sqrt{\pi}$ and $\Gamma(1) = 1$. The frequency of vibration is

$$\Omega = \frac{2\pi}{T} = c_\alpha. \tag{6.18}$$

(3) Coefficient of the nonlinearity c_α has also a significant influence on the period of vibration: the higher the value of the coefficient, the shorter the period of vibration. The influence of the parameter c_α is not of the same order for various types of oscillators. Considering the relation: $T = f(F_0, \alpha)(c_\alpha)^{-1/\alpha}$ it is evident that for the linear oscillator ($\alpha = 1$) the frequency of vibration directly depends on c_α, while for α tends to infinity the influence of the parameter c_α disappears.

(4) For $F_0 = 1$ and $c_\alpha^2 = 1$, the period of vibration increases with decreasing of the order of nonlinearity. Thus, the higher the order of nonlinearity, the higher the frequency of oscillation.

(5) For the case when the nonlinearity is of odd-integer order, i.e. $\alpha = 2n - 1$, where $n = 1, 2, 3, \ldots$, the differential equation (6.1) transforms into

$$\ddot{x} + c_{2n-1}^2 x^{2n-1} = F_0, \tag{6.19}$$

and the frequency of vibration is

$$\Omega = \sqrt{2\pi}\frac{(2n-1)}{(2n)^{1/2(2n-1)}} \left(c_{2n-1} F_0^{n-1}\right)^{1/(2n-1} \frac{\Gamma(\frac{n}{2n-1})}{\Gamma(\frac{1}{2(2n-1)})}. \tag{6.20}$$

6.1.1 Solution of the Odd-Integer Order Oscillator

For the case when $\alpha = 2n - 1$, where $n = 1, 2, 3, \ldots$, is an odd-integer the approximate solution for (6.1) i.e. (6.19) is assumed in the following form

$$x = F(1 - \cos \Omega t), \tag{6.21}$$

where F is an unknown value and Ω is the exact frequency of vibration (6.20). Substituting (6.21) into (6.1) we have

$$F\Omega^2 \cos \Omega t + c_\alpha^2 F^\alpha (1 - \cos \Omega t)^\alpha = F_0. \tag{6.22}$$

The determination of F using (6.22) is not an easy task. Namely, the harmonic balance method is not directly applicable due to the form of the relation (6.22). For mathematical reasons, let us introduce the series expansion of the power order function (see Gradstein and Rjizhik 1971; Mickens Mickens (2004)) into (6.22)

$$\begin{aligned} F_0 = F\Omega^2 \cos \Omega t + F^\alpha c_\alpha^2 [1 + \sum_{n=1} \frac{\alpha(\alpha-1)\ldots(\alpha-2n+1)}{(2n)!} \cos^{2n} \Omega t \\ - \sum_{n=1} \frac{\alpha(\alpha-1)\ldots(\alpha-(2n-1)+1)}{(2n-1)!} \cos^{2n-1} \Omega t)]. \end{aligned} \tag{6.23}$$

and the series expansion for functions $\cos^{2n}(\Omega t)$ and $\cos^{2n-1} \Omega t$

$$\begin{aligned} \cos^{2n}(\Omega t) &= \frac{1}{2^{2n}} \left(\sum_{k=0}^{n-1} 2 \binom{2n}{k} \cos 2(n-k)\omega t + \binom{2n}{2} \right), \\ \cos^{2n-1}(\Omega t) &= \frac{1}{2^{2n-2}} \sum_{k=0}^{n-1} \binom{2n-1}{k} \cos(2n-2k-1)(\Omega t). \end{aligned} \tag{6.24}$$

Using (6.23) with (6.24), and separating only the terms with $\cos \Omega t$ the following algebraic equation is obtained

$$F\Omega^2 - F^\alpha c_\alpha^2 \sum_{n=1} \frac{\alpha(\alpha-1)\ldots(\alpha-(2n-1)+1)}{(2n-1)!} \frac{1}{2^{2n-2}} \binom{2n-1}{n-1} = 0. \tag{6.25}$$

It follows that

$$F = \left(\frac{\Omega^2}{q c_\alpha^2} \right)^{1/(\alpha-1)}, \tag{6.26}$$

where

$$q = \sum_{n=1} \frac{\alpha(\alpha-1)\ldots(\alpha-(2n-1)+1)}{(2n-1)!} \frac{1}{2^{2n-2}} \binom{2n-1}{n-1}. \tag{6.27}$$

Substituting (6.14) into (6.26) we have

$$F = \left(\frac{1}{q} (\frac{1}{\alpha+1})^{1/\alpha} (\frac{\Gamma(\frac{1+\alpha}{2\alpha})}{\sqrt{2\pi}\alpha\Gamma(\frac{1}{2\alpha})})^2 \right)^{1/(\alpha-1)} (\frac{F_0}{c_\alpha^2})^{1/\alpha}. \tag{6.28}$$

For a certain oscillator the amplitude of vibration depends on the excitation value and coefficient of the nonlinear term: The higher the excitation F_0, the higher is the amplitude of vibration F. The influence of the coefficient of nonlinearity c_α on the amplitude of vibration F is the same as on the frequency of vibration (6.14).

1. For the linear oscillator ($\alpha = 1$) the amplitude of vibration is, according to (6.28), $F = F_0/c_\alpha^2$ and for $\Omega = c_\alpha^2$, and the exact solution

$$x = \frac{F_0}{c_\alpha^2}[1 - \cos(c_\alpha t)]. \tag{6.29}$$

2. For the pure cubic oscillator ($\alpha = 3$) the exact period of vibration is

$$T = \frac{\sqrt{2}}{3F_0^{1/3}} (\frac{(4)}{c_\alpha^2})^{1/6} \frac{\Gamma(\frac{1}{6})\Gamma(\frac{1}{2})}{\Gamma(\frac{2}{3})} = \frac{4.3274}{\sqrt[3]{F_0 c_\alpha}}, \tag{6.30}$$

and the corresponding angular frequency is

$$\Omega = 1.452\sqrt[3]{F_0 c_\alpha}. \tag{6.31}$$

The amplitude of vibration is according to (6.28)

$$F = 0.74981 \left(\frac{F_0}{c_\alpha^2}\right)^{1/3}. \tag{6.32}$$

In general the approximate solution of the oscillator with a pure cubic term is

$$x = 0.74981 \left(\frac{F_0}{c_\alpha^2}\right)^{1/3} (1 - \cos(1.452t\sqrt[3]{F_0 c_\alpha})). \tag{6.33}$$

Some numerical examples are considered and the analytical solutions are compared with numerical ones:

(a) For $F_0 = 1$ and $c_\alpha^2 = 1$ and the differential equation

$$\ddot{x} + x^3 = 1, \tag{6.34}$$

the approximate solution is

$$x_A = 0.74981(1 - \cos(1.452t)). \tag{6.35}$$

(b) For $F_0 = 0.5$ and $c_\alpha^2 = 1$ the mathematical model of the oscillator is

$$\ddot{x} + x^3 = 0.5, \tag{6.36}$$

the approximate analytical solution is

$$x_A = 0.59512(1 - \cos(1.1525t). \tag{6.37}$$

In Fig. 6.1 the analytical (x_A) and numerical (x_N) solutions for $F_0 = 0.5$ and $F_0 = 1$ are plotted.

It can be concluded that the approximate solution is on the top of the numerical one. The maximal difference is for the amplitudes, but even this difference is smaller than 5% and is negligible.

2. For the pure fifth order oscillator

$$\ddot{x} + x^5 = 0.5 \tag{6.38}$$

the approximate analytical solution is

$$x_A = 0.621\,19(1 - \cos(1.242\,9t)). \tag{6.39}$$

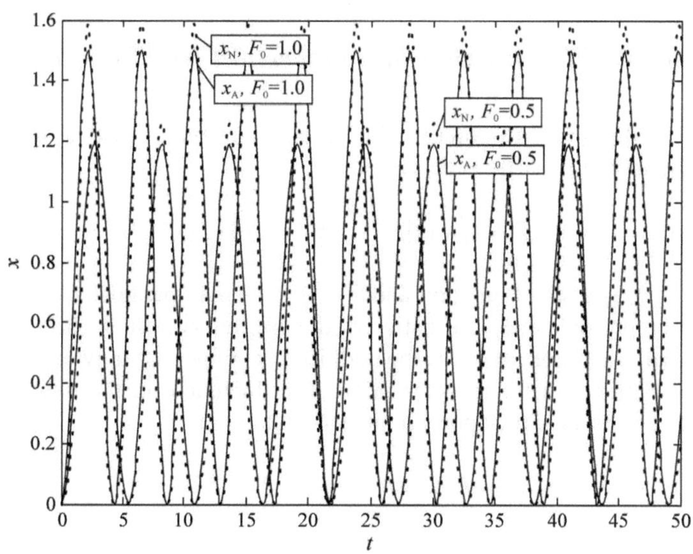

Fig. 6.1 Analytical $x_A - t$ (*full line*) and numerical $x_N - t$ (*dashed line*) solutions of pure cubic oscillator for $F_0 = 1$ and $F_0 = 0.5$

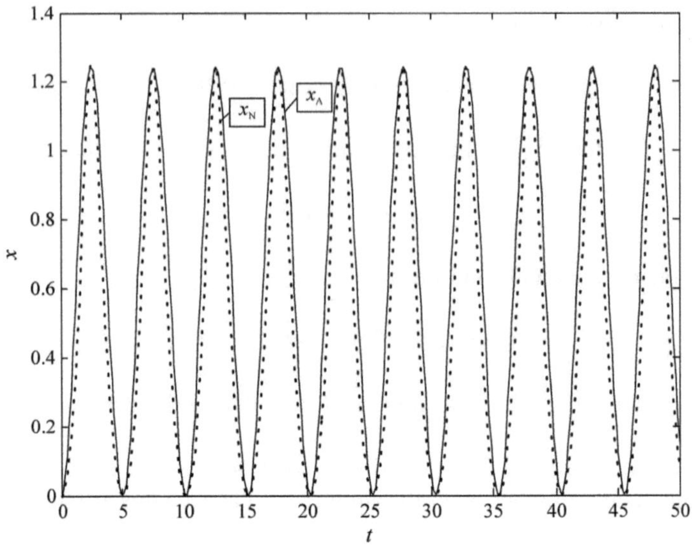

Fig. 6.2 Analytical $x_A - t$ (*full line*) and numerical $x_N - t$ (*dashed line*) solutions of pure fifth order oscillator

In Fig. 6.2 the analytical solution x_A, (6.39), is compared with numerical solution x_N to (6.38).

The difference between the solutions is seen to be minimal.

6.1.2 The Oscillator with Additional Small Nonlinearity

The model of the pure odd-order nonlinear oscillator with constant excitation and additional small nonlinearity is described as

$$\ddot{x} + c_\alpha^2 x \left|x\right|^{\alpha-1} = F_0 + \varepsilon f(x, \dot{x}), \tag{6.40}$$

where $\varepsilon f(x, \dot{x})$ is a small function and $\varepsilon \ll 1$ is a small parameter. The additional nonlinearity represents the perturbation of the oscillatory system (6.1). It is known that in the real oscillatory systems the additional nonlinearity may cause amplitude and phase variations. Such a phenomena should be included into the assumed solution of (6.40). Let us assume the approximate trial solution as

$$x = F - F^*(t) \cos \Psi(t) \equiv F - F^* \cos \Psi, \tag{6.41}$$

and its first time derivative

$$\dot{x} = F^* \Omega \sin \Psi, \tag{6.42}$$

where $F^*(t)$ and $\theta(t)$ are unknown time t functions and

$$\Psi = \Omega t + \theta(t) \equiv \Omega t + \theta. \tag{6.43}$$

The amplitude and phase of vibration, $F^*(t)$ and $\theta(t)$, are the perturbed values of the previously calculated one, F and θ.The solution (6.41) with (6.43) and the assumed first time derivative (6.42) have the same form as (6.21) for (6.1) but with time variable amplitude and phase.

Let us determine the first time derivative of (6.41) and compare it with the assumed one. It can be seen, that the assumed time derivative (6.42) and the calculated one

$$\dot{x} = F^*\Omega \sin\Psi + F^*\dot{\theta} \sin\Psi - \dot{F}^* \cos\Psi, \tag{6.44}$$

are equal if

$$F^*\dot{\theta} \sin\Psi - \dot{F}^* \cos\Psi = 0. \tag{6.45}$$

Substituting (6.41), the time derivative of (6.42) and its time derivative into (6.40) we obtain

$$\begin{aligned} &\dot{F}^*\Omega \sin\Psi + F^*\Omega^2 \cos\Psi + F^*\dot{\theta}\Omega \cos\Psi + c_\alpha^2 (F - F^* \cos\Psi)^\alpha \\ &+F_0 + \varepsilon f((F - F^* \cos\Psi), F^*\Omega \sin\Psi). \end{aligned} \tag{6.46}$$

The two first order differential equations (6.45) and (6.46) represent the rewritten version of the second order differential equation (6.40) in the new variables F^* and θ. Equations (6.45) and (6.46) are coupled and linear for $\dot{F}^*$ and $\dot{\theta}$. Separating the variables $\dot{F}^*$ and $\dot{\theta}$ in (6.45) and (6.46) the two uncoupled equations for $\dot{F}^*$ and $\dot{\theta}$ are obtained

$$\begin{aligned} F^*\dot{\theta}\Omega &= -c_\alpha^2 (F - F^* \cos\Psi)^\alpha \cos\Psi - F^*\Omega^2 \cos^2\Psi \\ &\quad + F_0 \cos\Psi + \varepsilon f((F - F^* \cos\Psi), F^*\Omega \sin\Psi) \cos\Psi, \end{aligned} \tag{6.47}$$

$$\begin{aligned} \dot{F}^*\Omega &= -c_\alpha^2 (F - F^* \cos\Psi)^\alpha \sin\Psi - F^*\Omega^2 \cos\Psi \sin\Psi \\ &\quad + F_0 \sin\Psi + \varepsilon f(F - F^* \cos\Psi, F^*\Omega \sin\Psi) \sin\Psi. \end{aligned} \tag{6.48}$$

In general, the two coupled Eqs. (6.47) and (6.48) can not be solved analytically for F^* and θ. Observing that the right-hand sides of (6.47) and (6.48) are periodic in Ψ with period 2π, the averaging procedure over this period is introduced. It gives the two averaged expressions in the new variables F^* and θ

$$\dot{F}^* = \frac{\varepsilon}{2\pi\Omega} \int_0^{2\pi} f((F - F^* \cos\Psi), F^*\Omega \sin\Psi) \sin\Psi d\Psi, \tag{6.49}$$

$$F^*\dot{\theta} = -\frac{c_\alpha^2}{2\pi\Omega} \int_0^{2\pi} (F - F^* \cos\Psi)^\alpha \cos\Psi d\Psi - \frac{1}{2} F^*\Omega$$
$$+ \frac{\varepsilon}{2\pi\Omega} \int_0^{2\pi} f((F - F^* \cos\Psi), F^*\Omega \sin\Psi) \cos\Psi d\Psi. \tag{6.50}$$

Solving the averaged differential equations (6.49) and (6.50) for the initial conditions (6.2), i.e.,

$$F^*(0) = F, \qquad \theta(0) = 0, \tag{6.51}$$

the approximate amplitude and phase of vibration are obtained.

Some special cases of additional terms are considered.

1. For the case when the small additional term is a function only of x the relation (6.49) transforms into

$$\dot{F}^* = 0, \tag{6.52}$$

i.e., for the initial conditions (6.51), it is

$$F^* = F. \tag{6.53}$$

Substituting (6.53) into (6.50) and integrating, it follows

$$\theta = -\frac{c_\alpha^2 F^{\alpha-1} t}{2\pi\Omega} \int_0^{2\pi} (1 - \cos\Psi)^\alpha \cos\Psi d\Psi - \frac{1}{2}\Omega t$$
$$+ \frac{\varepsilon t}{2\pi\Omega F} \int_0^{2\pi} f(F - F\cos\Psi) \cos\Psi d\Psi. \tag{6.54}$$

The approximate analytical solution is

$$x = F - F\cos(\frac{\Omega t}{2} - \frac{c_\alpha^2 F^{\alpha-1} t}{\Omega} \int_0^{2\pi} (1 - \cos\Psi)^\alpha \cos\Psi d\Psi + \frac{\varepsilon t}{\Omega F} \int_0^{2\pi} f(F - F\cos\Psi) \cos\Psi d\Psi). \tag{6.55}$$

The amplitude of vibration is independent of the value of the additional term, but the frequency of vibration is modified.

2. For the case when the small additional term is a function of damping $\dot{x}$

$$\ddot{x} + c_\alpha^2 x\,|x|^{\alpha-1} = F_0 + \varepsilon f(\dot{x}), \tag{6.56}$$

the averaged differential equations are

$$\dot{F}^* = \frac{\varepsilon}{\Omega}\int_0^{2\pi} f(F^*\Omega \sin\Psi)\sin\Psi d\Psi, \tag{6.57}$$

$$\dot{\theta} = -\frac{c_\alpha^2 F^\alpha}{\Omega F^*}\int_0^{2\pi} (1 - \frac{F^*}{F}\cos\Psi)^\alpha \cos\Psi d\Psi - \frac{1}{2}\Omega. \tag{6.58}$$

The amplitude of vibration is a function of the additional nonlinearity, but the phase angle is independent on that value.

Due to damping during the long time ($t \to \infty$) the oscillatory behavior of the system disappears and the motion stops at a fixed position which is the steady-state solution of the equation

$$c_\alpha^2 x^\alpha = F_0,$$

i.e.,

$$x = (\frac{F_0}{c_\alpha^2})^{1/\alpha}.$$

The coefficient of damping has no influence on the steady state position of the system. The higher the damping coefficient, the shorter the time for the system to get into the fixed position.

6.1.3 Examples

(1) To prove the correctness of the suggested procedure the result of CAM is applied for solving of a linear differential equation with a constant excitation and a small linear damping

$$\ddot{x} + c_\alpha^2 x = F_0 - \varepsilon b\dot{x}, \tag{6.59}$$

where $\varepsilon b \ll 0$ is the damping coefficient. According to (6.55) the approximate analytical solution is

$$x = \frac{F_0}{c_\alpha^2}(1 - \exp(-\frac{\varepsilon bt}{2})\cos(c_\alpha t)). \tag{6.60}$$

The exact analytical solution of (6.59) for initial conditions (6.2) is

$$x = \frac{F_0}{c_\alpha^2} + \frac{F_0}{2c_\alpha^2}\sqrt{1+\frac{(\varepsilon b)^2}{\omega^2}}\exp(-\frac{\varepsilon bt}{2})$$
$$[-\frac{2}{\sqrt{1+\frac{(\varepsilon b)^2}{\omega^2}}}\cos(\omega t) - \frac{\varepsilon b}{\sqrt{\omega^2+(\varepsilon b)^2}}\sin(\omega t)], \tag{6.61}$$

where

$$\omega = \sqrt{c_\alpha^2 - \frac{(\varepsilon b)^2}{4}}. \tag{6.62}$$

Comparing (6.60) with the exact analytical solution of (6.59) it can be concluded that for $(\varepsilon b)^2 \approx 0$ the expression (6.61) simplifies into (6.60). Due to the negligible difference between the relations (6.61) and (6.60), which describe the transient motion to the steady state position $x_{t\to\infty} = \frac{F_0}{c_\alpha^2}$, it is obvious that the averaging method presented in this section gives very accurate results.

(2) The suggested averaged procedure is applied for solving a pure cubic oscillator with constant excitation and small linear damping,too

$$\ddot{x} + x^3 = F_0 - \varepsilon b\dot{x}, \tag{6.63}$$

where εb is the constant coefficient. According to (6.57) the approximate amplitude is

$$F^* = F\exp(-\frac{\varepsilon b}{2}t), \tag{6.64}$$

and the approximate solution is

$$x = F - F\exp(-\frac{\varepsilon b}{2}t)\cos((\frac{1}{2}\Omega + \frac{3c_\alpha^2F^2}{2\Omega})t + \frac{3c_\alpha^2F^2}{8\varepsilon b\Omega}(1-\exp(-\varepsilon bt))). \tag{6.65}$$

For the numerical case when $\varepsilon b = 0.01,\ F_0 = 1$ and $c_\alpha^2 = 1$ the analytical solution is

$$x_A = 0.74981 - 0.74981\exp(-0.005t)\cos((1.31)t + 14.233(1-\exp(-0.01t)). \tag{6.66}$$

The relation (6.66) describes the transient motion to a fixed position $x_A = 0.999\,75$.

In Fig. 6.3 the analytical solution x_A and the numerical x_N for (6.63) are compared.

In Fig. 6.4 the $x - t$ diagrams obtained numerically for $\varepsilon b = 0.1; 0.5; 0.75,\ F_0 = 1$ and $c_\alpha^2 = 1$ for large value of t are plotted. It is evident that for large t the oscillator tends to a fixed position which is independent on the coefficient of damping. The value of the damping coefficient has no influence on the steady state position, as it was previously stated after the analytical analysis.

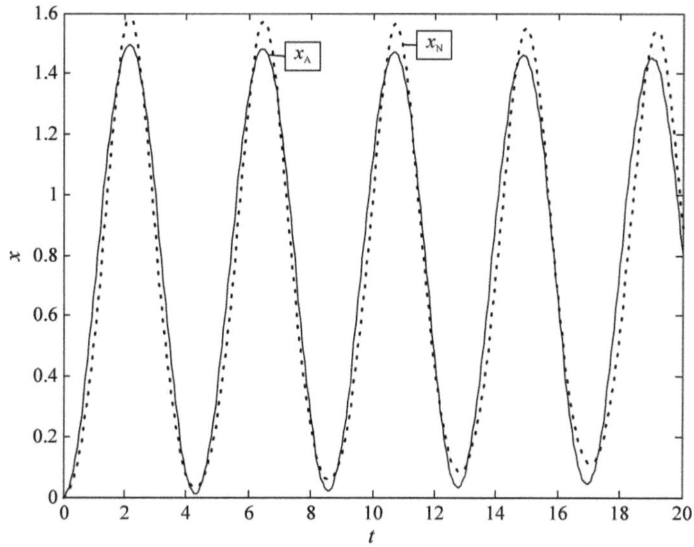

Fig. 6.3 Comparison of the analytical x_A (*full line*) and the numerical x_N (*dashed line*) solutions

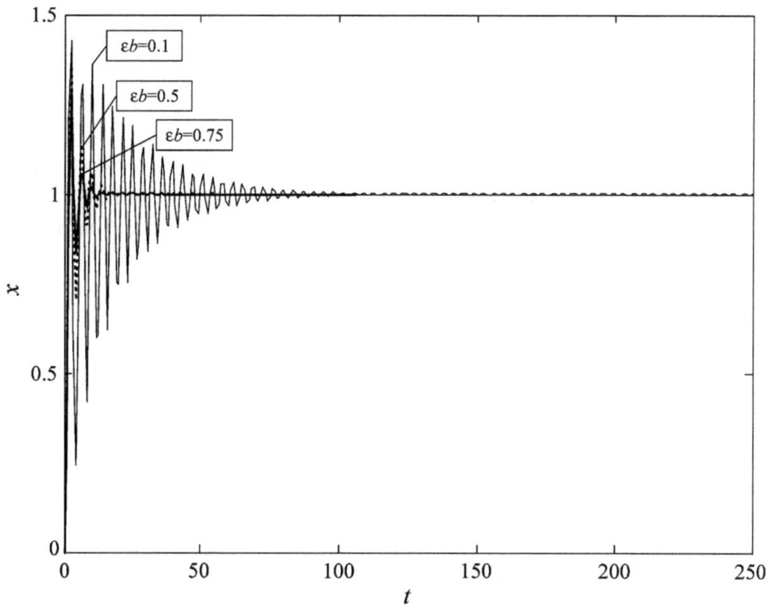

Fig. 6.4 The x-t diagrams for various values of the damping parameter

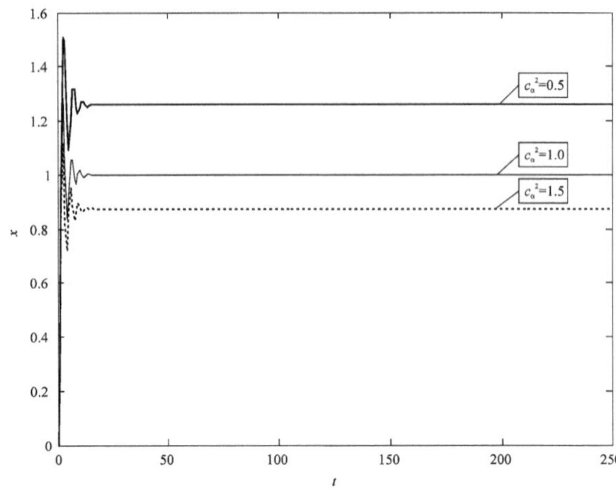

Fig. 6.5 The x-t diagrams for various values of the coefficient of nonlinearity: $c_a^2 = 0.5$, $c_a^2 = 1.0$ and $c_a^2 = 1.5$

In Fig. 6.5 the time-history diagrams for $\varepsilon b = 0.1,\ F_0 = 1$ and various values of coefficient of nonlinearity $c_\alpha^2 = 0.5; 1; 1.5$ are plotted. The result obtained for the steady-state solution is compared with analytically obtained one

$$x_A = \sqrt[3]{\frac{F_0}{c_\alpha^2}} = const. \tag{6.67}$$

Namely, for $c_\alpha^2 = 0.5$ the analytical value is 1.2596 and the numerically calculated one is 1.259921; for $c_\alpha^2 = 1$ we have $x_A = 0.999\,75$ and $x_N = 1$ and for $c_\alpha^2 = 1.5$ we have $x_A = 0.87336$ and $x_N = 0.873580$. Comparing the analytical and numerical values it can be seen that the difference is negligible.

6.1.4 Conclusion

The following is concluded:

1. For the pure nonlinear oscillator with constant excitation the period and angular frequency of vibration in the closed analytical form can be calculated. The period and also the angular frequency of vibration depend on the order of nonlinearity, value of excitation and the coefficient of nonlinearity.

For higher values of excitation the period of vibration is shorter and the frequency is higher. If the excitation tends to zero, the frequency is zero. If the excitation tends to infinity, the period of vibration tends to zero. For the linear oscillator the period of vibration is independent on the excitation.

Increasing the coefficient of nonlinearity the period of vibration decreases The influence of the parameter c_α is not of the same order for various types of oscillators. Namely, for the linear oscillator the angular frequency is directly dependent on coefficient. For higher order of nonlinearity, the influence of coefficient on the period of vibration is smaller and tends to disappear for large order (α tends to infinity).

2. The amplitude of vibration of the pure nonlinear oscillator with constant excitation also depends on the excitation value, order and coefficient of nonlinearity. In a certain oscillator the amplitude of vibration increases with increasing of the excitation value. For the oscillator with fixed excitation the amplitude of vibration is higher for smaller value of coefficient of nonlinearity when the order of nonlinearity is fixed.

3. For the case when additional small perturbation to the oscillator exists, the amplitude and frequency of vibration are time dependent.

4. If the perturbation of the pure nonlinear and excited oscillator is of damping type, for the long time t, the oscillatory motion stops, and the system gets into a constant steady state position. The distance from to zero is smaller for higher values of coefficient of nonlinearity, smaller excitation and higher order of nonlinearity. Besides, the value of the damping coefficient has no influence on the steady state position.

5. The suggested averaging method gives approximate analytical solutions which describe also the transient motion. The difference between analytical and numerical solutions is negligible.

6. The investigations reported in the paper are of special interest for engineers who are dealing with the problem of vibration reduction in the oscillator with constant excitation. It can be seen that the proper choice of material of oscillator (for example, with high coefficient of nonlinearity) decreases the vibration level and when even small damping exists the oscillations disappear.

6.2 Harmonically Excited Pure Nonlinear Oscillator

In general, the pure nonlinear oscillator with harmonic excitation is described as

$$\ddot{x} + c_{\alpha}^{2} x \left| x \right|^{\alpha - 1} = F_0 \cos(\Omega t), \tag{6.68}$$

where F_0 and Ω are the amplitude and frequency of the excitation force. The expression (6.68) is a strong nonlinear nonhomogeneous second order differential equation whose solution is not easy to be determined. Namely, due to strong nonlinearity the principle of superposition of the homogenous and nonhomogeneous solutions (which is valid for the linear differential equation) does not matter. So, the most important and complex problem is how to suggest the form of the solution. On the basis of the assertion of Burton (1984), that if the waveform of oscillation of a nonlinear oscillator is reasonably close to simple harmonic oscillation, the solution is assumed to have two parts: one, which considers the frequency of the excitation force and the other, where the frequency properties of the free oscillator are treated. The unknown parameters of the solution are determined by applying of the generalized harmonic balance method (Cveticanin 2011).

In this section, we treat the pure odd-integer order differential equation with periodical excitation (Cveticanin 2012), given as

$$\ddot{y} + c_{2n-1}^2 y^{2n-1} = F_0 \cos(\Omega t), \tag{6.69}$$

where $\alpha = 2n - 1$ ($n = 1, 2, \ldots$) and $c_\alpha^2 \equiv c_{2n-1}^2 = const. > 0$. The mathematical model (6.69) is a strong nonlinear ordinary second order differential equation with the time variable function y.

6.2.1 Pure Odd-Order Nonlinear Oscillator

As the motion of the oscillator (6.69) is caused by the periodical excitation force with amplitude F_0 and frequency Ω, it is assumed that the oscillator (6.69) vibrates with the frequency Ω, which is fixed by excitation (it is the excitation frequency), and also with unknown frequency ω, which is dependent on the specific characteristics of the nonlinear system. Therefore in this paper the solution of the nonlinear oscillator (6.69)is assumed in the following form

$$y = A\cos(\omega t) + Y\cos(\Omega t), \tag{6.70}$$

where Ω is the excitation frequency and A, Y and ω are unknown values. For zero initial displacement and velocity

$$y(0) = 0, \qquad \dot{y}(0) = 0,$$

we have $A = -Y$, and the relation (6.70) transforms into

$$y = Y(\cos\Omega t - \cos\omega t), \tag{6.71}$$

where Y and ω are the amplitude and the frequency of the excited vibration. The form of the solution (6.71) is exactly the same as for the linear differential equation.

Substituting the assumed solution (6.71) into (6.69) we have

$$c_{2n-1}^2 Y^{2n-1}(\cos\omega t - \cos\Omega t)^{2n-1} + Y(\omega^2\cos\omega t - \Omega^2\cos\Omega t) - F_0\cos\Omega t = 0. \tag{6.72}$$

Introducing the series expansion of $(\cos\omega t - \cos\Omega t)^{2n-1}$ into (6.72) and also the expressions of the even and odd order cosine functions (Gradstein and Rjizhik 1971)

$$\cos^{2n}(\omega t) = \frac{1}{2^{2n}}\left(\sum_{k=0}^{n-1} 2\binom{2n}{k}\cos 2(n-k)\omega t + \binom{2n}{2}\right), \tag{6.73}$$

$$\cos^{2n-1}(\omega t) = \frac{1}{2^{2n-2}} \sum_{k=0}^{n-1} \binom{2n-1}{k} \cos(2n-2k-1)\omega t, \tag{6.74}$$

and using the harmonic balance method to eliminate the terms with the same order of trigonometric function, it yields

$$\cos \Omega t : c_{2n-1}^2 Y^{2n-1} s - Y\Omega^2 - F_0 = 0, \tag{6.75}$$

$$\cos \omega t : \omega^2 Y - c_{2n-1}^2 Y^{2n-1} s = 0, \tag{6.76}$$

where s is the coefficient of the nonlinear term calculated according to (6.73) and (6.74). The coefficient with nonlinear term s in both relations is equal due to symmetry. The order of the nonlinear algebraic equation (6.75) is equal to that of the nonlinear differential equation (6.69), i.e.$(2n-1)$. The only unknown variable in the Eq. (6.75) is Y which depends on c_{2n-1}^2, F_0 and Ω. Solving (6.75), the amplitude of vibration is obtained. Mathematically, the number of real solutions of (6.75) varies from 1 to $(2n-1)$ by changing the parameter values c_{2n-1}^2, F_0 and Ω. It is important to say that Eq. (6.75) and its solutions are the basic ones for obtaining the frequency response function (as achieved by Worden (1996) for the cubic Duffing oscillator). Equation (6.76) is a new one which introduces the restriction to the number and properties of the solution Y based on the physical characteristics of the oscillator. The main objective is to give restrictions for the excited frequency of vibration to be real and positive ($\omega^2 > 0$).

Substituting the solution of (6.75) into (6.76) the frequency ω is calculated. Comparing the relations (6.75) and (6.76) we obtain

$$\omega^2 - \Omega^2 = \frac{F_0}{Y}, \tag{6.77}$$

and the approximate analytic solution (6.71) transforms into

$$y = \frac{F_0}{\omega^2 - \Omega^2}(\cos \Omega t - \cos \omega t). \tag{6.78}$$

In spite of the fact that (6.77) and (6.78) have the same form as for the linear oscillator, the physical sense of the problem differs:

1. In general, for the nonlinear oscillator when $n \neq 1$, the excited frequency ω is not constant but depends on the amplitude Y (see (6.76)). Only for the linear system, when $n = 1$ is the excited frequency constant ($\omega = \sqrt{c_{2n-1}^2}$).

2. By analyzing Eq. (6.77) it is obvious that the difference $(\omega^2 - \Omega^2)$ is never zero for a nonlinear oscillator when $n \neq 1$. The frequency of excited vibration and of the excitation force may be equal only for the linear oscillator when $n = 1$. This frequency is called the 'resonant' one in the linear theory of oscillations.

3. Due to the statement given in 2., it follows that the amplitude of vibration of the nonlinear oscillator ($n \neq 1$)

$$Y = \frac{F_0}{\omega^2 - \Omega^2}, \tag{6.79}$$

does not increases indefinitely. For $n \neq 1$ the amplitude of vibration is bounded. Due to the equality of the frequencies of excitation and the excited system, the amplitude of vibration tends to infinity only for the linear oscillator when $n = 1$. In the theory of linear vibrations this state is called the 'resonant' one and the corresponding amplitude is the 'resonance response'.

4. The differential equation (6.69), i.e., the algebraic equation (6.75) has 1, 2, ..., $(2n - 1)$ real solutions, but the fact that ω^2 has to be a positive value gives some limitation to the obtained solutions Y. For $F_0 > 0$ and when only one real solution for Y exists, its sign is positive. For a positive value of Y the relation (6.77) i.e., (6.79) is satisfied for $(\omega^2 - \Omega^2) > 0$, i.e., $\omega > \Omega$. For $F_0 > 0$ and $\Omega > \omega$ relation (6.79) requires only the negative values of amplitude Y. It means, that among all of the real solutions of Y the only one which satisfies the relation (6.79) is that which is negative and with smallest absolute value. In the real mechanical system the amplitude of vibration increases by increasing of the excitation frequency up to the value which is smaller than the excited frequency of the system. By further increasing of the frequency of excitation which is higher than the excited frequency of mechanical system the amplitude jumps to a small value which tends to zero for infinitely high value of the excitation frequency.

5. According to the previous statement 4., two regions for oscillations are detected: one, for $\omega > \Omega$, when the excited amplitude Y needs to be positive (6.77), and the second for $\Omega > \omega$ when Y needs to be negative. For the low values of excitation frequency, when $\omega - \Omega > 0$, the averaged solution is

$$y^* = Y \cos(\Omega t), \tag{6.80}$$

which describes the long period vibrations on which the excited vibrations with frequency ω are added. For high values of excitation frequency when $\Omega - \omega > 0$, the excited vibrations are with long period $2\pi/\omega$ and the averaged solution has the form

$$y^* = Y \cos(\omega t). \tag{6.81}$$

6. Using the relations (6.75) and (6.76) the approximate solution for (6.69), has the form

$$y = -\frac{2F_0}{\omega^2 - \Omega^2} \sin(\frac{\Omega - \omega}{2} t) \sin(\frac{\Omega + \omega}{2} t), \tag{6.82}$$

which is the same as for the linear oscillator. In the region where the difference $(\omega^2 - \Omega^2)$ changes the sign the effect of flutter appears and the analytical shape of the envelope amplitude is approximately described as

$$y^* = \pm \frac{2F_0}{\omega^2 - \Omega^2} \sin(\frac{\Omega - \omega}{2} t). \tag{6.83}$$

6.2.2 Bifurcation in the Oscillator

In general, the changes in the oscillation are connected with bifurcation in the oscillatory system. Namely, for certain parameter values the number of real solutions of (6.69) is changing or the solutions change their properties. It is of interest to determine the relation between the excitation parameters and the parameters of the oscillatory system for which these changes appear.

Due to (6.77), the excitation frequency is expressed as

$$\Omega = \sqrt{\frac{c_{2n-1}^2 Y^{2n-1} s - F_0}{Y}}. \tag{6.84}$$

and the excitation amplitude

$$F_0 = c_{2n-1}^2 s Y^{2n-1} - \Omega^2 Y. \tag{6.85}$$

Calculating the partial derivatives $\partial\Omega/\partial Y$ of (6.84) and $\partial F_0/\partial Y$ of (6.85) and equating them to zero the conditions for an extreme value which is a function of amplitude Y are obtained

$$(2n-2)\gamma s Y^{2n-1} + F_0 = 0, \tag{6.86}$$

and

$$c_{2n-1}^2 s(2n-1) Y^{2n-2} - \Omega^2 = 0. \tag{6.87}$$

Eliminating Y from relations (6.86) and (6.87), the connection between F_0 and Ω follows

$$\frac{\Omega^{2n-1}}{F_0^{n-1}} = \frac{\sqrt{\gamma s}(2n-1)^{\frac{2n-1}{2}}}{(2n-2)^{n-1}}. \tag{6.88}$$

Condition (6.88) expresses the bifurcation condition for the differential equation (6.69) and gives the 'jump' in the oscillatory motion. The condition (6.88) expresses the relation between the excitation frequency and amplitude for oscillators with various orders and coefficients of nonlinearity. For the pure cubic oscillator, where $n = 2$ relation (6.88) gives

$$\frac{\Omega^3}{F_0} = 3.8971\sqrt{c_{2n-1}^2}. \tag{6.89}$$

For $F_0 = 1$ Eq. (6.89) transforms into $\Omega = 1.5737\sqrt[6]{c_{2n-1}^2}$. This result is obtained as the condition for jump-up by Brennan et al. (2008). For the pure fifth-order oscillator, where $n = 3$, we have

$$\frac{\Omega^5}{F_0^2} = 8.7346\sqrt{c_{2n-1}^2}. \tag{6.90}$$

For $F_0 = 1$ the Eq. (6.90) gives $\Omega = 1.5426\sqrt[10]{c_{2n-1}^2}$. Comparing (6.89) and (6.90) for $F_0 = 1$ and $c_{2n-1}^2 \le 1$ it is concluded that the bifurcation frequency is smaller for higher order of nonlinearity and for the small value of coefficient of nonlinearity ($c_{2n-1}^2 \ll 1$), the bifurcation point (6.88) tends to zero.

The bifurcation value (6.88) separates two motion regions. If

$$\frac{\Omega^{2n-1}}{F_0^{n-1}} < \frac{\sqrt{sc_{2n-1}^2}(2n-1)^{\frac{2n-1}{2}}}{(2n-2)^{n-1}}, \tag{6.91}$$

the algebraic equation (6.75) has only one real solution which decreases from the value $Y_1 = \sqrt[2n-1]{F_0/(c_{2n-1}^2 s)}$ to a boundary value which is reached for (6.88). Using these values for Y_1, according to (6.76), the excited frequency ω is larger than the excitation frequency Ω. The period of vibration of the excitation force $2\pi/\Omega$ is higher than for excited vibration $2\pi/\omega$ and the appropriate averaged solution is $y = Y\cos(\Omega t)$.

For the boundary condition (6.91) Eq. (6.75) has three real solutions: one, which is a positive and corresponds to the previously mentioned value for Y_1, and an additional two which are equal and negative ($Y_2 = Y_3$, $Y_2 < 0$, $Y_3 < 0$). The other solutions, if they exist, are complex and conjugate and have nothing to do with the motion properties.

For

$$\frac{\Omega^{2n-1}}{F_0^{n-1}} > \frac{\sqrt{sc_{2n-1}^2}(2n-1)^{\frac{2n-1}{2}}}{(2n-2)^{n-1}}, \tag{6.92}$$

the positive solution Y_1 increases and the two previously equal negative solutions differ: one, Y_2, tends to decrease and to have the absolute value as the positive solution, and the other negative solution Y_3 tends to zero. Only the last mentioned solution Y_3 satisfies the conditions for the real value of excited vibration ω and represents the realized amplitude of vibration. For the negative value of solution Y_3 the difference between ω^2 and Ω^2 is negative and means that the excited frequency is smaller than the excitation frequency. The period of excited vibration $2\pi/\omega$ is higher than the period of excitation force $2\pi/\Omega$. Due to that fact the averaged solution has the form (6.81).

Brennan et al. (2008) suggest that the other bifurcation point is for Y_1 maximal. The positive real solution Y_1 of (6.75) increases from zero to infinitely high value. However, due to condition (6.76) the solution Y_1 has a real physical meaning only for the values which correspond to excitation with frequency $\Omega > 0$ to a boundary value (6.88). This value is the maximal value for Y_1 and the 'jump down' phenomena occurs. The conclusion is that the jump-down and the jump-up phenomena are the same in the pure nonlinear oscillator described with differential equation (6.69).

It is of interest to obtain the conditions for which the excited oscillation tends to zero in spite of the fact that the excitation force acts. This phenomenon occurs when the previously mentioned real solution Y_3 is marginally nonzero and the difference between the absolute value of the other negative solution Y_2 and of the positive solution Y_1 is negligible ($|Y_2| \approx Y_3$). For the small value of excited amplitude $Y_3 \approx \Delta \ll 1$, the frequency of excited vibration, calculated according to (6.76), is also very small $\omega \ll 1$. Then, only the component of vibration at this excitation frequency appears, and the approximate motion of the oscillator is

$$y = -\Delta(\cos \Omega t - 1) = 2\Delta \sin^2(2\Omega t). \tag{6.93}$$

For that case the averaged relation is

$$y^* = -\Delta. \tag{6.94}$$

6.2.3 Harmonically Forced Pure Cubic Oscillator

As a special case, let us assume the harmonically excited Duffing oscillator with pure cubic nonlinearity

$$\ddot{y} + c_3^2 y^3 = F_0 \cos(\Omega t), \tag{6.95}$$

Using (6.75) and (6.76) the characteristic algebraic equations are for $n = 2$

$$\frac{9}{4} c_3^2 Y^3 - Y\Omega^2 - F_0 = 0, \tag{6.96}$$

$$\omega^2 - \frac{9}{4} c_3^2 Y^2 = 0. \tag{6.97}$$

Solving the third order equation (6.96) the values for Y are obtained. The number of real solutions of the Eq. (6.96) depends on the sign of the discriminant D

$$D = \frac{16}{81(c_3^2)^2} \left[\left(\frac{F_0}{2} \right)^2 - \left(\frac{2\Omega^3}{9\sqrt{3c_3^2}} \right)^2 \right]. \tag{6.98}$$

For $D > 0$, i.e.,

$$\left(\frac{F_0}{2}\right) - \left(\frac{2\Omega^3}{9\sqrt{3c_3^2}}\right) > 0, \tag{6.99}$$

only one real solution for (6.96) exists

$$Y_1 = \sqrt[3]{\frac{2F_0}{9c_3^2} + \sqrt{D}} + \sqrt[3]{\frac{2F_0}{9c_3^2} - \sqrt{D}}. \tag{6.100}$$

For $D < 0$ three real solutions for (6.96) are calculated

$$Y_1 = \frac{4\Omega}{3\sqrt{3c_3^2}} \cos\frac{\varphi}{3}, \qquad Y_2 = \frac{4\Omega}{3\sqrt{3c_3^2}} \cos\frac{\varphi + 2\pi}{3}, \qquad Y_3 = \frac{4\Omega}{3\sqrt{3c_3^2}} \cos\frac{\varphi + 4\pi}{3}, \tag{6.101}$$

where

$$\cos\varphi = \frac{9F_0\sqrt{3c_3^2}}{4\Omega^3}. \tag{6.102}$$

If $D = 0$, i.e.,

$$\left(\frac{F_0}{2}\right) - \left(\frac{2\Omega^3}{9\sqrt{3c_3^2}}\right) = 0, \tag{6.103}$$

and

$$\Omega = 1.5737\sqrt[3]{F_0\sqrt{c_3^2}}, \tag{6.104}$$

the two real solutions are

$$Y_1 = \frac{4\Omega}{3\sqrt{3c_3^2}}, \qquad Y_{2,3} = -\frac{2\Omega}{3\sqrt{3c_3^2}}. \tag{6.105}$$

Hence, for $\Omega = 1.5737\sqrt[3]{F_0\sqrt{c_3^2}}$ the number of real solutions of Eq. (6.96) changes from one to three and Eq. (6.104) represents the bifurcation condition.

The same result is obtained by assuming that the two equal solutions are negative ($Y_{2,3} = Y_u < 0$) and the third solution is positive ($Y_1 > 0$). Equation (6.96) is rewritten as

$$(Y + Y_u)^2(Y - Y_1) = 0. \tag{6.106}$$

Comparing (6.106) with (6.96) we have

$$y^2 : Y_1 = 2\,|Y_u|\,, \tag{6.107}$$

$$y : Y_u = -\sqrt[3]{\frac{2F_0}{9c_3^2}},\; Y_1 = 2\sqrt[3]{\frac{2F_0}{9c_3^2}}, \tag{6.108}$$

$$y^0 : Y_u = -\frac{2\Omega}{3\sqrt{3c_3^2}},\; Y_1 = \frac{4\Omega}{3\sqrt{3c_3^2}}. \tag{6.109}$$

By equating (6.108) and (6.109) we prove the existence of the previously obtained $\Omega(F_0, c_3^2)$ relation (6.103). Simplifying the result for the response-frequency relation obtained by Nayfeh and Mook (Nayfeh and Mook 1979) by introducing the assumption that the linear term is zero ($\alpha = 0$) in (6.68) we have the solution $(6.108)_1$.

Using the relations (6.96) and (6.97), and substituting the values for Y_1 and Y_u the following relations are obtained

$$\omega^2 - \Omega^2 = \frac{F_0}{Y} = \frac{\Omega^2}{3} > 0, \tag{6.110}$$

$$\Omega^2 - \omega^2 = -\frac{F_0}{Y} = \frac{2\Omega^2}{3} > 0. \tag{6.111}$$

Due to (6.110) the motion is oscillatory if

$$\omega^2 - \frac{4}{3}\Omega^2 > 0, \tag{6.112}$$

and the solution Y is positive and satisfies the relation

$$0 < Y \leq \frac{3F_0}{\Omega^2}. \tag{6.113}$$

The motion is also oscillatory if

$$\frac{1}{3}\Omega^2 - \omega^2 > 0, \tag{6.114}$$

and the solution Y is negative and satisfies the relation

$$|Y| < \frac{3F_0}{2\Omega^2}, \qquad Y < 0. \tag{6.115}$$

The previous relations give the restriction to the mathematically obtained solutions. The amplitude of excited oscillations Y increases to a boundary value

$$Y_1 = 2\sqrt[3]{\frac{2F_0}{9c_3^2}} = 1.2114\sqrt[3]{\frac{F_0}{c_3^2}}, \tag{6.116}$$

for (6.104) and after that decreases to the amplitude Y_3 which is the negative solution of (6.96). The 'boundary' value of amplitude corresponds to the excited frequency

$$\omega = \frac{3}{2}Y\sqrt{c_3^2} = \frac{2\Omega}{\sqrt{3}} = 1.1547\Omega. \tag{6.117}$$

To obtain the condition for which the vibration disappears, let us rewrite Eq. (6.96) as

$$(Y - a)(Y + a - \Delta_1)(Y + \Delta_2) = 0, \tag{6.118}$$

where a is the positive solution, $a + \Delta_1$ is the assumed negative solution which differs from the positive solution for a small value Δ_1 and Δ_2 is the third solution which differs from zero for a small value. Comparing (6.118) with (6.96) the system of algebraic equations follows

$$\Delta_1 = \Delta_2, \tag{6.119}$$

$$\Delta^2 - a\Delta + a^2 - \frac{\Omega^2}{sc_3^2} = 0, \tag{6.120}$$

$$a\Delta^2 - a^2\Delta + \frac{F_0}{sc_3^2} = 0. \tag{6.121}$$

Assuming that $\Delta^2 \approx 0$ and omitting the terms with Δ^2 the solution of (6.120) and (6.121) for Δ_2 is

$$\Delta_2 \approx -\frac{F_0}{\Omega^2}. \tag{6.122}$$

The solution Δ_2 is valid if

$$\frac{F_0}{\Omega^2} \ll 1. \tag{6.123}$$

The relation (6.122) with (6.97) gives the excited frequency

$$\omega = \frac{3F_0}{2\Omega^2}. \tag{6.124}$$

The oscillatory motion described analytically is

$$y \approx -\frac{F_0}{\Omega^2}(\cos(\Omega t) - 1) = \frac{2F_0}{\Omega^2}\sin^2(2\Omega t), \tag{6.125}$$

with amplitude of vibration $\left|2F_0/\Omega^2\right|$.

Analyzing the above relations the following is concluded:

1. The boundary amplitude of vibration Y_1 given with (6.116) and the corresponding frequency depend on the properties of the excitation force: amplitude F_0 and frequency Ω, but also on the coefficient of nonlinearity c_3^2. For higher values of the amplitude of excitation, the higher the excited amplitude of vibration. The smaller the coefficient of nonlinearity, the higher the amplitude of the excited vibration.

2. For specific excitation frequency of (6.104), for which the boundary amplitude Y_1 (6.116) appears, the ratio of the excited and excitation frequency of vibration is constant: $\omega/\Omega = 2/\sqrt{3} = 1.1547$. Comparing the obtained value with that for linear case, when $\omega/\Omega = 1$, it can be concluded that the critical frequency ratio for the boundary amplitude of vibration is higher for the cubic nonlinear oscillator than for the linear case. It is necessary to mention that the excited frequency ω is not a constant value but depends on the properties of the excitation force and on the coefficient of nonlinearity.

3. Substituting (6.104) into (6.117) the excited frequency for maximal amplitude is obtained

$$\omega = 1.8171\sqrt[3]{F_0}\sqrt[6]{c_3^2}. \tag{6.126}$$

The frequency is higher for higher values of excitation amplitude and coefficient of nonlinearity.

4. Comparing the pure cubic oscillator with the linear one it is evident that the boundary amplitude for the pure nonlinear oscillator is not unbounded (as for undamped linear oscillator (at resonance), but has a bounded value.

5. For $\Omega = 0$ the amplitude of the excited vibration is $Y = \sqrt[3]{4F_0/9c_3^2}$ and the approximate analytical solution is

$$y = \sqrt[3]{\frac{4F_0}{9c_3^2}}(1 - \cos(\omega t) = 2\sqrt[3]{\frac{4F_0}{9c_3^2}} \sin^2(2\omega t), \tag{6.127}$$

where

$$\omega = \frac{3}{2}\sqrt[3]{\frac{4F_0}{9c_3^2}}. \tag{6.128}$$

6. For

$$0 < \omega^2 - \Omega^2 < \frac{\Omega^2}{3}, \tag{6.129}$$

the convenient averaging solution is $y = Y\cos(\Omega t)$ and for

$$0 < \Omega^2 - \omega^2 < \frac{2\Omega^2}{3}, \tag{6.130}$$

the calculated analytical averaging solution (6.81) is sufficiently accurate. For

$$\frac{\Omega^2}{3} < \left|\omega^2 - \Omega^2\right| < \frac{2\Omega^2}{3}, \tag{6.131}$$

the averaged overall envelope curve is

$$y = \pm 2Y \sin(\frac{\Omega - \omega}{2} t). \tag{6.132}$$

7. For $F_0/\Omega^2 \ll 1$ the amplitude of excited vibration is extremely small. The oscillatory motion is given with (6.125) with the constant averaged value $y^* = F_0/\Omega^2 = const.$

8. To improve the approximate averaged solution $y = Y\cos(\Omega t)$, the relation by introducing the Jacobi elliptic *cn* function (Byrd and Friedman 1954). As is now well known (see Cveticanin 2009) the exact solution of the free vibration ($F_0 = 0$) of the pure cubic oscillator is in general

$$y = Acn(\omega_f t, k^2), \tag{6.133}$$

where A is the amplitude which depends on the initial conditions, and the parameters of the Jacobi elliptic function are

$$\omega_f = A\sqrt{c_3^2}, k^2 = 1/2. \tag{6.134}$$

We assume that the modulus of the Jacobi elliptic function for the excited vibrations remain constant and has the value $k^2 = 1/2$. The frequency of the elliptic function of the forced vibration is calculated applying (6.134). Using the previous values the period of vibration is

$$T = \frac{2\pi}{\omega} = \frac{4K(k)}{\omega_f}, \tag{6.135}$$

where $4K(k)$ is the period of the *cn* Jacobi elliptic function and $K(k)$ is the total Jacobi integral of the first kind. For $K(1/2) = 1.854075$ (see Byrd and Friedman 1954) we rewrite (6.135) as

$$\omega_f = \frac{4K(1/2)}{2\pi}\omega = 1.1803\omega. \tag{6.136}$$

Substituting (6.136) into (6.71) the approximate solution is

$$y = Y(\cos\Omega t - cn(1.1803\omega t, 1/2). \tag{6.137}$$

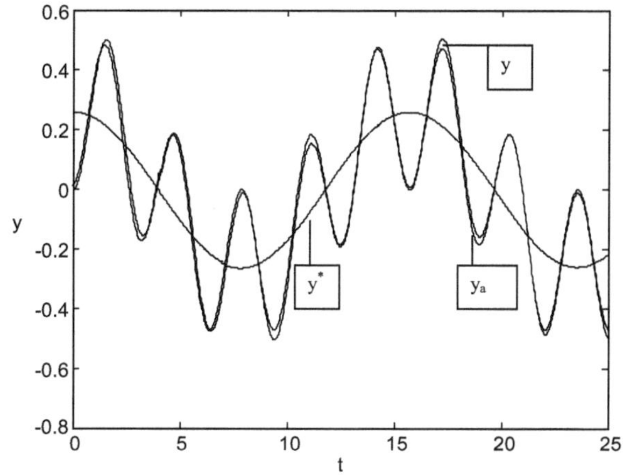

Fig. 6.6 The time history diagrams obtained numerically (y_N), analytically (y) and the averaged curve (y^*) for $\gamma = 1$, $F_0 = 1$ and $\Omega = 2$

The suggested approximation (6.137) is more accurate than (6.71) since the cn function includes all harmonics of odd order and $\cos(\omega t)$ is just the first term in the Fourier expansion of the cn function.

6.2.4 Numerical Simulation and Discussion

In this section the above stated results are illustrated with some numerical examples. The Runge-Kutta method is applied to obtain the numerical solutions of the problem which are compared with analytically obtained approximate ones. The influence of various values of excitation parameters F_0, Ω and also of the coefficient of the nonlinearity c_3^2 is analyzed.

1. In Fig. 6.6 the analytically and numerically obtained solutions are compared.

For the differential equation (6.95), where the parameter values are $c_3^2 = 1$, $F_0 = 1$ and $\Omega = 2$, the approximate analytical solution (6.71) has the form

$$y_a = 0.25987(\cos(0.4t) - \cos(2t)). \tag{6.138}$$

Comparing the approximate analytical solution y_a (6.138) with an accurate numerical solution y obtained by the integration of Eq. (6.95) it is seen that the difference is negligible. Besides, in this figure the averaged curve

$$y^* = 0.25987\cos(0.4t), \tag{6.139}$$

is also plotted in Fig. 6.6. It is shown the vibration with excitation frequency is around the averaged curve.

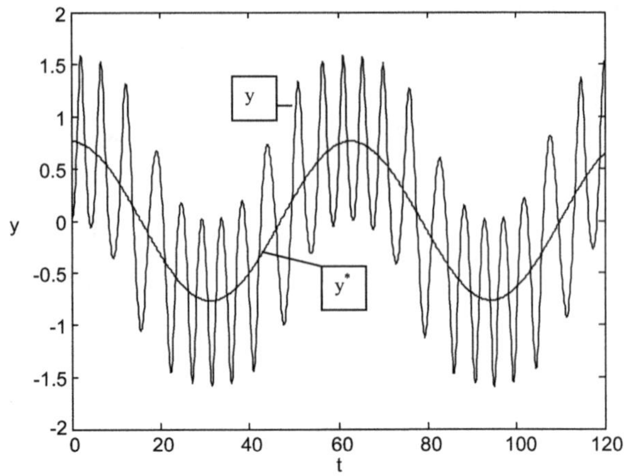

Fig. 6.7 Numerically obtained $y - t$ and analytically obtained $y^* - t$ diagrams for $F_0 = 1, c_3^2 = 1$ and $\Omega = 0.1$

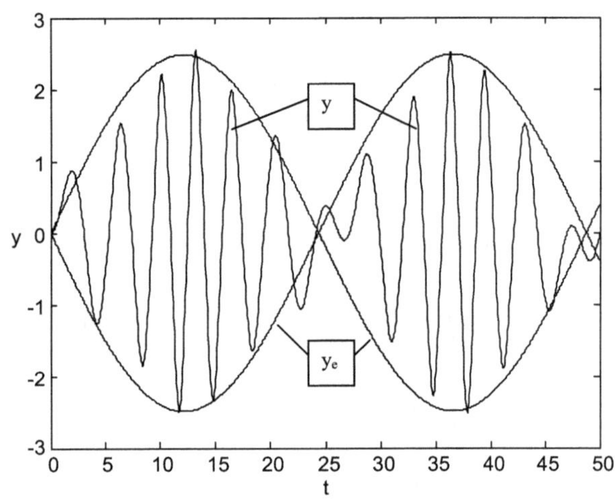

Fig. 6.8 The numerically obtained $y - t$ and the envelope curve $y^* - t$ diagrams for $F_0 = 1, c_3^2 = 1$ and $\Omega = 1.4$

2. In Figs. 6.7, 6.8, 6.9, 6.10, 6.11 the $y - t$ and $y^* - t$ diagrams obtained numerically and the analytically for $c_3^2 = 1,\ F_0 = 1$ and various values of excitation frequency Ω are plotted.

In Fig. 6.7, the numerically obtained curve for $\Omega = 0.1$ and the analytical solution

$$y^* = 0.76508 \cos(0.1t), \tag{6.140}$$

are plotted. It is evident that (6.140) represents the averaged curve for the excited vibration where the frequency is $\omega = 1.1476$.

In Fig. 6.8 the vibration curve for $\Omega = 1.4$ is plotted.

As the frequency of excited vibration ($\omega = 1.6878$) is close to the frequency of the excitation force the phenomena of fluttering occurs. We calculate the envelope amplitude function

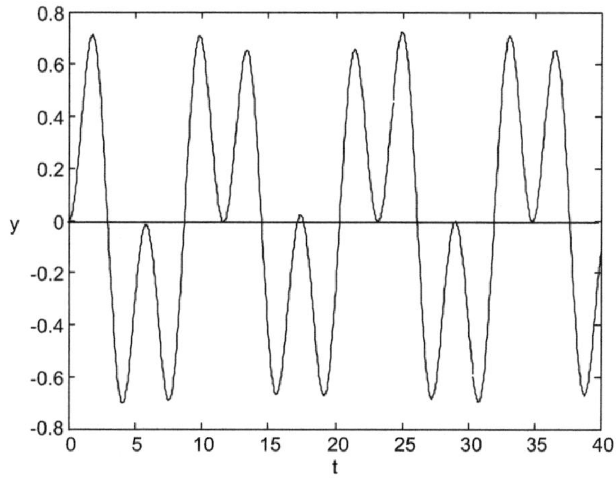

Fig. 6.9 The numerically obtained $y - t$ diagram for $\Omega_b = 1.627$

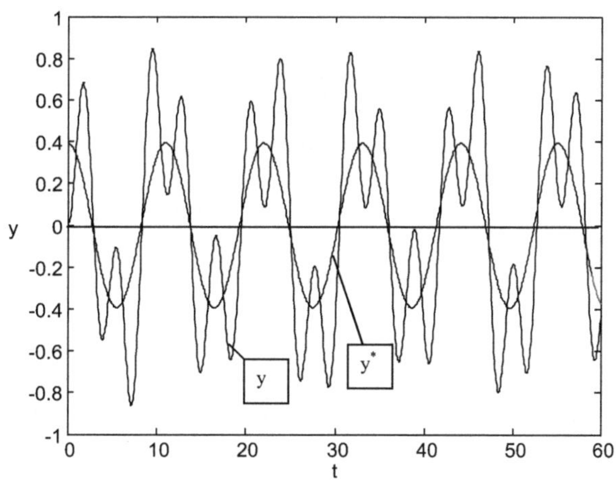

Fig. 6.10 Numerically obtained $y - t$ and analytically obtained $y^* - t$ diagrams for $F_0 = 1, \gamma = 1$ and $\Omega = 1.7$

$$y^* = 1.46 \sin(0.1439t), \tag{6.141}$$

and the obtained $y^* - t$ function is plotted. The analytically obtained solution (6.141) is accurate over a long time period.

In Fig. 6.9 the numerically obtained $y - t$ diagram for $\Omega_b = 1.627$ is plotted.

This is the boundary curve between two types of oscillatory motions. According to (6.104) the analytically obtained critical value is $\Omega_{ba} = 1.574$ and it differs from the exact one by 3.4365%.

In Fig. 6.10 the vibration obtained numerically for $\Omega = 1.7$ is plotted.

As the frequency of excited vibration is $\omega = 0.59015$, i.e., $\Omega > \omega$ the averaged analytical solution is with period $T = 2\pi/\omega = 10.647$, and has the form

$$y^* = 0.39343 \cos(0.59015t). \tag{6.142}$$

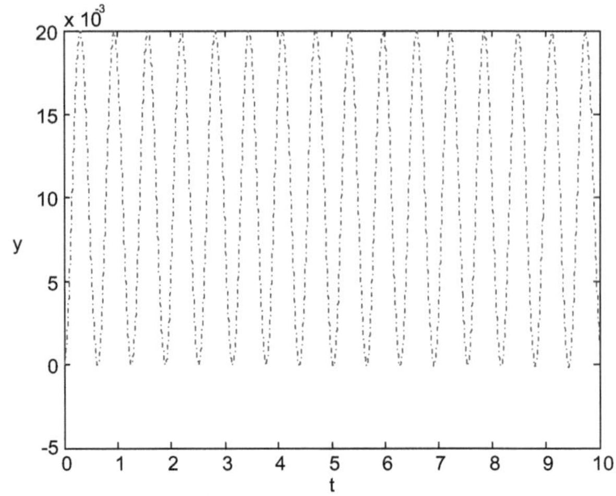

Fig. 6.11 The $y-t$ diagram for $F_0 = 1$, $c_3^2 = 1$ and $\Omega = 10$

The system oscillates around the suggested curve with the frequency $\Omega = 1.7$.

In Fig. 6.11 the time history diagram for $\Omega = 10$ obtained numerically is plotted. The calculated analytical solution (6.125) transforms into

$$y = 0.02 \sin^2(20t), \tag{6.143}$$

and the averaged value (6.94) is

$$y^* = 0.01 = const. \tag{6.144}$$

The best resume of the previously obtained results is seen in Fig. 6.12, where the $Y-\Omega$ and $\omega-\Omega$ diagrams are plotted.

Increasing the excitation frequency Ω the amplitude Y and frequency ω of excited vibration increase to a certain value which corresponds to the boundary frequency Ω_b and jumps to a smaller value. Increasing further the excitation frequency the amplitude and frequency of the excited vibration decrease and tend to zero (see Friswell and Penny 1994; Brennan et al. 2008).

3. In Fig. 6.13 the influence of excitation amplitude F_0 on the system where $c_3^2 = 1$ and $\Omega = 0.1$ is shown. The $y-t$ numerically obtained curves, the averaged solution

$$y^* = 0.3584 \cos(0.1t), \tag{6.145}$$

for $F_0 = 0.1$ and excited frequency $\omega = 0.5376$ and also the so called 'steady state' solution

$$y^* = 1.645 \cos(0.1t). \tag{6.146}$$

for $F_0 = 10$ and $\omega = 2.4675$ are plotted.

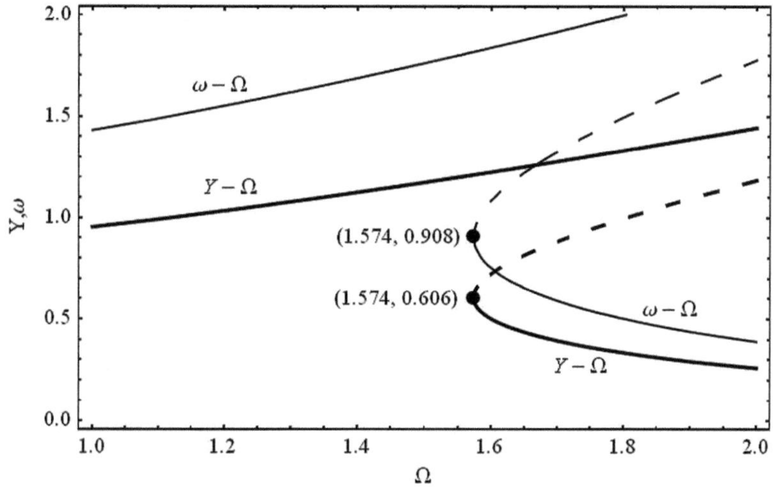

Fig. 6.12 The $Y - \Omega$ and the $\omega - \Omega$ diagrams obtained analytically for the pure cubic oscillator with $F_0 = 1$ and $\gamma = 1$

Fig. 6.13 The numerically calculated $y - t$ and averaged $y^* - t$ diagrams for $\Omega = 0.1$, $c_3^2 = 0.1$ and: $F_0 = 0.1$ and $F_0 = 10$

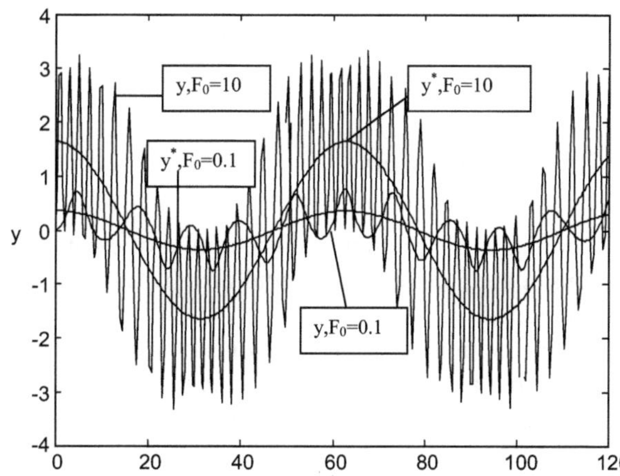

The figure shows that for higher values of excitation amplitude F_0 the excited amplitude Y is higher and also the excited frequency ω.

4. In Fig. 6.14, the numerical $y - t$ solutions for $F_0 = 0.1$ and $\Omega = 0.1$, when c_3^2 has the values 0.1 and 100, are compared with the calculated usually applied steady state solutions

$$y^* = 0.78255 \cos(0.1t), \tag{6.147}$$

and

$$y^* = 0.0765 \cos(0.1t), \tag{6.148}$$

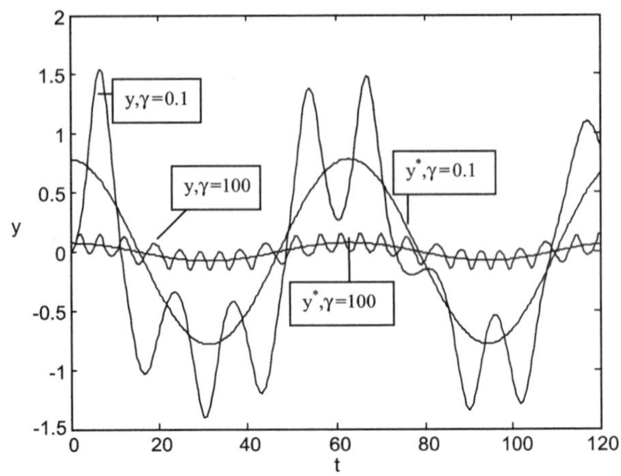

Fig. 6.14 The numerically calculated $y - t$ and averaged $y^* - t$ diagrams for $F_0 = 0.1$, $\Omega = 0.1$ and: $c_3^2 = 0.1$ and $c_3^2 = 100$

respectively.

Namely, for $c_3^2 = 0.1$, the frequency of excited vibration is $\omega = 0.37120$ and for $c_3^2 = 100$, it is $\omega = 1.1475$. For both parameter values it is $\Omega < \omega$ and the averaged Eqs. (6.147) and (6.148) have the frequency $\Omega = 0.1$.

6.2.5 *Conclusion*

According to the previous consideration of excited vibrations with zero initial conditions of the pure odd-order nonlinear harmonically excited oscillator the following is concluded:

1. The motion of the pure nonlinear harmonically excited oscillator has a periodic wave-form in spite of the fact that the parameters of the system and of the excitation force are not small values. Approximately, the oscillatory motion is a sum of two periodical motions: one, at the frequency of the excitation force and the second, at the frequency of the excited vibration. The amplitude of the excited vibration depends on the parameters of the system (order and coefficient of nonlinearity) and parameters of the excitation force (amplitude and frequency).

2. The frequency of the excited oscillation is not a constant value (as is the case for the linear oscillator) but is a function of the amplitude of the excited vibration and the order and coefficient of the nonlinearity.

3. For all parameter values of the pure harmonically excited oscillator the frequency of the excited vibration differs from the frequency of excitation force except for the linear oscillator. Besides, the amplitude of vibration is bounded independently on the parameter values of the oscillator, except that of the linear oscillator where the amplitude tends to infinity for the resonant case. Thus, in spite of the fact that there is no damping in the system the amplitude of vibration is never infinitely large:

the higher the order of nonlinearity, the smaller the extreme amplitude of vibration. Finally, in all working regimes of the oscillator the amplitude of vibration is bounded and the frequency of excitation and excited vibration differ.

4. In the pure nonlinear harmonically excited oscillator for certain excitation parameters the well known jump phenomenon occurs which depends not only on the coefficient but also on the order of nonlinearity. The higher the order of the nonlinearity the extreme jump response is smaller.

5. If the excitation frequency and amplitude ratio is a large number the oscillations tend to zero.

6. In the pure cubic harmonic oscillator for a higher value of the coefficient of nonlinearity the amplitude of vibration is smaller. If the amplitude of excitation force is higher, the amplitude of vibration is higher. Increasing the excitation frequency the amplitude of vibration increases and tends to the bifurcation value after which it decreases and tends to zero if the excitation frequency and amplitude ratio is high.

7. The harmonic balance method is applied for approximate analytical solving of the mathematical oscillator model, described by a second order pure nonlinear differential equation, and gives very accurate qualitative and quantitative description of the oscillations. The so called 'steady state solution' represents the averaged solution. The frequency of the averaged solution depends on the correlation between the excitation frequency and excited one. If the excited frequency is higher than the excitation frequency, the averaged oscillatory solution has the excited frequency of the system. For the case of an excited frequency which is smaller that the excitation frequency, the averaged solution is equal to the excitation frequency. This is a new finding in comparison to the previous results in the literature.

6.3 Forced Vibrations of the Pure Nonlinear Oscillator

In this section a procedure for designing the external excitation of the purely nonlinear oscillator to yield the closed-form solution in terms of the Ateb or Jacobi cn function is developed. Based on the known exact solution of the purely nonlinear oscillator the procedure for forced vibration is developed. Frequency-response relations have to be formed to illustrate the response in the frequency domain and the forced vibrations in the time domain.

6.3.1 Design of Excitation and Derivation of Amplitude-Frequency Equation

Let us consider the purely nonlinear oscillators externally excited by the force F:

$$\ddot{x} + c_\alpha^2 x \left| x \right|^{\alpha-1} = F\left(t\right). \tag{6.149}$$

The question of interest here is the way how F should change with time explicitly so that Eq. (6.149) has the exact closed-form solution for the corresponding forced response. To answer this, the excitation is assumed as

$$F = \frac{F_0}{A\,|A|^{\alpha-1}} x\,|x|^{\alpha-1}\,, \tag{6.150}$$

as a result of which the equation of motion (6.149) turns into

$$\ddot{x} + \left(c_\alpha^2 - \frac{F_0}{A\,|A|^{\alpha-1}}\right) x\,|x|^{\alpha-1} = 0. \tag{6.151}$$

Comparing it with Eq. (3.2), whose solution is given by Eq. (3.29), one concludes that Eq. (6.151) has the solution

$$x = A\text{ca}\,(\alpha, 1, \omega_r t)\,, \tag{6.152}$$

where

$$\omega_r = |A|^{(\alpha-1)/2}\sqrt{\frac{\alpha+1}{2}\left(c_\alpha^2 - \frac{F_0}{A\,|A|^{\alpha-1}}\right)}, \tag{6.153}$$

under the condition that the radicand is positive.

This implies that the required form of the force (6.150) is given by:

$$F = F_0\text{ca}\,(\alpha, 1, \omega_r t)\,|\text{ca}\,(\alpha, 1, \omega_r t)|^{\alpha-1}\,, \tag{6.154}$$

Based on the results presented Sect. 3.2, one can interpret this excitation as a multi-term harmonic excitation with odd harmonics whose amplitudes and frequencies are mutually related.

The sought autonomous equation of motion is, then:

$$\ddot{x} + c_\alpha^2 x\,|x|^{\alpha-1} = F_0\text{ca}\,(\alpha, 1, \omega_r t)\,|\text{ca}\,(\alpha, 1, \omega_r t)|^{\alpha-1}\,. \tag{6.155}$$

The expression (6.153) can further be transformed into the following amplitude-frequency equation:

$$-\frac{2}{\alpha+1}\omega_r^2 A + c_\alpha^2 \text{sgn}\,(A)\,|A|^\alpha = F_0. \tag{6.156}$$

For given values of F_0 and c_α^2, Eq. (6.156) enables one to find the amplitude(s) corresponding to different values of the frequency ω_r and to plot frequency-response (amplitude-frequency) curves. The branches of this curve will be located around the backbone curves. The relationship for it is obtained from Eq. (6.156) with $F_0 = 0$:

$$\omega_{bc} = |A|^{(\alpha-1)/2}\sqrt{c_\alpha^2\frac{\alpha+1}{2}}. \tag{6.157}$$

To find the value of frequency at which the number of possible solutions changes, i.e. where the saddle-node (SD) point occurs, one can differentiate Eq. (6.156) with respect to A to derive

$$\omega_r^2 = |A|^{\alpha-1}\,c_\alpha^2\frac{\alpha(\alpha+1)}{2}. \tag{6.158}$$

Substituting this back into Eq. (6.156) and then into Eq. (6.158), one can derive the following expressions for the amplitude $\left|\hat{A}\right|$ and frequency $\hat{\omega}_r$ at which the SD point exists

$$\left|\hat{A}\right| = \left[\frac{F_0}{c_\alpha^2\,|\alpha-1|}\right]^{1/\alpha}, \tag{6.159}$$

$$\hat{\omega}_r = \sqrt{\frac{c_\alpha^2\alpha(\alpha+1)}{2}}\left[\frac{F_0}{c_\alpha^2\,|\alpha-1|}\right]^{\frac{\alpha-1}{2\alpha}}. \tag{6.160}$$

Figure 6.15 shows the rescaled SD amplitudes ($|A^*| = \left|\hat{A}\right| / \left[\frac{F_0}{c_\alpha^2}\right]^{1/\alpha}$) and frequency ($\omega_r^* = \hat{\omega}_r / \left(c_\alpha\left[\frac{F_0}{c_\alpha^2}\right]^{\frac{\alpha-1}{2\alpha}}\right)$) versus the order of nonlinearity α.

These graphs can be used to estimate where the point SD occurs depending on the order α.

6.3.2 Special Cases

Case $\alpha = 1$

In the linear case, Eq. (6.155) reduces the harmonically excited linear oscillator

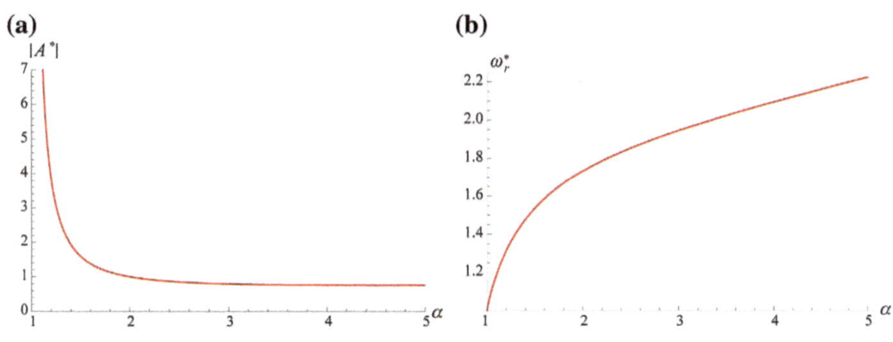

Fig. 6.15 Saddle-node: **a** amplitudes and **b** frequencies versus the order of nonlinearity α

$$\ddot{x} + c_1^2 x = F_0 \cos(\omega_r t), \tag{6.161}$$

where

$$\omega_r = \sqrt{c_1^2 - \frac{F_0}{A}}, \tag{6.162}$$

with the forced response being

$$x = A \cos(\omega_r t). \tag{6.163}$$

Indeed, it is easy to check that Eq. (6.163) satisfies Eq. (6.161).

Case $\alpha = 2$
For the purely nonlinear oscillator with odd quadratic nonlinearity, Eq. (6.155) transforms into

$$\ddot{x} + c_2^2 x\,|x| = F_0 \mathrm{ca}(2, 1, \omega_r t)\,|\mathrm{ca}(2, 1, \omega_r t)|. \tag{6.164}$$

It solution is

$$x = A \mathrm{ca}(2, 1, \omega_r t), \tag{6.165}$$

where

$$\omega_r = \sqrt{\frac{3}{2}|A|\left(c_2^2 - \frac{F_0}{A\,|A|}\right)}. \tag{6.166}$$

Equation (6.156) is used to plot frequency-response branches in Fig. 6.16a for $c_2 = 1$ and $F_0 = 1/2$.

The backbone curve (6.158) is also presented and the SD point labelled by the asterix. The frequency-response branches are bent to the right-hand side and there are frequency regions in which a multi-valued response exists.

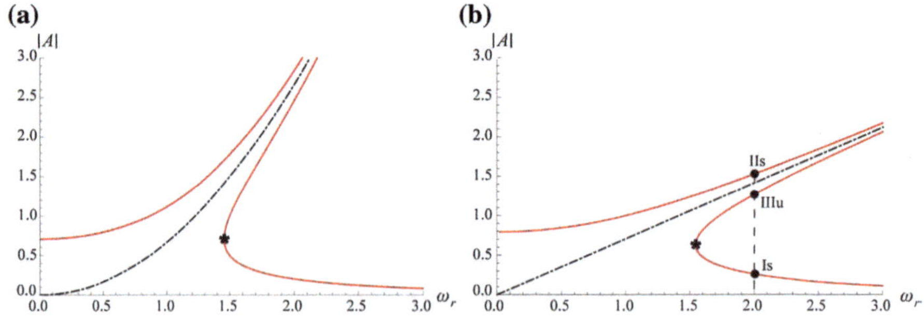

Fig. 6.16 Frequency response (*full line*) and backbone (*dotted line*) curves for: **a** $\alpha = 2$, $c_2 = 1$ and $F_0 = 1/2$, **b** $\alpha = 3$, $c_3 = 1$ and $F_0 = 1/2$

Case $\alpha = 3$
For the pure cubic case, Eq. (6.155) becomes

$$\ddot{x} + c_3^2 x^3 = F_0 \mathrm{ca}^3\,(3, 1, \omega_r t)\,. \tag{6.167}$$

The exact closed-form solution for the forced response is

$$x = A\,\mathrm{ca}\,(3, 1, \omega_r t)\,, \tag{6.168}$$

with

$$\omega_r = |A|\sqrt{2\left(c_3^2 - \frac{F_0}{A^3}\right)}. \tag{6.169}$$

The corresponding frequency-response curve is plotted in Fig. 6.16b for $c_3 = 1$ and $F_0 = 1/2$. To illustrate the time history, the case $\omega_r = 2$ is considered. Three values of the amplitude are calculated from Eq. (6.156) and shown in Fig. 6.16b with asterix. Two stable ones are labelled by Is and IIs: $A_{\mathrm{Is}} = -0.258652$ and $A_{\mathrm{IIs}} = 1.52569$. The third one is unstable: $A_{\mathrm{IIIu}} = -1.26704$. The stability can be cheked by introducing the perturbation into the equation of motion (6.149) and deriving the linear variational equation (see for example, [?]).

For the determined stable amplitudes the forced response is obtained in the form of the Fourier series (see Appendix B):

$$\begin{aligned} x_{\mathrm{Is}} &= -0.25865\,\mathrm{ca}\,(3, 1, 2t) \\ &\approx -0.24701\cos(1.19814t) - 0.01113\cos(3.59442t) \\ &\quad - 0.00048\cos(5.9907t) - 0.00002\cos(8.38698t)\,, \end{aligned} \tag{6.170}$$

and

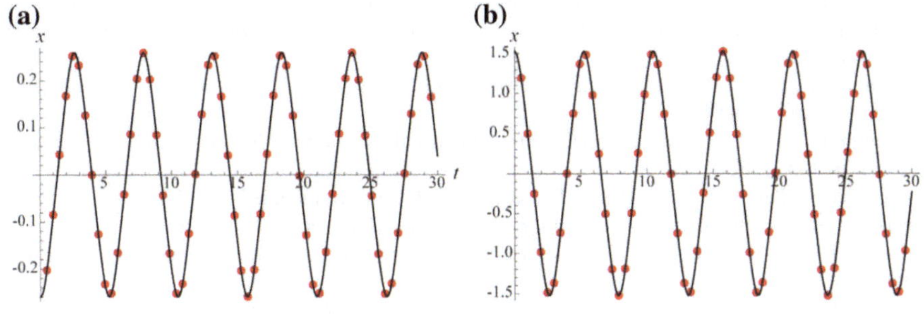

Fig. 6.17 Time history diagrams for: **a** $A_{\mathrm{Is}} = -0.258652$, **b** $A_{\mathrm{IIs}} = 1.52569$ obtained analytically (*full line*) and numerically (*dotted line*)

$$\begin{aligned} x_{\mathrm{IIs}} &= 1.52569\,\mathrm{ca}\,(3, 1, 2t) \\ &\approx 1.45704\cos(1.19814t) + 0.06568\cos(3.59442t) \\ &\quad + 0.00284\cos(5.9907t) + 0.00012\cos(8.38698t)\,. \end{aligned} \tag{6.171}$$

The solutions (6.170) and (6.171) are shown in Fig. 6.17a, b together with numerical solutions of Eqs. (6.167), (6.169) to demonstrate their mutual agreement.

Note that given the transformation of the Ateb function to the Jacobi cn function for the case of cubic nonlinearity (see Sect. 3.2.3), one can represent the equation of motion in the form

$$\ddot{x} + c_3^2 x^3 = F_0 \mathrm{cn}^3\left(\omega_r t \middle| \frac{1}{2}\right). \tag{6.172}$$

with the exact solution for the forced response

$$x = A\mathrm{cn}\left(\omega_r t \middle| \frac{1}{2}\right), \tag{6.173}$$

where now

$$\omega_r = |A|\sqrt{\left(c_3^2 - \frac{F_0}{A^3}\right)}. \tag{6.174}$$

References

Brennan, M. J., Kovacic, I., Carrella, A., & Waters, T. P. (2008). On the jump-up and jump-down frequencies of the Duffing oscillator. *Journal of Sound and Vibration*, *318*, 1250–1261.

Burton, T. D. (1984). A perturbation method for certain non-linear oscillators. *International Journal of Non-Linear Mechanics*, *19*, 397–407.

Byrd, P. F., & Friedman, M. D. (1954). *Handbook of elliptic integrals for engineers and physicists*. Berlin: Springer.

Cveticanin, L. (2009). The approximate solving methods for the cubic Duffing equation based on the Jacobi elliptic functions. *International Journal of Nonlinear Sciences and Numerical Simulation*, *10*, 1491–1516.

Cveticanin, L. (2011). Pure odd-order oscillators with constant excitation. *Journal of Sound and Vibration*, *330*, 976–986.

Cveticanin, L. (2012). Forced pure nonlinear symmetrical oscillators. *Mathematical and Computer Modelling*, *55*, 1580–1593.

Friswell, M., & Penny, J. E. T. (1994). The accuracy of jump frequencies in series solutions of the response of a Duffing oscillator. *Journal of Sound and Vibration*, *169*, 261–269.

Gradstein, I. S., & Rjizhik, I. M. (1971). *Tablici integralov, summ, rjadov i proizvedenij*. Moscow: Nauka.

Mickens, R. E. (2004). *Mathematical Methods for the Natural and Engineering Sciences*. New Jersey: World Scientific.

Nayfeh, A.H., & Mook, D.T. (1979). *Nonlinear oscillations*. New York: John Wiley & Sons.

Worden, K. (1996). On jump frequencies in the response of the Duffing oscillator. *Journal of Sound and Vibration*, *198*, 522–525.

Chapter 7
Two-Degree-of-Freedom Oscillator

Nowadays, an enormous interest in investigating of vibrations of two mass systems with two degrees of freedom is shown. This is mainly due to the fact that the model of two coupled oscillators is suitable for analysis the processes and devices in various fields of science and engineering. For example, the two degree of freedom system is assumed to describe the properties of micro-electro-mechanical systems (MEMS) (Jones and Nenadic 2013) like sensors in automobile airbag systems and actuators in printer cartridges (Welte et al. 2013), electrostatic gap closing actuators (Cortopassi and Englander 2010), valves, gears, gyroscopes and microactuators of electrostrictive, magnetostrictive, photostrictive, mechanochemical type, artificial muscle actuator, nano-electro-mechanical devices (Polo et al. 2009), micromirrors (Burns and Bright 1997), nanometer switches (Polo et al. 2010) which are applied in aerospace, optics, electronics and medical engineering, in piezoelectric actuators which employ piezoelectric materials and are appropriate to be applied in atomic force microscopes, mobile phones, digital camera lens drive etc., in smart sensors and actuators with shape memory alloys (Machado et al. 2013). For all of MEMS it is common that have the mechanical structure which is rather simple and the number of independent motions is usually limited. In the papers (Kucab et al. 2013; Snamina and Orkisz 2014) the energy harvesting from vibrations of a two-degree-of-freedom mechanical oscillator system and the energy harvesting in the two-masses piezoelastic system is investigated, respectively. Coupled oscillators have a significant importance in electronic communication in generating spectrally pure signal, combining power of signals from different sources, coherent modulation and detection processes (Sarkar et al. 2013). Model of two coupled oscillators is applied for explaining phenomena in vibration-, acoustic- and impact isolators (Bondar 1978; Pilipchuk 2007, 2010; Afsharfard and Farshidianfar 2012), ship-isolator system, which can protect the ships from the sea waves vibrations (Xiong et al. 2005), etc. The concept of the passive vibration isolator is based on the model of two masses and three elastic springs fixed at one end and connected to each other at the other end (Alabudzev et al. 1989; Carela et al. 2007; Kovacic et al. 2008; Gatti et al. 2010). Such model describes the two

L. Cveticanin, *Strong Nonlinear Oscillators*, Mathematical Engineering,
DOI 10.1007/978-3-319-58826-1_7

mass system with permanent magnets, too (Xing et al. 2005). The passive vibration isolators are important for protecting of the buildings during the earthquake (Araky et al. 2010). The phenomena of earthquake is often simulated with a two-degree-of-freedom spring-block model (Abe and Kato 2013; Sugiura et al. 2014). Recently, Pham et al. (2011) investigated the vibration of cable-stayed-bridges as a system of two coupled oscillators where the connection between the bridge with deck or with pile is taken into consideration. Support-machine system, cutting machines with periodical motion of the cutting tool, press-working piece system, vehicle hanging system (Zhu and Ishitoby 2004) are also modeled as two degrees of freedom ones.

In this Chapter the asymptotic solving methods for the oscillatory motion, described with a system of two coupled strong nonlinear differential equations, are presented. Two groups of oscillatory systems are considered: a two-mass system with two-degrees-of-freedom and a two-degrees of freedom system described with a complex function.

The vibration of a two-mass system with two-degrees-of-freedom is described with a system of two coupled differential equations. The connection between the masses is assumed to be nonlinear. Various analytical solution procedures for oscillatory motion of the system of two strong nonlinear and coupled differential equations are presented. The methods are based on the already published results in Vakakis and Rand (1992a, b), Cveticanin (2001b, c, 2002, 2003, 2004c).

The vibration of the one-mass system with two-degrees-of-freedom is described with a second order differential equation with a complex function. The procedures for solving these strong nonlinear differential equations are presented. Besides the adopted Krylov–Bogolubov method (1943), the procedure based on the first integrals of the oscillatory system is discussed. Some numerical examples are calculated.

7.1 System with Nonlinear Viscoelastic Connection

Vibration of a two mass system with two degrees of freedom is considered. Two equal harmonic oscillators are coupled with a strong nonlinear viscoelastic connection. Mathematical model of the system are two coupled second order strong nonlinear differential equations. Based on the Ateb function the approximate solution of the problem is obtained (Cveticanin 2015). The procedure is applied for investigation of the vibration of a vocal cord, which is modeled with two harmonic oscillators with strong nonlinear non-integer order viscoelastic connection.

7.1.1 Model with Strong Nonlinear Viscoelastic Connection

The model of the system is shown in Fig. 7.1.

The model contains two harmonic oscillators with mass m and linear stiffness k, and the strong nonlinear viscoelastic connection. Mathematical description of such

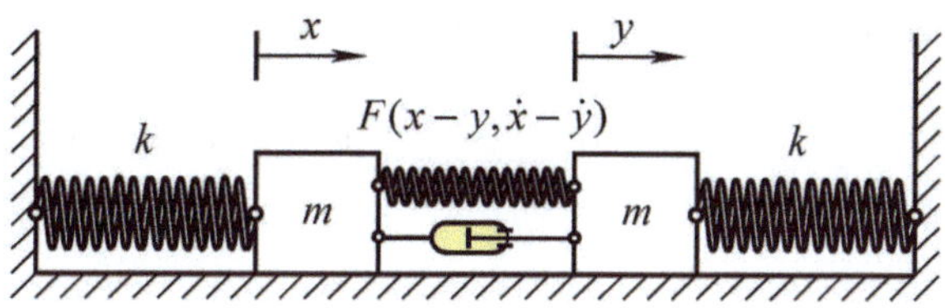

Fig. 7.1 Model of the two-body system

oscillatory system is as follows

$$m\ddot{x} + kx + F(x - y, \dot{x} - \dot{y}) = 0, \tag{7.1}$$

$$m\ddot{y} + ky - F(x - y, \dot{x} - \dot{y}) = 0, \tag{7.2}$$

with initial conditions

$$x(0) = x_0, \quad y(0) = y_0, \quad \dot{x}(0) = \dot{x}_0, \quad \dot{y}(0) = \dot{y}_0, \tag{7.3}$$

where $F(x - y, \dot{x} - \dot{y})$ correspond to viscoelastic connection. Introducing the new variables Eqs. (7.1) and (7.2) can be rewritten in the more convenient form.

Thus, for new variables

$$u = x + y, \quad z = x - y, \tag{7.4}$$

and the notation

$$\omega_0^2 = \frac{k}{m}, \quad F^* = \frac{F}{m}, \tag{7.5}$$

Equations (7.1) and (7.2) transform into

$$\ddot{u} + \omega_0^2 u = 0, \tag{7.6}$$

$$\ddot{z} + \omega_0^2 z + 2F^*(z, \dot{z}) = 0, \tag{7.7}$$

with initial conditions

$$u(0) = u_0 \equiv x_0 + y_0, \quad z(0) = z_0 \equiv x_0 - y_0, \tag{7.8}$$
$$\dot{u}(0) = \dot{u}_0 \equiv \dot{x}_0 + \dot{y}_0, \quad \dot{z}(0) = \dot{z}_0 \equiv \dot{x}_0 - \dot{y}_0.$$

Equations (7.6) and (7.7) are two uncoupled second order differential equations: Eq. (7.6) is a linear, while Eq. (7.7) is strong nonlinear.

Otherwise, using new variables

$$w = x, \qquad z = x - y, \tag{7.9}$$

Equations (7.1) and (7.2) give the relations

$$\ddot{w} + \omega_0^2 w + F^*(z, \dot{z}) = 0, \tag{7.10}$$

$$\ddot{w} - \ddot{z} + \omega_0^2 w - \omega_0^2 z - F^*(z, \dot{z}) = 0, \tag{7.11}$$

i.e.,

$$\ddot{w} + \omega_0^2 w + F^*(z, \dot{z}) = 0, \tag{7.12}$$

$$\ddot{z} + \omega_0^2 z + 2F^*(z, \dot{z}) = 0, \tag{7.13}$$

with initial conditions

$$w(0) = w_0 \equiv x_0, \quad z(0) = z_0 \equiv x_0 - y_0, \tag{7.14}$$

$$\dot{w}(0) = \dot{w}_0 \equiv \dot{x}_0, \quad \dot{z}(0) = \dot{z}_0 \equiv \dot{x}_0 - \dot{y}_0. \tag{7.15}$$

For both transformed systems of differential equations (7.1), (7.2) it is common that one of equations in the system (7.6), (7.7) or (7.12), (7.13) is a one-variable function. In the next section a solving procedure for such a system where one of differential equations is a function of only one variable will be developed. This requirement represents the limitation of the suggested method.

7.1.2 Solution Procedure

For both transformations of (7.1) and (7.2) the existence of an uncoupled strong nonlinear Eq. (7.7) i.e. (7.13) and an additional differential equation (7.6) or (7.12) is evident. Based on the form of equations the suggested procedure is divided into three steps: (1) the solution of (7.7) i.e. (7.13) is obtained, (2) the solution of (7.6) or (7.12) is determined and (3) the solution of (7.1), (7.2) is follows. In spite of the fact that (7.1) and (7.2) form a system of coupled strong nonlinear differential equations but due to limitations given to the model, the superposition procedure is applicable for solving such a system of differential equations.

1. For the viscoelastic connection with the dominant elastic force the differential equation (7.7) is rewritten into the form

$$\ddot{z} + F_1(z) = \varepsilon F_2(z, \dot{z}), \tag{7.16}$$

where $F_1(z)$ is a strong nonlinear function, $\varepsilon F_2(z, \dot{z})$ is a weak nonlinear function, while $\varepsilon \ll 1$ is a small parameter. To solve the strong nonlinear equation perturbed with additional small function is not an easy task. In the text a method for obtaining of the approximate solution of (7.16) will be developed.

The procedure suggested in this paper is based on the solution of a strong nonlinear so called "generating equation" of (7.16) for $\varepsilon = 0$:

$$\ddot{z} + F_1(z) = 0. \tag{7.17}$$

The function $F_1(z)$ is an odd one: $F_1(-z) = -F_1(z)$. The generating equation (7.17) describes the oscillatory motion. The oscillator (7.17) is conservative and the elastic force depends on the deflection z. The generating solution has to satisfy the Eq. (7.17) and to describe the periodical motion. The solution of (7.17) depends on the type of the function $F_1(z)$. For example, if the nonlinearity is quadratic or cubic nonlinearity, the solution is a Jacobi elliptic function. Nevertheless, due to generality and to eliminate the specification of the type of the nonlinearity, we introduce the generating solution in the form of a periodical function

$$z = Z_0 cp(\psi), \tag{7.18}$$

where cp is a notation for a periodical function, Z_0 is the amplitude, ψ is the phase

$$\psi = \Omega t + \theta, \tag{7.19}$$

Ω is the frequency and θ is the initial phase angle. The parameters Z_0 and θ are integration constants and have to satisfy initial conditions $(7.8)_2$ and $(7.8)_4$. Due to our knowledge in nonlinear vibrations, the frequency Ω depends on the amplitude Z_0. Namely, for the real nonlinear oscillators the experiments show that the frequency and the period of vibration depend on the initial deflection (Bogolubov and Mitropolski 1963; Nayfeh and Mook 1979). Let us determine the first time derivative of (7.18)

$$\dot{z} = Z_0 \left(cp(\psi)\right)^{\cdot} = Z_0 \Omega \left(cp(\psi)\right)', \tag{7.20}$$

where $(\bullet)^{\cdot} \equiv d(\bullet)/dt$ and $(\bullet)' \equiv d(\bullet)/d\psi$.

As the parameter ε is a small one, it is supposed that (7.16) is the perturbed version of the Eq. (7.17). According to this assumption it is suggested that the solution of (7.16) has to be the perturbed version of (7.18). Namely, the solution of (7.16) and its first time derivative are assumed in the form (7.18) and (7.20), but with time variable amplitude and phase

$$z = Z_0(t) cp(\psi(t)), \quad \dot{z} = Z_0(t) \Omega \left(cp(\psi(t))\right)', \tag{7.21}$$

where

$$\psi(t) = \int \Omega(Z_0) dt + \theta(t). \tag{7.22}$$

and $(\bullet)' \equiv d(\bullet)/d\psi$. Otherwise, the first derivative of $(7.21)_1$ is

$$\dot{z} = \dot{Z}_0(t)cp(\psi(t)) + Z_0(t)\Omega\left(cp(\psi(t))\right)' + Z_0(t)\dot{\theta}(t)\left(cp(\psi(t))\right)'. \tag{7.23}$$

Equation (7.23) is equal to $(7.21)_2$ if the following relation is satisfied

$$\dot{Z}_0(t)cp(\psi(t)) + Z_0(t)\dot{\theta}(t)\left(cp(\psi(t))\right)' = 0. \tag{7.24}$$

The second time derivative of $(7.21)_2$ is

$$\begin{aligned}\ddot{z} = &\dot{Z}_0(t)\Omega\left(cp(\psi(t))\right)' + Z_0(t)\dot{Z}_0(t)\Omega'\left(cp(\psi(t))\right)' \\ &+ Z_0(t)\Omega^2\left(cp(\psi(t))\right)'' + Z_0(t)\Omega\dot{\theta}(t)\left(cp(\psi(t))\right)'',\end{aligned} \tag{7.25}$$

where $(\bullet)'' \equiv d^2(\bullet)/d\psi^2$ and $\Omega' = d\Omega/dZ_0$. Substituting (7.21) and (7.25) into (7.16) we obtain

$$\begin{aligned}&\dot{Z}_0(t)\Omega\left(cp(\psi(t))\right)' + Z_0(t)\dot{Z}_0(t)\Omega'\left(cp(\psi(t))\right)' + Z_0(t)\Omega\dot{\theta}(t)\left(cp(\psi(t))\right)'' \\ &= \varepsilon F_2(Z_0(t)cp(\psi(t)), Z_0(t)\Omega\left(cp(\psi(t))\right)').\end{aligned} \tag{7.26}$$

Equation (7.24) and (7.26) represent the system of two coupled first order differential equations which correspond to (7.16). After some transformation the following system of equations is obtained

$$\dot{Z}_0\left(\left(\Omega + Z_0\Omega'\right)(cp)'^2 - \Omega cp\,(cp)''\right) = \varepsilon F_2(Z_0 cp, Z_0\Omega\,(cp)')\,(cp)', \tag{7.27}$$

$$Z_0\dot{\theta}\left(\left(\Omega + Z_0\Omega'\right)(cp)'^2 - \Omega cp\,(cp)''\right) = -\varepsilon F_2(Z_0 cp, Z_0\Omega cp')cp \tag{7.28}$$

where $cp(\psi(t) = cp$, $Z_0(t) = Z_0$, $\theta(t) = \theta$.

It is worth to say that Eqs. (7.27), (7.28) are exactly the same as (7.16) rewritten in new variables. This transformation is done to obtain a new form of (7.16) which will be more convenient for solving. In the Eqs. (7.27) and (7.28) the periodic function cp and its derivatives, which are also periodic, appear. Here, the well known suggestion given by Bogolubov and Mitropolski (1963) for averaging of periodical functions for solving weak nonlinear oscillatory systems, is applied. Namely, averaging Eqs. (7.27) and (7.28) over the period of the cp function, we obtain

$$\dot{Z}_0\left((\Omega' Z_0 - \Omega)\left\langle(cp')^2\right\rangle - \Omega\left\langle cp\,(cp)''\right\rangle\right) = \varepsilon\left\langle F_2(Z_0 cp, Z_0\Omega cp')cp'\right\rangle, \tag{7.29}$$

$$Z_0\dot{\theta}\left((\Omega' Z_0 - \Omega)\left\langle(cp')^2\right\rangle - \Omega\left\langle cp\,(cp)''\right\rangle\right) = -\varepsilon\left\langle F_2(Z_0 cp, Z_0\Omega cp')cp\right\rangle, \tag{7.30}$$

where $\langle\bullet\rangle = \int_0^T (\bullet)\,d\psi$ and T is the period of the periodical function. Averaged Eqs. (7.29), (7.30) are more simple than (7.27), (7.28) but convenient to be solved and discussed. Equation (7.29) depends only on Z_0 and has to be solved first. Substituting Z_0 with Ω into (7.30) and integrating for $\theta(t)$ i.e. $\psi(t)$, relation for the phase angle is obtained. In general, the solution of (7.29) is

$$z = z(t, C_1, C_2), \tag{7.31}$$

where integration constants C_1 and C_2 are calculated according to initial conditions $(7.8)_2$ and $(7.8)_4$.

2. Now, the second step of the procedure follows. The Eqs. (7.6) or (7.12) is solved. For simplicity, let us consider the Eq. (7.6). For initial conditions (7.8) the solution of the linear equation (7.6) is

$$u = u_0 \cos(\omega_0 t) + \frac{\dot{u}_0}{\omega_0} \sin(\omega_0 t), \tag{7.32}$$

i.e.,

$$u = (x_0 + y_0) \cos(\omega_0 t) + \frac{\dot{x}_0 + \dot{y}_0}{\omega_0} \sin(\omega_0 t). \tag{7.33}$$

3. Based on (7.32) and (7.31) the body motions (7.4) are

$$x = \frac{u + z}{2} = \frac{1}{2}(u_0 \cos(\omega_0 t) + \frac{\dot{u}_0}{\omega_0} \sin(\omega_0 t) + z(t, C_1, C_2)), \tag{7.34}$$

$$y = \frac{u - z}{2} = \frac{1}{2}(u_0 \cos(\omega_0 t) + \frac{\dot{u}_0}{\omega_0} \sin(\omega_0 t) - z(t, C_1, C_2)), \tag{7.35}$$

i.e., sum and difference of a harmonical u and a periodical function z

$$x = \frac{x_0 + y_0}{2} \cos(\omega_0 t) + \frac{\dot{x}_0 + \dot{y}_0}{2\omega_0} \sin(\omega_0 t) + z(t, x_0, y_0, \dot{x}_0, \dot{y}_0), \tag{7.36}$$

$$y = \frac{x_0 + y_0}{2} \cos(\omega_0 t) + \frac{\dot{x}_0 + \dot{y}_0}{2\omega_0} \sin(\omega_0 t) - z(t, x_0, y_0, \dot{x}_0, \dot{y}_0). \tag{7.37}$$

Motions of masses can be treated as complex ones: a periodic motion is added or subtracted from a harmonic motion with the amplitude $(x_0 + y_0)/2$ and angular frequency ω_0.

Remark 4 It is important to note that the solution of (7.1), (7.2) is obtained as a sum of an exact solution of a linear differential equation and an approximate solution of a strongly nonlinear differential equation. The approximate solution is the exact solution of the averaged nonlinear equation. The form of the averaged nonlinear equation deeply depends on the type of the nonlinear connection.

7.1.3 Pure Nonlinear Viscoelastic Connection

If the viscoelastic connection is purely nonlinear (there is no linear stiffness term) and the nonlinear function is

$$F(x-y,\dot{x}-\dot{y}) = c_{\alpha}^{2}(x-y)\,|x-y|^{\alpha-1} + \sum \varepsilon d_{\beta}^{*}(\dot{x}-\dot{y})\,|\dot{x}-\dot{y}|^{\beta-1}, \quad (7.38)$$

we obtain the function

$$F^{*}(z,\dot{z}) = k_{\alpha}^{2} z\,|z|^{\alpha-1} + \sum \varepsilon d_{\beta}\dot{z}\,|\dot{z}|^{\beta-1}, \quad (7.39)$$

and the corresponding differential equation (7.16)

$$\ddot{z} + k_{\alpha}^{2} z\,|z|^{\alpha-1} = -\varepsilon k_{1}^{2} z - \sum \varepsilon d_{\beta}\dot{z}\,|\dot{z}|^{\beta-1}, \quad (7.40)$$

where $\alpha \geqslant 1$ and $\beta \geqslant 1$ are orders of nonlinearity, c_{α}^{2} and $\varepsilon d_{\beta} \ll 1$ are parameters of elasticity and damping, $k_{\alpha}^{2} = 2c_{\alpha}^{2}/m$ and $\varepsilon d_{\beta} = 2\varepsilon d_{\beta}^{*}/m$, $\varepsilon k_{1}^{2} = \omega_{0}^{2}$ and ε is a small parameter, i.e., $\varepsilon \ll 1$.

Based on the first step of the suggested procedure given in the previous section we have to obtain the solution of (7.40).

Neglecting the right-hand side term with the small parameter ε we obtain a pure strong nonlinear differential equation

$$\ddot{z} + k_{\alpha}^{2} z\,|z|^{\alpha-1} = 0. \quad (7.41)$$

For initial conditions $(7.8)_2$ and $(7.8)_4$, there is an exact solution of the Eq. (7.41) in the form of a *ca*-cosine Ateb function (Rosenberg 1963)

$$z = Z_0 ca(\alpha, 1, \psi), \quad (7.42)$$

where the phase angle is

$$\psi = \Omega t + \theta, \quad (7.43)$$

with $\theta = const$, Z_0 is the amplitude of the oscillatory function and Ω is the frequency of the Ateb function, which depends on the amplitude of vibration Z_0 as

$$\Omega = \sqrt{\frac{k_{\alpha}^{2}\,|Z_0|^{\alpha-1}(\alpha+1)}{2}}. \quad (7.44)$$

The *ca* -cosine Ateb function is the inverse Beta function with the properties given in the Appendix. Using to the first derivative of the ca Ateb function (Cveticanin and Pogany 2012)

$$\frac{d}{d\psi}\left(ca(\alpha, 1, \psi)\right) \equiv \left(ca(\alpha, 1, \psi)\right)' = -\frac{2}{\alpha+1} sa(1, \alpha, \psi), \tag{7.45}$$

the time derivative of the solution (7.42) is

$$\dot{z} = -\frac{2Z_0\Omega}{\alpha+1} sa(1, \alpha, \psi) \equiv -Z_0^{(\alpha+1)/2}\sqrt{\frac{2k_\alpha^2}{\alpha+1}} sa(1, \alpha, \psi), \tag{7.46}$$

where sa is the sine Ateb function.

According to (7.21), we assume the solution of (7.40) and the first time derivative in the form which corresponds to (7.42) and (7.46), but with time variable amplitude and phase. Thus, the trial solution and the assumed first derivative trial solution are

$$z = Z_0(t) ca(\alpha, 1, \psi(t)), \tag{7.47}$$

$$\dot{z} = -\frac{2Z_0(t)\Omega(Z_0)}{\alpha+1} sa(1, \alpha, \psi(t)) = -(Z_0(t))^{\frac{\alpha+1}{2}}\sqrt{\frac{2k_\alpha^2}{\alpha+1}} sa(1, \alpha, \psi(t)), \tag{7.48}$$

where

$$\psi(t) = \int \Omega(Z_0) dt + \theta(t), \qquad \Omega(Z_0) = \sqrt{\frac{k_\alpha^2 \left|Z_0(t)\right|^{\alpha-1}(\alpha+1)}{2}}, \tag{7.49}$$

and $\theta(t)$ is an unknown function. Nevertheless, the first time derivative of (7.47) is

$$\dot{z} = \dot{Z}_0(t) ca(\alpha, 1, \psi(t)) - \frac{2Z_0(t)\Omega(Z_0)}{\alpha+1} sa(1, \alpha, \psi(t)) - \frac{2Z_0(t)}{\alpha+1}\dot{\theta}(t) sa(1, \alpha, \psi(t)), \tag{7.50}$$

and it differs from the assumed relation (7.48). The relations (7.46) and (7.50) are equal if

$$\dot{Z}_0 ca - \frac{2Z_0}{\alpha+1}\dot{\theta} sa = 0, \tag{7.51}$$

where for simplicity the following notation is introduced: $Z_0(t) = Z_0$, $\theta(t) = \theta$, $\psi(t) = \psi$, $\Omega(Z_0) = \Omega$, $ca(\alpha, 1, \psi(t)) = ca$, $sa(1, \alpha, \psi(t)) = sa$. This mathematical transaction gives us an opportunity to rewrite the second order differential equation (7.40) into the two first order coupled differential equations.

Namely, applying the time derivative of sa Ateb function

$$\frac{d}{d\psi}\left(sa(1, \alpha, \psi)\right) = \left(sa(1, \alpha, \psi)\right)' = ca^\alpha(\alpha, 1, \psi), \tag{7.52}$$

the time derivative of (7.48) is

$$\ddot{z} = -Z_0^{(\alpha-1)/2}\dot{Z}_0\sqrt{\frac{(\alpha+1)k_\alpha^2}{2}}sa - k_\alpha^2 Z_0 \left|Z_0\right|^{\alpha-1} ca^\alpha - Z_0^{(\alpha+1)/2}\dot{\theta}\sqrt{\frac{2k_\alpha^2}{\alpha+1}}ca^\alpha. \tag{7.53}$$

Substituting (7.21) and (7.27) into (7.32) we obtain

$$\dot{Z}_0 sa + \frac{2}{\alpha+1}Z_0\dot{\theta}ca^\alpha = \varepsilon k_1^2 Z_0 \left|Z_0\right|^{(1-\alpha)/2}\sqrt{\frac{2}{(\alpha+1)k_\alpha^2}}ca \tag{7.54}$$
$$-\sum \varepsilon d_\beta Z_0\sqrt{\frac{2}{(\alpha+1)k_\alpha^2}}\left(\sqrt{\frac{2k_\alpha^2}{\alpha+1}}\right)^\beta sa\left|Z_0^{(\alpha+1)/2}sa\right|^{\beta-1}.$$

After some modification of Eqs. (7.51) and (7.54), and using the relation

$$sa^2(1,\alpha,\psi(t)) + ca^{\alpha+1}(\alpha,1,\psi(t)) = 1, \tag{7.55}$$

two first order equations follow which represent the rewritten second order Eq. (7.40) with new variables θ and Z_0

$$\dot{Z}_0 = \varepsilon k_1^2 Z_0 \left|Z_0\right|^{(1-\alpha)/2}\sqrt{\frac{2}{(\alpha+1)k_\alpha^2}}saca \tag{7.56}$$
$$-\sum \varepsilon d_\beta Z_0\sqrt{\frac{2}{(\alpha+1)k_\alpha^2}}\left(\sqrt{\frac{2k_\alpha^2}{\alpha+1}}\right)^\beta sa^2\left|Z_0^{(\alpha+1)/2}sa\right|^{\beta-1},$$

$$Z_0\dot{\theta} = \varepsilon k_1^2 Z_0 \left|Z_0\right|^{(1-\alpha)/2}\sqrt{\frac{\alpha+1}{2k_\alpha^2}}ca^2 \tag{7.57}$$
$$-\sum \varepsilon d_\beta\sqrt{\frac{\alpha+1}{2k_\alpha^2}}\left(\sqrt{\frac{2k_\alpha^2}{\alpha+1}}\right)^\beta saZ_0\left|Z_0^{(\alpha+1)/2}sa\right|^{\beta-1}ca.$$

The system of Eqs. (7.56) and (7.57) is exactly transformed into two coupled first order equations. Our task is to solve Eqs. (7.56) and (7.57) for Z_0 and ψ i.e. θ. Due to complexity of the problem, a simplification is introduced. Averaging Ateb functions in (7.56) and (7.57) over their period $2\Pi_\alpha$ which is

$$\Pi_\alpha = B\left(\frac{1}{\alpha+1},\frac{1}{2}\right), \tag{7.58}$$

averaged equations follow

$$\dot{Z}_0 = -\sum \varepsilon d_\beta Z_0 \left|Z_0\right|^{(\alpha+1)(\beta-1)/2} S_{\beta+1}\sqrt{\frac{2}{(\alpha+1)k_\alpha^2}}\left(\sqrt{\frac{2k_\alpha^2}{\alpha+1}}\right)^\beta, \tag{7.59}$$

$$Z_0\dot{\theta} = \varepsilon k_1^2 Z_0 \left|Z_0\right|^{(1-\alpha)/2} C\sqrt{\frac{\alpha+1}{2k_\alpha^2}}. \tag{7.60}$$

where B is the Beta function and

$$\langle saca\rangle = \frac{1}{2\Pi_\alpha}\int_0^{2\Pi_\alpha} sacad\psi = 0, \quad \left\langle sa\left|sa\right|^{\beta-1} ca\right\rangle = \frac{1}{2\Pi_\alpha}\int_0^{2\Pi_\alpha} sa\left|sa\right|^{\beta-1} cad\psi = 0,$$

$$S_{\beta+1} = \left\langle sa^2\left|sa\right|^{\beta-1}\right\rangle = \frac{1}{2\Pi_\alpha}\int_0^{2\Pi_\alpha} sa^2\left|sa\right|^{\beta-1} d\psi,$$

$$C = \left\langle ca^2\right\rangle = \frac{1}{2\Pi_\alpha}\int_0^{2\Pi_\alpha} ca^2 d\psi = \left(\frac{2}{3+\alpha}\right)^{2/(\alpha+1)}. \tag{7.61}$$

Equation (7.59) is an uncoupled one, and after integration it is

$$\int \frac{dZ_0}{\sum \varepsilon d_\beta Z_0\left|Z_0\right|^{(\alpha+1)(\beta-1)/2} S_{\beta+1}\sqrt{\frac{2}{(\alpha+1)k_\alpha^2}}\left(\sqrt{\frac{2k_\alpha^2}{\alpha+1}}\right)^\beta} = -t + C_1, \tag{7.62}$$

where C_1 is an arbitrary constant. Substituting (7.62) into (7.60) and solving, we obtain

$$\theta = \varepsilon k_1^2 C\sqrt{\frac{\alpha+1}{2k_\alpha^2}}\int \left|Z_0\right|^{(1-\alpha)/2} dt + C_2, \tag{7.63}$$

where C_2 is also an arbitrary constant. Using (7.62), (7.63) and the corresponding relation (7.49), the solution (7.47) which satisfies the initial conditions $(7.8)_2$ and $(7.8)_4$ is determined. Equations (7.62) and (7.63) represent the solutions of averaged Eqs. (7.49) and (7.50).

It can be concluded that the amplitude and the frequency of vibration are time variable. Amplitude of vibration depends on the orders and coefficients of nonlinearity of elastic and damping terms and is independent on the elasticity of the harmonic oscillators. The phase angle ψ depends on the rigidity of the connecting springs, too.

7.1.4 Special Case

For the special case when only one damping term exists the solution of (7.62) is

$$\frac{1}{(Z_0(0))^{\frac{(\alpha+3)(\beta-1)}{2}-1}}-\frac{1}{Z_0^{\frac{(\alpha+3)(\beta-1)}{2}-1}}$$
$$=-\left(\frac{(\alpha+3)(\beta-1)}{2}-1\right)\varepsilon d_\beta S_{\beta+1}\sqrt{\frac{2}{(\alpha+1)k_\alpha^2}}\left(\sqrt{\frac{2k_\alpha^2}{\alpha+1}}\right)^\beta t, \tag{7.64}$$

where $Z_0(0)$ is the initial value for Z_0. If the damping parameter d_β is positive, independently on the values of α and β the amplitude of vibration decreases, while for negative coefficient of damping d_β the amplitude of vibrations increases. It is constant for $d_\beta = 0$.

7.1.5 'Steady-State' Solution

(a) If two damping terms exist, a positive $\varepsilon d_{\beta 1}$ and a negative $\varepsilon d_{\beta 2}$, the Eq. (7.40) has the form

$$\ddot{z}+k_\alpha^2 z\,|z|^{\alpha-1}=-\varepsilon k_1^2 z-\varepsilon d_{\beta 1}\dot{z}\,|\dot{z}|^{\beta_1-1}+\varepsilon d_{\beta 2}\dot{z}\,|\dot{z}|^{\beta_2-1}. \tag{7.65}$$

After the transformations suggested in the previous text and using the relation (7.59), it is

$$\dot{Z}_0=-Z_0\,|Z_0|^{\frac{(\alpha+1)\beta 1}{2}}\varepsilon D\left(1-Q\frac{\varepsilon d_{\beta 2}}{\varepsilon d_{\beta 1}}\,|Z_0|^{\frac{(\alpha+1)(\beta 2-\beta 1)}{2}}\right), \tag{7.66}$$

where

$$\varepsilon D=\sqrt{\frac{2}{(\alpha+1)k_\alpha^2}}\left(\sqrt{\frac{2k_\alpha^2}{\alpha+1}}\right)^{\beta 1}\varepsilon d_{\beta 1}S_{\beta 1+1},\qquad Q=\frac{S_{\beta 2+1}}{S_{\beta 1+1}}\left(\sqrt{\frac{2k_\alpha^2}{\alpha+1}}\right)^{\beta 2-\beta 1}. \tag{7.67}$$

Analyzing the relation (7.66) it is obvious that the amplitude of vibration decreases if $Q\frac{\varepsilon d_{\beta 2}}{\varepsilon d_{\beta 1}}Z_0(0)\,|Z_0(0)|^{\frac{(\alpha+1)(\beta 2-\beta 1)}{2}}<1$, and increases for $Q\frac{\varepsilon d_{\beta 2}}{\varepsilon d_{\beta 1}}Z_0(0)\,|Z_0(0)|^{\frac{(\alpha+1)(\beta 2-\beta 1)}{2}}>1$. The behavior of the system depends not only the relation between the damping parameters, but also on the order of nonlinearity of the damping terms, β_1 and β_2, and on the order of nonlinearity of the stiffness term α.

Equating the left side of Eq. (7.66) with zero, the steady state motion with constant amplitude is obtained as

$$1 - Q\frac{\varepsilon d_{\beta 2}}{\varepsilon d_{\beta 1}} Z_0 \left| Z_0 \right|^{\frac{(\alpha+1)(\beta 2-\beta 1)}{2}} = 0, \tag{7.68}$$

i.e.,

$$Z_0 \left| Z_0 \right|^{\frac{(\alpha+1)(\beta 2-\beta 1)}{2}} = \frac{1}{Q}\frac{\varepsilon d_{\beta 1}}{\varepsilon d_{\beta 2}}. \tag{7.69}$$

Such motion occurs when the initial value satisfies the relation (7.69), i.e., $Z_0(0) = (\frac{1}{Q}\frac{\varepsilon d_{\beta 1}}{\varepsilon d_{\beta 2}})^{\frac{2}{(\alpha+1)(\beta 2-\beta 1)}}$.

For the steady state motion the averaged differential equation (7.60) transforms into

$$\dot{\theta} = \varepsilon k_1^2 \left| Z_0(0) \right|^{(1-\alpha)/2} C\sqrt{\frac{\alpha+1}{2k_\alpha^2}}, \tag{7.70}$$

and after integration

$$\theta = C_2 + \varepsilon k_1^2 \left| Z_0(0) \right|^{(1-\alpha)/2} Ct\sqrt{\frac{\alpha+1}{2k_\alpha^2}}, \tag{7.71}$$

while the frequency function is

$$\Omega = \left| Z_0(0) \right|^{(\alpha-1)/2} \sqrt{\frac{k_\alpha^2(\alpha+1)}{2}}. \tag{7.72}$$

Using (7.71) and (7.72) frequency of the solution is

$$\Omega^* = \left| Z_0(0) \right|^{(\alpha-1)/2} \sqrt{\frac{k_\alpha^2(\alpha+1)}{2}} + \varepsilon k_1^2 \left| Z_0(0) \right|^{(1-\alpha)/2} C\sqrt{\frac{\alpha+1}{2k_\alpha^2}}. \tag{7.73}$$

The corresponding period of vibration is

$$T = \frac{2\Pi_\alpha}{\Omega^*}, \tag{7.74}$$

where Π_α is given with (7.58). Finally, the $z - t$ relation is approximately

$$z = Z_0(0) ca\left(\alpha, 1, \left((Z_0(0))^{(\alpha-1)/2}\sqrt{\frac{k_\alpha^2(\alpha+1)}{2}} + \varepsilon k_1^2 \left| Z_0(0) \right|^{(1-\alpha)/2} C\sqrt{\frac{\alpha+1}{2k_\alpha^2}}\right) t + C_2\right). \tag{7.75}$$

Due to additional small terms in (7.40), the frequency of the *ca* - Ateb function differs from (7.44). Based on obtained solutions (7.32) and (7.75), motions of both masses are

$$x = \frac{1}{2}\Bigg(u_0 \cos t\sqrt{\varepsilon k_1^2} + \frac{\dot{u}_0}{\omega_0}\sin t\sqrt{\varepsilon k_1^2} \tag{7.76}$$

$$+Z_0(0) ca\Bigg(\alpha, 1, \Bigg((Z_0(0))^{(\alpha-1)/2}\sqrt{\frac{k_\alpha^2(\alpha+1)}{2}} + \varepsilon k_1^2 \left|Z_0(0)\right|^{\frac{1-\alpha}{2}} C\sqrt{\frac{\alpha+1}{2k_\alpha^2}}\Bigg)t + C_2\Bigg)\Bigg),$$

$$y = \frac{1}{2}\Bigg(u_0 \cos t\sqrt{\varepsilon k_1^2} + \frac{\dot{u}_0}{\omega_0}\sin t\sqrt{\varepsilon k_1^2} \tag{7.77}$$

$$-Z_0(0) ca\Bigg(\alpha, 1, \Bigg((Z_0(0))^{(\alpha-1)/2}\sqrt{\frac{k_\alpha^2(\alpha+1)}{2}} + \varepsilon k_1^2 \left|Z_0(0)\right|^{\frac{1-\alpha}{2}} C\sqrt{\frac{\alpha+1}{2k_\alpha^2}}\Bigg)t + C_2\Bigg)\Bigg).$$

(b) For the case when the positive damping is linear and the differential equation (7.40) is

$$\ddot{z} + k_\alpha^2 z\left|z\right|^{\alpha-1} = -\varepsilon k_1^2 z - \varepsilon d_1 \dot{z} + \varepsilon d_{\beta 2}\dot{z}\left|\dot{z}\right|^{\beta_2 - 1}.$$

the Eq. (7.59) transforms into

$$\dot{Z}_0 + q_1 Z_0 - q_2 Z_0 \left|Z_0\right|^r = 0, \tag{7.78}$$

where

$$q_1 = \varepsilon d_1 S_2 \frac{2}{\alpha+1}, \quad q_2 = \varepsilon d_\beta S_{\beta+1}\sqrt{\frac{2}{(\alpha+1)k_\alpha^2}}\left(\sqrt{\frac{2k_\alpha^2}{\alpha+1}}\right)^\beta,$$

$$r = \frac{(\alpha+1)(\beta-1)}{2}, \quad S_2 = \langle sa^2\rangle = \frac{1+\alpha}{3+\alpha}. \tag{7.79}$$

Equation (7.78) is a Bernoulli one with the solution

$$Z_0 = \left(\frac{q_2}{q_1} + \left|(Z_0(0))^{-r} - \frac{q_2}{q_1}\right| \exp((r-1)q_1 t)\right)^{-\frac{1}{r}}. \tag{7.80}$$

The value of $Z_0 \equiv Z_{0p}$ remains constant, if the initial value $Z_0(0)$ satisfies the relation

$$Z_0(0) = \left(\frac{q_1}{q_2}\right)^{\frac{1}{r}}, \tag{7.81}$$

and the vibration is a steady state one. Specifying the relation (7.69), when $\beta_1 = 1$, the result (7.81) is obtained. For the steady state motion (7.81), the $\theta - t$ function corresponds to (7.71) and the oscillations in x and y direction are given in the form (7.76) and (7.77).

Remark 5 Due to the suggested procedure, the motion of bodies with purely nonlinear viscoelastic connection is given with (7.36) and (7.37)

$$x = u(t, x_0, y_0, \dot{x}_0, \dot{y}_0) + Z_0 ca\left(\alpha, 1, \int \sqrt{\frac{k_\alpha^2 \left|Z_0\right|^{\alpha-1}(\alpha+1)}{2}} dt + \theta\right), \tag{7.82}$$

$$y = u(t, x_0, y_0, \dot{x}_0, \dot{y}_0) - Z_0 ca\left(\alpha, 1, \int \sqrt{\frac{k_\alpha^2 \left|Z_0\right|^{\alpha-1}(\alpha+1)}{2}} dt + \theta\right), \tag{7.83}$$

where $u(t, x_0, y_0, \dot{x}_0, \dot{y}_0) = \frac{x_0+y_0}{2}\cos(\omega_0 t) + \frac{\dot{x}_0+\dot{y}_0}{2\omega_0}\sin(\omega_0 t)$, $Z_0 \equiv Z_0(t, x_0, y_0, \dot{x}_0, \dot{y}_0)$ and $\theta \equiv \theta(t, x_0, y_0, \dot{x}_0, \dot{y}_0)$ are the solutions of the averaged first order differential equations (7.62) and (7.63).

Remark 6 The obtained appropriate analytical result has a physical sense and based on the solution we give conclusions about the amplitude, phase and frequency of vibration, about the influence of the order of nonlinearity, stiffness and damping parameters, initial conditions on the oscillatory properties of the system.

7.1.6 Mechanical Vibration of the Vocal Cord

Ishizaka and Flangan (1972) were the first to introduce the mathematical model of the vocal cord. After that, a significant number of investigation on the dynamics of vocal cords is done (Cveticanin 2012) with the aim to explain the cause of vocalization. They assumed that the left and the right vocal cords are bilaterally symmetric. Therefore, properties of only one cord are discussed and the same being implied for the opposing cord. The vocal cord is supposed as a two body model (Fig. 7.2).

Namely, the vocal cord is divided in the depth in two parts: upper and lower. Each part represents a simple mechanical oscillator having a mass and spring which permitted only lateral motions, x and y, of masses. The two oscillators are connected with a spring and a damper which represent the effect of flexural stiffness and damping of the vocal cord. In general, (7.1) and (7.2) model of the vocal cord where m is the mass of the upper and lower part. Recently, experimental investigation show that the force F of the vocal fold lamina propria varies as a function of depth relative to the epithelial surface (Heris et al. 2013) as

$$Fe = c_\alpha^2(x-y)^\alpha + r_1(\dot{x}-\dot{y}) - r_\beta(\dot{x}-\dot{y})\left|\dot{x}-\dot{y}\right|^{\beta-1}. \tag{7.84}$$

α and β are the orders of elastic and damping nonlinearities, r_1 is the coefficient of linear damping while c_α^2 and r_β are coefficients which depend on the stiffness

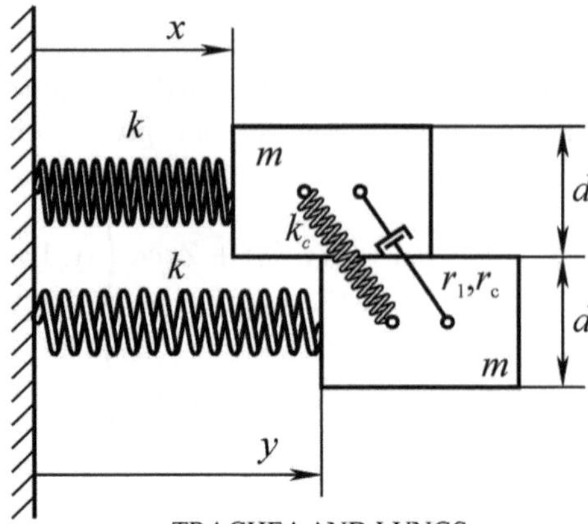

Fig. 7.2 Model of the vocal cord

k_c and damping parameter r_c of the tissue, i.e., $c_{\alpha}^2 \approx 2k_c/(d_1 + d_2)^{\alpha-1}$ and $r_{\beta} \approx 2r_c/(d_1 + d_2)^{\beta-1}$ where d_1 and d_2 are the thickness of the upper and lower parts of the vocal cord. Usual value of nonlinearity order of the human vocal cord tissue is (see Mergel et al. 2000): $\alpha = \beta = 1.5$, while the mass is $2m = 0.2g$, depth of the vocal cord $d = 0.3cm$, the lateral rigidity coefficient $k = 0.01\text{g/ms}^2$, flexural rigidity parameter $k_c = 0.001369\,3\text{g/ms}^2$, damping coefficients $r_1 = 0.0025$ g/ms and $r_c = 2.738\,6 \times 10^{-4}$ g/ms. Substituting the numerical values into the Eqs. (7.1), (7.2) and (7.84) and assuming the equality of masses of both parts of the vocal cord we have

$$\ddot{x} + 0.1x + 0.5(x - y)\,|x - y|^{\frac{1}{2}} = -0.025(\dot{x} - \dot{y}) + 0.1(\dot{x} - \dot{y})\,|\dot{x} - \dot{y}|^{\frac{1}{2}}\,, \tag{7.85}$$

$$\ddot{y} + 0.1y - 0.5(x - y)\,|x - y|^{\frac{1}{2}} = 0.025(\dot{x} - \dot{y}) - 0.1(\dot{x} - \dot{y})\,|\dot{x} - \dot{y}|^{\frac{1}{2}}\,. \tag{7.86}$$

The rewritten version of equations is

$$\ddot{u} + 0.1u = 0, \tag{7.87}$$

$$\ddot{z} + z\,|z|^{1/2} = -0.1z - 0.05\dot{z} + 0.2\dot{z}\,|\dot{z}|^{1/2}\,. \tag{7.88}$$

The Eq. (7.88) describes the flexural vibration of the vocal cord, while the Eq. (7.87) gives the sum of lateral motions of the upper and lower part of the vocal fold whose minimal value affects the pressures inside the vocal fold and the flow volume velocity (Wan et al. 2013).

In Fig. 7.3a, $u - t$ and $z - t$ diagrams of (7.87) and (7.88) for initial conditions

$$u(0) = 0.3, \quad \dot{u}(0) = 0, \tag{7.89}$$

$$z(0) = 0.12, \quad \dot{z}(0) = 0, \tag{7.90}$$

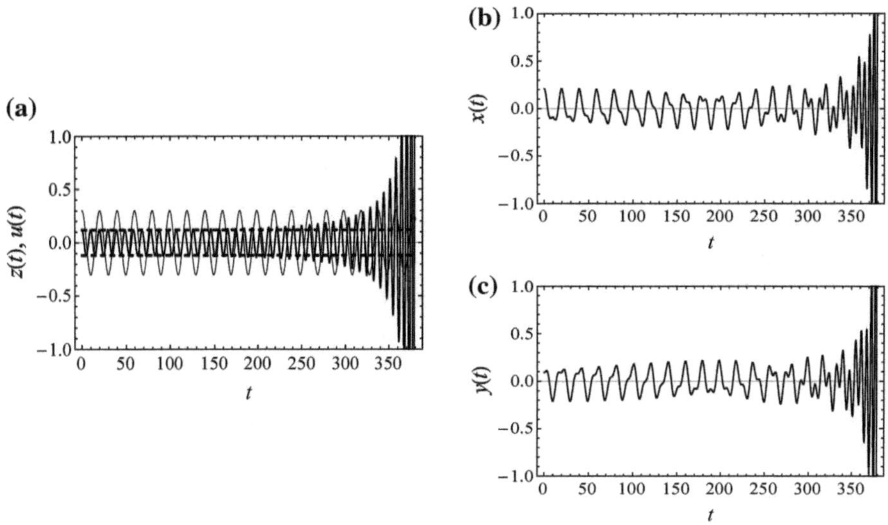

Fig. 7.3 **a** $u - t$ (*thin line*) and $z - t$ (*thick line*) diagrams; **b** $x - t$ and **c** $y - t$ diagrams for $u(0) = 0.3$ and $z(0) = 0.12$

are plotted. The solution of the uncoupled differential equation (7.87) is periodical

$$u = 0.3\cos(0.316\,23t). \tag{7.91}$$

the $z - t$ diagram has the tendency of increase. According to these results the amplitude of vibration of both masses increases in time due to priority of the negative damping in the connection (Fig. 7.3b, c). The flexural vibration of the vocal cord is with increasing amplitude and it causes the increase of amplitude of vibration for both parts (upper and lower) of the vocal cord.

For initial conditions (7.89) and

$$z(0) = 0.1, \quad \dot{z}(0) = 0, \tag{7.92}$$

$u - t$ and $z - t$ diagrams are plotted (Fig. 7.4a).

The $u - t$ diagram is periodical, while the $z - t$ diagram has the tendency of decrease. Flexural vibration of the vocal cord disappears and the both parts of the vocal cord have the same vibration with equal amplitude and period (Fig. 7.4b, c). Namely, for $z \longrightarrow 0$ motion of masses is harmonical and independent on the flexural properties of the tissue, i.e.,

$$x = y = \frac{u}{2} = 0.15\cos(0.316\,23t). \tag{7.93}$$

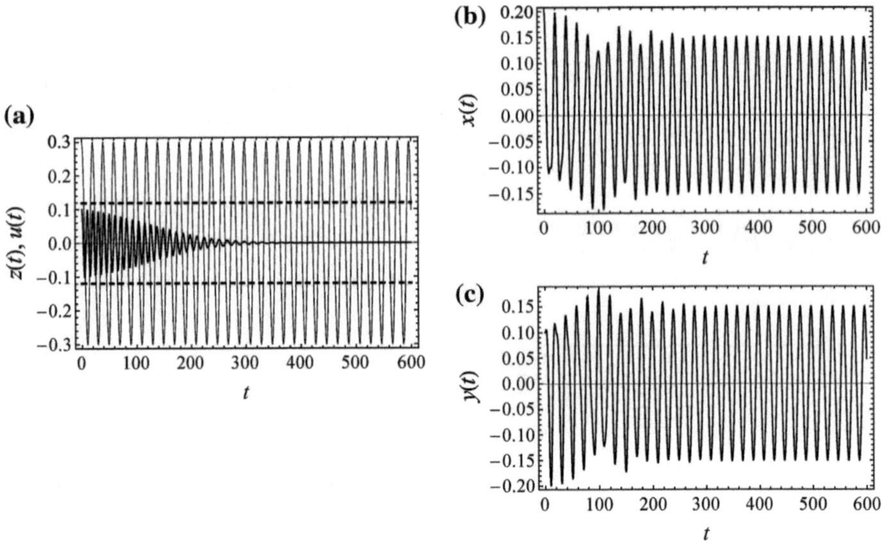

Fig. 7.4 **a** $u - t$ (*thin line*) and $z - t$ (*thick line*) diagrams; **b** $x - t$ and **c** $y - t$ diagrams for $u(0) = 0.3$ and $z(0) = 0.10$

The value of the amplitude of vibration is half of the sum of initial displacements of masses. The frequency is independent on the properties of the connection between bodies.

Comparing the analytical results (7.93) with those obtained experimentally (see Wan et al. 2013), it can be concluded that the model of the system is correctly defined as the analytical results $x - t$ and $y - t$ are in good agreement with measured ones for the same values of vocal cord parameters.

According to (7.79) and (7.88) the parameter values $q_1 = 0.02222$, $q_2 = 0.084065$, $r = 0.625$, $S_2 = 0.55555$ and $S_{5/2} = 0.55556$, the constant amplitude of the periodical solution of (7.88) is obtained

$$Z_0(0) = \left(\frac{q_1}{q_2}\right)^{1/r} = 0.11896. \tag{7.94}$$

For the initial amplitude

$$z(0) = 0.11896, \quad \dot{z}(0) = 0, \tag{7.95}$$

the steady solution (7.75) is

$$z = 0.11896.ca\,(1.5, 1, 0.756\,10t)\,. \tag{7.96}$$

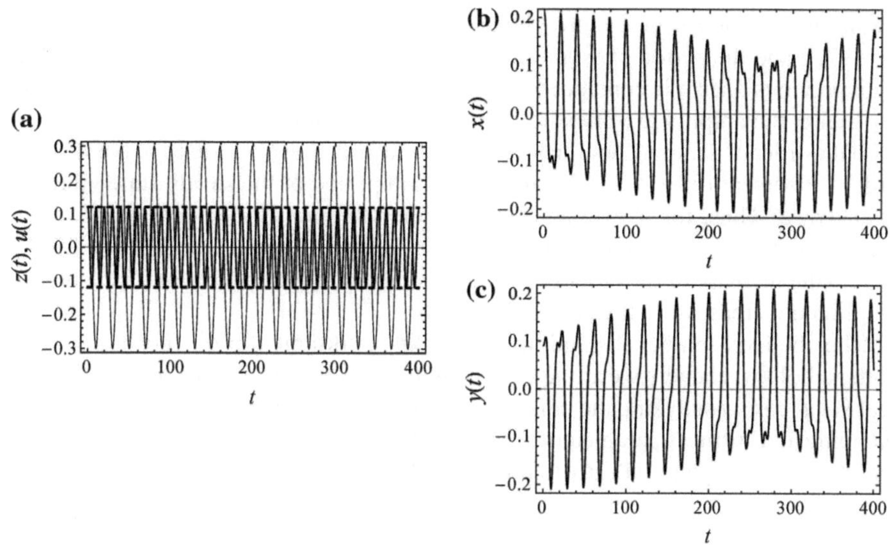

Fig. 7.5 **a** $u - t$ (*thin line*) and $z - t$ (*thick line*) diagrams; **b** $x - t$ and **c** $y - t$ diagrams for $u(0) = 0.3$ and $z(0) = 0.11894$

For $\Pi = 3.67909$ and $\Omega^* = 0.75610$ the period of vibration is $T = 9.731\,86$. The relation (7.96) gives the steady-state flexural vibrations of the vocal cord with constant amplitude and period. For initial conditions (7.95) and (7.89), Eqs. (7.87) and (7.88) are solved numerically, too (see Fig. 7.5a).

Comparing the analytical and numerical solutions it is obvious that the difference is negligible. Namely, the numerically calculated amplitude is $Z_{0p} = 0.11894$, and the period of vibration $T = 9.75775$. Based on (7.91) and (7.96), motions of bodies (7.76) and (7.77) are as follows

$$x = 0.15\cos(0.316\,23t) + 0.05948ca\,(1.5, 1, 0.756\,10t)\,, \quad (7.97)$$

$$y = 0.15\cos(0.316\,23t) - 0.05948ca\,(1.5, 1, 0.756\,10t)\,. \quad (7.98)$$

In Fig. 7.5b, c, the $x - t$ and $y - t$ diagrams are plotted. The motion of masses is bounded.

As it is well known the stiffness and damping properties for all human vocal cords are not equal. Even the left and the right vocal cord may differ. Besides, the same vocal cord changes their properties during the life time of a person or due to some defects or illness. In the following text the influence of elasticity and damping nonlinearity order on the amplitude and period of vibration of the vocal cord is analyzed.

If the order of the elasticity is $\alpha = 2$, i.e., higher than the previous one, the equation of flexural vibration is

$$\ddot{z} + z\,|z| = -0.1z - 0.05\dot{z} + 0.2\dot{z}\,|\dot{z}|^{1/2}\,. \quad (7.99)$$

Using the mentioned procedure, the steady-state amplitude of vibration is $Z_{0p} = 0.159736$ and the period of vibration $T = 6.48817$. Comparing the obtained values with previous ones it is concluded that the higher the order of stiffness nonlinearity, the higher the amplitude of flexural vibrations and shorter the period.

If the order of damping nonlinearity in flexural direction is varying and the stiffness order is constant ($\alpha = 3/2$), the amplitude and period of the steady state vibration is analyzed. Thus, for $\beta = 2$ amplitude of the vibration is $Z_{0p} = 0.3745215$ and the period $T = 7.4559$, while for $\beta = 4/3$ it is $Z_{0p} = 0.036700$ and $T = 11.9719$. Due to the increase of the order of the negative damping in connection, the vibration amplitude increases and the period decreases.

The obtained analytical solutions give us an opportunity for diagnostics of vocal cord pathology. Namely,we arrange the analytical solution to a scope of vibration of a real vocal cord. Analyzing the obtained parameters we conclude about the change in damping and stiffness parameters of the model, i.e., of the vocal cord. The parameter variation offers us a possibility to conclude about the type of vocal cord disease.

7.1.7 Conclusions

The following is concluded:

1. For the case when two masses are connected with strong pure nonlinear viscoelastic connection (without linear term) the motion is suitable to be described as a sum of a harmonic and a periodic function. Due to nonlinearity the most convenient form of the periodic solution is the Ateb function whose parameters are obtained by adopting of the method of time variable amplitude and phase. The amplitude depends on the properties of the connection (damping and stiffness parameters and the order of nonlinearity). Depending on the initial values the amplitude of periodic function increases or decreases, while for certain values it is constant. The phase and frequency of the periodical Ateb function include also the stiffness characteristics of the connection to the fixed body.

2. Based on the obtained solution which is the sum of a harmonic and a periodic motion it follows:

(a) If the amplitude of the periodical solution increases, the amplitude of motion of both masses tends to infinity.

(b) If the amplitude of the periodical solution decreases, the motion of both masses tend to have equal oscillatory motions: masses oscillate harmonically with amplitude, which is the half sum of initial deflections of masses, and the frequency of vibration depends directly on the mass and indirectly on the rigidity of connection to the fixed body, but is independent on properties of the connection and initial values.

(c) If the periodical solution is 'steady state' with constant amplitude and phase, the total motion of one mass is a sum of the harmonic motion and of the periodic solution, while the motion of the other mass is a difference of these two functions. In general, motion of both bodies are bounded, but not periodical.

3. The results of theoretical consideration of the two mass oscillatory model with strong nonlinear viscoelastic connection can be applied for mechanical vibration of a human vocal fold and for analyzing of the influence of the stiffness and damping coefficients and the order of nonlinearity properties of the vocal cord tissue on the vocalization. Namely, the lower is the order of elasticity in the vocal cord between the upper and the lower part, the amplitude and the frequency of vibration of the vocal cord are lower. The higher is the order of damping property of the tissue along the vocal cord, the amplitude of vibration of the vocal cord decreases, while the period of vibration increases. Due to decreasing of the elasticity of the vocal fold and changing the damping properties of the tissue, the fundamental frequency decreases. Finally, the suggested model and the obtained results are suitable for simulation of defects of the vocal cord.

4. The analytical method suggested in this paper gives approximate solutions whose difference in comparison with numerical solutions is negligible.

7.2 System with Nonlinear Elastic Connection

In this section the model of the system which contains two masses m_1 and m_2 connected with an elastic spring with strong nonlinear properties is considered. The additional small forces, which depend on the position and velocity of masses, may also be introduced. The mathematical model of the oscillator is in general

$$\begin{aligned} m_1\ddot{x} + k_\alpha^2(x-y)\,|x-y|^{\alpha-1} &= \varepsilon f_1(x,\dot{x},y,\dot{y}), \\ m_2\ddot{y} + k_\alpha^2(y-x)\,|y-x|^{\alpha-1} &= \varepsilon f_2(x,y,\dot{x},\dot{y}), \end{aligned} \tag{7.100}$$

where m_1 and m_2 are two different masses, k_α^2 is the rigidity coefficient of the connecting spring, εf_1 and εf_2 are small functions dependent on deflections (x and y) and velocities ($\dot{x}$ and $\dot{y}$) and α is a rational number (integer or non-integer) not smaller than 1 which represents the order of nonlinearity. Initial conditions are assumed to be

$$x(0) = A_1, \qquad y(0) = A_2, \qquad \dot{x}(0) = 0, \qquad \dot{y}(0) = 0. \tag{7.101}$$

Introducing the new variable

$$Y = y - x, \tag{7.102}$$

the system of differential equations (7.100) is rewritten as

$$\begin{aligned} m_1\ddot{x} - k_\alpha^2 Y\,|Y|^{\alpha-1} &= \varepsilon f_1(x,\dot{x},Y,\dot{Y}), \\ m_2(\ddot{Y}+\ddot{x}) + k_\alpha^2 Y\,|Y|^{\alpha-1} &= \varepsilon f_2(x,\dot{x},Y,\dot{Y}), \end{aligned} \tag{7.103}$$

with initial conditions

$$x(0) = A_1, \qquad \dot{x}(0) = 0, \tag{7.104}$$

$$Y(0) = A, \qquad \dot{Y}(0) = 0, \tag{7.105}$$

where $A = A_2 - A_1$. Introducing the notation

$$\frac{1}{\mu} = \frac{1}{m_1} + \frac{1}{m_2},$$

$$c_\alpha^2 = \frac{k_\alpha^2}{\mu}, \quad \frac{1}{\lambda} = \frac{\mu}{m_1},$$

$$\varepsilon F_Y(x, \dot{x}, Y, \dot{Y}) \equiv \varepsilon F_Y = \frac{1}{m_2} f_2 - \frac{1}{m_1} f_1,$$

$$\varepsilon F_x(x, \dot{x}, Y, \dot{Y}) \equiv \varepsilon F_x = \frac{1}{m_2 \lambda}(f_1 + f_2), \tag{7.106}$$

the Eq. (7.103) transform into

$$\ddot{Y} + c_\alpha^2 Y \,|Y|^{\alpha-1} = \varepsilon F_Y(x, \dot{x}, Y, \dot{Y}), \quad \ddot{x} = -\frac{1}{\lambda}\ddot{Y} + \varepsilon F_x(x, \dot{x}, Y, \dot{Y}). \tag{7.107}$$

For $\varepsilon = 0$, the second differential equation $(7.107)_2$ simplifies to $\ddot{x} = -\frac{1}{\lambda}\ddot{Y}$ which, after integration, gives the relation

$$x = -\frac{1}{\lambda} Y. \tag{7.108}$$

Substituting (7.108) and its first time derivative into $(7.107)_1$, it follows

$$\ddot{Y} + c_\alpha^2 Y \,|Y|^{\alpha-1} = \varepsilon F_Y(Y, \dot{Y}). \tag{7.109}$$

The differential equation (7.109) is a single variable nonlinear differential equation which is already studied in Chap. 4. As is discussed in this Chap. 4, various solving procedures are developed. The approximate solution of (7.109) is in general

$$Y = Y_a(t, C_1, C_2), \tag{7.110}$$

where C_1 and C_2 are arbitrary constants which have to satisfy the initial conditions (7.105). Substituting (7.110) into $(7.107)_2$ we have

$$\ddot{x} = -\frac{1}{\lambda}\ddot{Y}_a + \varepsilon F_x(Y_a, \dot{Y}_a). \tag{7.111}$$

After two times of integrations of (7.111), it follows

$$x = -\frac{Y_a(t, C_1, C_2)}{\lambda} + \int \left(\int \varepsilon F_x(Y_a(t, C_1, C_2), \dot{Y}_a(t, C_1, C_2))dt \right) dt + C_3 t + C_4, \tag{7.112}$$

where C_3 and C_4 are arbitrary integration constants which satisfy the initial conditions (7.104). Substituting (7.110) and (7.112) into (7.102) it is

$$y = \frac{\lambda - 1}{\lambda} Y_a(t, C_1, C_2) + \int \left(\int \varepsilon F_x(Y_a(t, C_1, C_2), \dot{Y}_a(t, C_1, C_2))dt \right) dt + C_3 t + C_4. \tag{7.113}$$

Using (7.106) and the relations (7.112) and (7.113), the motion of the masses is in general

$$\begin{aligned} x &= -\frac{m_2 Y_a}{m_1 + m_2} + \frac{\varepsilon}{m_1 + m_2} \int \left(\int (f_1(Y_a, \dot{Y}_a) + f_2(Y_a, \dot{Y}_a))dt \right) dt + C_3 t + C_4, \\ y &= \frac{m_1 Y_a}{m_1 + m_2} + \frac{\varepsilon}{m_1 + m_2} \int \left(\int (f_1(Y_a, \dot{Y}_a) + f_2(Y_a, \dot{Y}_a))dt \right) dt + C_3 t + C_4. \end{aligned} \tag{7.114}$$

Analyzing the relations (7.114), it is obvious that the motion of the both masses has an oscillatory and a non-oscillatory part. The non-oscillatory part is equal for the both masses and represents the straightforward motion with the constant velocity C_3. For the arbitrary constants $C_3 = C_4 = 0$, the motion of the both masses is pure oscillatory.

7.3 Two-degree-of-freedom Van der Pol Oscillator

If the connection between two masses is with a negative damping function of Van der Pol type, the differential equations of motion are

$$\begin{aligned} m_1 \ddot{x} + k_\alpha^2 (x - y) \, |x - y|^{\alpha - 1} &= d(\dot{x} - \dot{y})(1 - |x - y|^{\beta - 1}), \\ m_2 \ddot{y} + k_\alpha^2 (y - x) \, |y - x|^{\alpha - 1} &= d(\dot{y} - \dot{x})(1 - |y - x|^{\beta - 1}), \end{aligned} \tag{7.115}$$

where d is the dissipation coefficient and β is a rational number (integer or non-integer). The considered Van der Pol nonlinearity is of general type. Introducing the new variable (7.102) and the notation (7.106) and assuming that the dissipation coefficient is small, i.e., $d = \varepsilon \ll 1$, the Eq. (7.115) are transformed into

$$\ddot{Y} + c_\alpha^2 Y \, |Y|^{\alpha - 1} = \frac{\varepsilon}{\mu} \dot{Y}(1 - |Y|^{\beta - 1}), \qquad \ddot{x} = -\frac{1}{\lambda} \ddot{Y}. \tag{7.116}$$

Integration of the relation $(7.116)_2$, reads

$$x = -\frac{1}{\lambda} Y + C_1 t + C_2, \tag{7.117}$$

where C_1 and C_2 are constants of integration. Due to the initial conditions (7.104) and (7.105) the integration constants are

$$C_1 = 0, \qquad C_2 = \frac{A_1 m_1 + A_2 m_2}{m_1 + m_2}. \tag{7.118}$$

Then, the expression (7.117) is in general

$$x = -\frac{m_2}{m_1 + m_2} Y + \frac{A_1 m_1 + A_2 m_2}{m_1 + m_2}. \tag{7.119}$$

The Eq. $(7.116)_1$ is an one variable uncoupled second order nonlinear differential equation with strong nonlinear deflection and damping terms of rational (integer or non-integer) order. Using the solution of the differential equation $(7.116)_1$, the solution of $(7.116)_2$ follows.

The averaged equations which correspond to $(7.116)_1$ are according to (4.43)

$$\begin{aligned} \dot{a} &= \frac{\varepsilon a}{\Pi_\alpha \mu(\alpha+1)} \int_0^{2\Pi_\alpha} sa(1 - |a\, ca|^{\beta-1}) sa d\psi, \\ \dot{\theta} &= \frac{\varepsilon}{2\Pi_\alpha \mu} \int_0^{2\Pi_\alpha} sa(1 - |a\, ca|^{\beta-1}) ca d\psi. \end{aligned} \tag{7.120}$$

After averaging and using the relations

$$\int_0^{2\Pi_\alpha} saca d\psi = 0, \qquad \int_0^{2\Pi_\alpha} saca\, |ca|^{\beta-1}\, d\psi = 0,$$

the Eq. (7.120) are transformed into

$$\dot{a} = \frac{2\varepsilon\, a}{\mu(\alpha+1)} \left(a_3 - a^{\beta-1} a_4\right), \qquad \dot{\theta} = 0. \tag{7.121}$$

where, using the procedure of averaging of the periodic Ateb functions (see Appendix C), we have

$$a_3 = \frac{1}{2\Pi_\alpha} \int_0^{2\Pi_\alpha} sa^2 d\psi = \frac{\alpha+1}{\alpha+3}, \qquad a_4 = \frac{1}{2\Pi_\alpha} \int_0^{2\Pi_\alpha} sa^2\, |ca|^{\beta-1}\, d\psi. \tag{7.122}$$

The parameter a_3 depends on the value of α, while a_4 is a function of α and β. Substituting $(7.122)_1$ into $(7.121)_1$ and integrating $(7.121)_2$ we obtain

$$\dot{a} = \frac{2\varepsilon a}{\mu}(\frac{1}{\alpha+3} - \frac{a_4 a^{\beta-1}}{\alpha+1}), \qquad \theta = C_3, \tag{7.123}$$

where C_3 is the constant of integration.

7.3.1 Transient Motion

Introducing the new variable $u = a^{1-\beta}$ the Eq. $(7.123)_1$ transforms into the well known Bernoulli differential equation

$$\dot{u} - \frac{2\varepsilon a_3(1-\beta)}{\mu(\alpha+1)}u + \frac{2\varepsilon a_4(1-\beta)}{\mu(\alpha+1)} = 0. \tag{7.124}$$

The solution of (7.124) is

$$u = \frac{a_4}{a_3} + \left(C_4 - \frac{a_4}{a_3}\right)\exp\left(\frac{2\varepsilon a_3(1-\beta)}{\mu(\alpha+1)}t\right), \tag{7.125}$$

where C_4 is the constant of integration. Then, the transient amplitude $a(t)$ varies from the initial value A to the steady state one a_S:

$$a(t) = \left(\frac{a_3}{a_4}\right)^{1/(\beta-1)} \frac{1}{\left[1 + \left(\frac{a_3}{a_4}C_4 - 1\right)\exp\left(-\frac{2\varepsilon a_3(\beta-1)}{\mu(\alpha+1)}t\right)\right]^{1/(\beta-1)}}. \tag{7.126}$$

Namely, for the initial condition $a(0) = A$ the integration constant is $C_4 = A^{1-\beta}$. For $t \to \infty$ the relation (7.126) tends to a constant value a_S.

Substituting (7.123) and (7.126) into $(4.36)_2$, i.e.,

$$\psi(t) := \theta + \sqrt{\frac{c_\alpha^2(\alpha+1)}{2}} \int [a(t)]^{(\alpha-1)/2} dt, \tag{7.127}$$

and after integration the phase angle is

$$\psi(t) = C_3 - \frac{({}_2F_1)}{s_3}\sqrt{\frac{c_\alpha^2(\alpha+1)}{2}} \frac{\left[s_1 + s_2\exp(s_3 t)\right]^{1+[(\alpha-1)/2(1-\beta)]}}{s_1\left(1 + [(\alpha-1)/2(1-\beta)]\right)}, \tag{7.128}$$

where ${}_2F_1$ is the hypergeometric function

$${}_2F_1 \equiv {}_2F_1\left(1 + \frac{\alpha-1}{2(1-\beta)}, 1, 2 + \frac{\alpha-1}{2(1-\beta)}, \frac{s_1 + s_2\exp(s_3 t)}{s_1}\right), \tag{7.129}$$

with $s_1 = \frac{a_4}{a_3}$, $s_2 = C_4 - s_1$, $s_3 = -\frac{2\varepsilon a_3(1-\beta)}{\alpha+1}$. Substituting (7.126) and (7.128) into the assumed solution (4.36) and using the initial conditions (3.3) the integration constants C_3 and C_4 are calculated

$$C_3 = \frac{{}_2F_1\left(\left(1+\frac{\alpha-1}{2(1-\beta)}, 1, 2+\frac{\alpha-1}{2(1-\beta)}, \frac{s_1+s_2}{s_1}\right)\right.}{s_3}\sqrt{\frac{c_\alpha^2(\alpha+1)}{2}}\frac{[s_1+s_2]^{1+[(\alpha-1)/2(1-\beta)]}}{s_1\left(1+[(\alpha-1)/2(1-\beta)]\right)}. \tag{7.130}$$

Based on (7.130) the amplitude variation during the time is determined as

$$a(t) = \left(\frac{a_3}{a_4}\right)^{1/(\beta-1)} \frac{1}{\left[1+\left(\frac{a_3}{a_4}A^{1-\beta}-1\right)\exp\left(-\frac{2\varepsilon a_3(\beta-1)}{\mu(\alpha+1)}t\right)\right]^{1/(\beta-1)}}. \tag{7.131}$$

Finally, the approximate solution of $(7.116)_1$ is

$$Y = \frac{\left(\frac{a_3}{a_4}\right)^{1/(\beta-1)}}{\left[1+\left(\frac{a_3}{a_4}A^{1-\beta}-1\right)\exp\left(-\frac{2\varepsilon a_3(\beta-1)}{\mu(\alpha+1)}t\right)\right]^{1/(\beta-1)}} ca(\alpha, 1, \psi(t)). \tag{7.132}$$

Substituting (7.132) into (7.102) and (7.119) the motions of the masses 1 and 2 are

$$x = -\frac{\left(\frac{m_2}{m_1+m_2}\right)\left(\frac{a_3}{a_4}\right)^{1/(\beta-1)} ca(\alpha, 1, \psi)}{\left[1+\left(\frac{a_3}{a_4}A^{1-\beta}-1\right)\exp\left(-\frac{2\varepsilon a_3(\beta-1)}{\mu(\alpha+1)}t\right)\right]^{1/(\beta-1)}} + \frac{A_1 m_1 + A_2 m_2}{m_1+m_2}, \tag{7.133}$$

$$y = \frac{\left(\frac{m_1}{m_1+m_2}\right)\left(\frac{a_3}{a_4}\right)^{1/(\beta-1)} ca(\alpha, 1, \psi)}{\left[1+\left(\frac{a_3}{a_4}A^{1-\beta}-1\right)\exp\left(-\frac{2\varepsilon a_3(\beta-1)}{\mu(\alpha+1)}t\right)\right]^{1/(\beta-1)}} + \frac{A_1 m_1 + A_2 m_2}{m_1+m_2}. \tag{7.134}$$

Analyzing (7.133) and (7.134) it is evident that the motions of the both masses, 1 and 2, are around a fixed position $\frac{A_1 m_1 + A_2 m_2}{m_1+m_2}$. It is of interest to analyze the steady state motion of the system.

7.3.2 *Steady State Motion*

For the steady state motion when the amplitude and phase variation is zero ($\dot{a} = 0$ and $\dot{\theta} = 0$) the relations (7.121) transform into

$$a_S = \left(\frac{a_3}{a_4}\right)^{1/(\beta-1)}, \qquad \omega = a_S^{(\alpha-1)/2}\sqrt{\frac{c_\alpha^2(\alpha+1)}{2}}. \tag{7.135}$$

Based on these solutions the steady state amplitude of the both masses has to be determined. Using the relations (7.133) and (7.134), the steady state motion of both masses is obtained as

$$x_S = \frac{A_1 m_1 + A_2 m_2}{m_1 + m_2} \tag{7.136}$$
$$- \left(\frac{m_2}{m_1+m_2}\right)\left(\frac{a_3}{a_4}\right)^{\frac{1}{\beta-1}} ca\left(\alpha, 1, \left(\frac{a_3}{a_4}\right)^{\frac{\alpha-1}{2(\beta-1)}} t\sqrt{\frac{c_\alpha^2(\alpha+1)}{2}}\right),$$

$$y_S = \frac{A_1 m_1 + A_2 m_2}{m_1 + m_2} \tag{7.137}$$
$$+ \left(\frac{m_1}{m_1+m_2}\right)\left(\frac{a_3}{a_4}\right)^{\frac{1}{\beta-1}} ca\left(\alpha, 1, \left(\frac{a_3}{a_4}\right)^{\frac{\alpha-1}{2(\beta-1)}} t\sqrt{\frac{c_\alpha^2(\alpha+1)}{2}}\right).$$

It is obvious that the amplitude and frequency of vibration for the both masses are constant. The amplitudes and the intervals of the steady state motions of the masses 1 and 2 are, respectively, as follows:

$$A_{1S} = \left(\frac{m_2}{m_1+m_2}\right)\left(\frac{a_3}{a_4}\right)^{\frac{1}{\beta-1}}, \text{ i.e.,}$$
$$x_S \in \left[\frac{A_1 m_1 + A_2 m_2}{m_1 + m_2} \pm \left(\frac{m_2}{m_1+m_2}\right)\left(\frac{a_3}{a_4}\right)^{\frac{1}{\beta-1}}\right], \tag{7.138}$$

$$A_{2S} = \left(\frac{m_1}{m_1+m_2}\right)\left(\frac{a_3}{a_4}\right)^{\frac{1}{\beta-1}}, \text{ i.e.,}$$
$$y_S \in \left[\frac{A_1 m_1 + A_2 m_2}{m_1 + m_2} \pm \left(\frac{m_1}{m_1+m_2}\right)\left(\frac{a_3}{a_4}\right)^{\frac{1}{\beta-1}}\right]. \tag{7.139}$$

7.3.3 Examples

The influence of the order β of the Van der Pol nonlinearity on the motion of the two-mass system is analyzed. The masses of the system are $m_1 = 2$ and $m_2 = 4$, the rigidity coefficient is $k_\alpha^2 = 1$ and the coefficient of the Van der Pol damping $d = 0.01$. The mathematical model of the system is

$$2\ddot{x} + (x-y)\,|x-y|^{2/3} = 0.01(\dot{x}-\dot{y})(1-|x-y|^{\beta-1}),$$
$$4\ddot{y} + (y-x)\,|y-x|^{2/3} = 0.01(\dot{y}-\dot{x})(1-|y-x|^{\beta-1}), \tag{7.140}$$

with initial conditions

$$x(0) = 0.5, \qquad \dot{x}(0) = 0,$$
$$y(0) = 0.25, \qquad \dot{y}(0) = 0. \tag{7.141}$$

Three values of parameter β are considered: 3, 4 and 4/3.

(a) For $\beta = 3$, i.e., the usual applied model of Van der Pol force, the model of the system is

$$2\ddot{x} + (x - y)\,|x - y|^{2/3} = 0.01(\dot{x} - \dot{y})(1 - |x - y|^2),$$
$$4\ddot{y} + (y - x)\,|y - x|^{2/3} = 0.01(\dot{y} - \dot{x})(1 - |y - x|^2). \tag{7.142}$$

According to the previous consideration the steady state motion is

$$x_S = 0.33333 - 1.3075 ca\left(\frac{5}{3}, 1, 1.2517t\right),$$
$$y_S = 0.33333 + 0.65375 ca\left(\frac{5}{3}, 1, 1.2517t\right),$$

with parameter values (see Appendix C)

$$a_3 = \frac{4}{7}, \qquad a_4 = \frac{1}{2\Pi_{5/3}} \int_0^{2\Pi_{5/3}} sa^2 ca^2 d\psi = 0.148567,$$
$$\Pi_{5/3} := B\left(\frac{3}{8}, \frac{1}{2}\right) = 3.85581$$

As the amplitudes of vibration of the mass 1 and 2 are 1.3075 and 0.65375, respectively, the limits for steady state motion are [−0.97417, 1.6408] and [−0.32042, 0.98708]. In Fig. 7.6 the numerically obtained solutions $x(t)$ and $y(t)$ of (7.142) and the analytically calculated limit values are plotted.

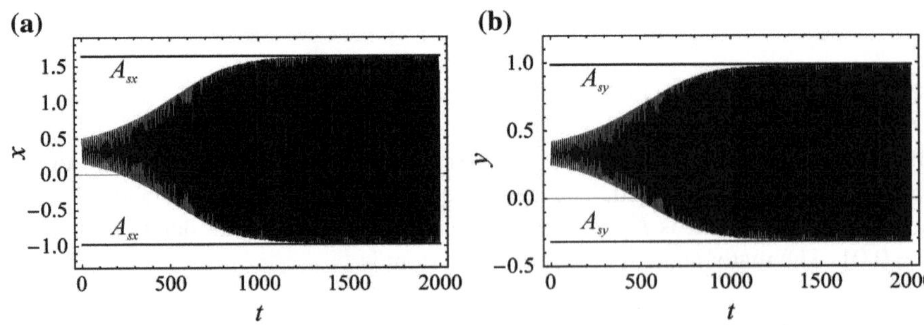

Fig. 7.6 **a** $x - t$ diagram and the limit steady state values; **b** $y - t$ diagram and the limit steady state values for $\beta = 3$

The analytically obtained values are on the top of the numerically obtained ones.

(b) For $\beta = 4$, the differential equations of motion are

$$2\ddot{x} + (x - y)\,|x - y|^{2/3} = 0.01(\dot{x} - \dot{y})(1 - |x - y|^3),$$
$$4\ddot{y} + (y - x)\,|y - x|^{2/3} = 0.01(\dot{y} - \dot{x})(1 - |y - x|^3). \tag{7.143}$$

The relations of the steady-state motions are

$$x_S = 0.33333 - 1.1846ca\left(\frac{5}{3}, 1, 1.2112t\right),$$
$$y_S = 0.33333 + 0.5923ca\left(\frac{5}{3}, 1, 1.2112t\right),$$

for the parameter value

$$a_4 = \frac{1}{2\Pi_{5/3}} \int_0^{2\Pi_{5/3}} sa^2\,|ca|^3\,d\psi = 0.101846.$$

In Fig. 7.7 the numerically obtained solutions x and y of the differential equations (7.125) and the limits of steady state motions $[-0.85127, 1.5179]$ and $[-0.25897, 0.92563]$ are plotted.

It can be seen that the analytical solutions are in good agreement with numerical ones.

(c) For $\beta = 4/3$, the differential equations of motion are

$$2\ddot{x} + (x - y)\,|x - y|^{2/3} = 0.01(\dot{x} - \dot{y})(1 - |x - y|^{1/3}),$$
$$4\ddot{y} + (y - x)\,|y - x|^{2/3} = 0.01(\dot{y} - \dot{x})(1 - |y - x|^{1/3}). \tag{7.144}$$

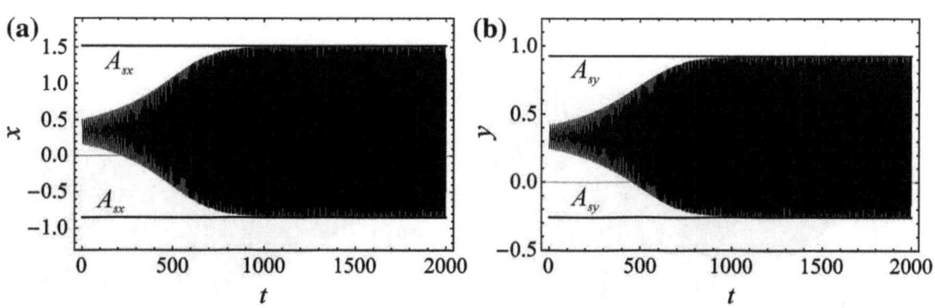

Fig. 7.7 **a** $x - t$ diagram and the limit steady state values; **b** $y - t$ diagram and the limit steady state values for $\beta = 4$

The analytically calculated steady state motions of the masses are given as

$$x_S = 0.33333 - 1.8399ca\left(\frac{5}{3}, 1, 1.4027t\right),$$
$$y_S = 0.33333 + 0.91993ca\left(\frac{5}{3}, 1, 1.4027t\right), \tag{7.145}$$

where

$$a_4 = \frac{1}{2\Pi_{5/3}}\int_0^{2\Pi_{5/3}} sa^2\,|ca|^{1/3}\,d\psi = 0.407385$$

In Fig. 7.8 the $x-t$ and $y-t$ diagrams obtained numerically, by solving of (7.140), and the limit steady state values obtained analytically, are plotted. The steady state oscillations are inside the limits $[-1.5066, 2.1732]$ and $[-0.5866, 1.2533]$, respectively.

The first time derivatives of (7.145) are according to (A9) (see Appendix A) are

$$\dot{x}_S = \frac{1}{2}\left(\frac{a_3}{a_4}\right)^4 sa\left(1, \frac{5}{3}, \left(\frac{a_3}{a_4}\right)t\right),$$
$$\dot{y}_S = -\frac{1}{4}\left(\frac{a_3}{a_4}\right)^4 sa\left(1, \frac{5}{3}, \left(\frac{a_3}{a_4}\right)t\right). \tag{7.146}$$

Substituting (7.145) and (7.146) into (A7) (Appendix A), the time t is eliminated and the $\dot{x} - x$ and $\dot{y} - y$ relations are obtained

$$4\left(\frac{a_4}{a_3}\right)^8 \dot{x}_S^2 + 2.9483(\frac{1}{3} - x_S)^{8/3}\left(\frac{a_4}{a_3}\right)^8 = 1,$$
$$16\dot{y}_S^2\left(\frac{a_4}{a_3}\right)^8 + 18.721(y_S - \frac{1}{3})^{8/3}\left(\frac{a_4}{a_3}\right)^8 = 1. \tag{7.147}$$

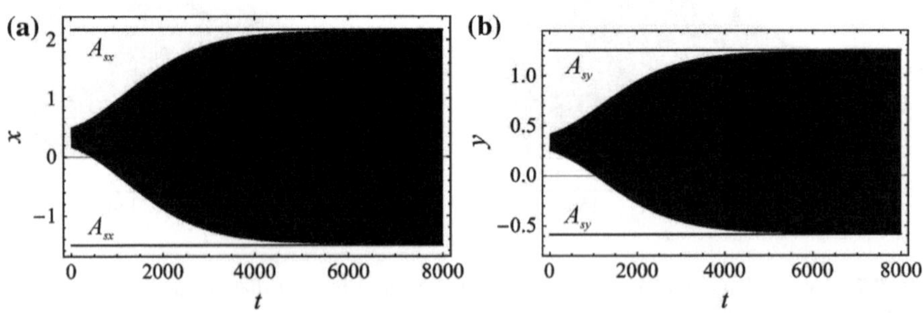

Fig. 7.8 **a** $x - t$ diagram and the limit steady state values; **b** $y - t$ diagram and the limit steady state values for $\beta = 4/3$

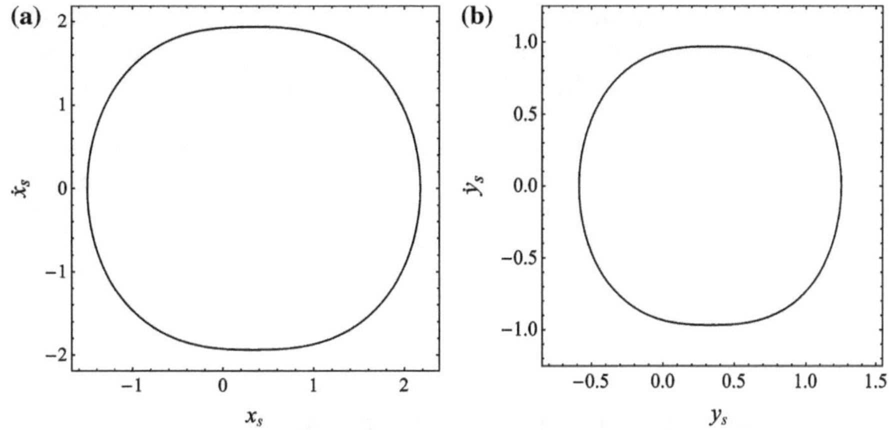

Fig. 7.9 Limit cycle diagrams in phase planes: **a** $\dot{x} - x$ and **b** $\dot{y} - y$ for $\beta = 4/3$

Based on (7.147) motion of masses in the phase plane are shown in Fig. 7.9. It is seen that the phase plane diagrams for steady state motion are closed curves.

7.3.4 Conclusion

From the previous consideration it can be concluded:

1. The free motion of masses connected to each other with an elastic spring is in general not a pure oscillatory one. Namely, the motion of the both masses is a sum of two terms: one, which is an oscillatory one and the second, which may be a straightforward one with a constant velocity or is a constant (including zero) which depend on the initial conditions. The latter motion is equal for the both masses, while the oscillatory motion is with different amplitudes and in opposite direction. The relation between the amplitudes of vibration depends on the mass ratio of the connected masses. Besides, the amplitude and frequency of vibration depend on the properties of the nonlinearity (order and coefficient).
2. The solution procedure based on the Ateb function is appropriate for solving of the coupled second order nonlinear differential equations with the damping of Van der Pol type. The analytically obtained solutions agree with those obtained numerically.
3. For the case of the pure elastic connection between masses, both masses oscillate around the same position which is the averaged value of the initial deflections. For both masses the amplitudes and periods of vibration are equal, but the motion is in opposite direction. Amplitude of vibration directly depends on the difference between initial deflections of masses. The frequency of vibration is a nonlinear function (of integer or non-integer order) of the difference between initial deflections of masses.

4. The steady state vibrations of the masses in the two-mass system with damping of Van der Pol type are with constant amplitude and frequency. The properties of the steady state vibrations are independent on the initial amplitudes, but depend on the mass values and the order of nonlinearities.

5. The order of Van der Pol nonlinearity has an influence on the limit value of the vibration motion: the higher the order the nonlinearity, the limits of steady state vibration are smaller and the interval of steady state motion is narrower. The frequency of vibration decreases by increasing of the order of nonlinearity. The time necessary to reach the limit cycle motion is shorter for higher values of nonlinearity of the Van der Pol type.

7.4 Complex-Valued Differential Equation

The special type of the two coupled differential equations is described with a complex function. Namely, the deflection in two directions are connected with an imaginary unit and described with a complex variable. This mathematical model corresponds to an one mass two-degree-of-freedom system, for example rotor. Dimentberg (1959) and Vance (1988) analyzed the vibrations of the linear rotor. Various methods are developed for solving linear differential equation with complex function and additional small nonlinearities. Let mention some of them: Krylov–Bogolubov method (1943) and Bogolubov and Mitropolski (1963) procedure adopted for the differential equation with complex function (Cveticanin 1992b, 1993), the generalized averaging solving method introduced by Mahmoud (1998), Mahmoud and Shaban (2000), etc. Unfortunately, the suggested procedures in this form are not applicable for solving the strong nonlinear differential equation with complex function.

In this Chapter the oscillator modeled with a second order differential equation with complex function and pure nonlinearity is shown. Some constraints to the form of the equation, initial conditions and type of nonlinearity are introduced. These limitations give additional requirements to the solution procedures of differential equations. Presented methods are based on the results published in papers of Cveticanin (1988, 1992b, 1998a, b, 2001a, c, 2004a, b, 2005) and Maccari (2003), which are developed for rotors with Van der Pol or Rayleigh nonlinearities (Manasevich et al. 1999) and also Duffing and Liénard nonlinearities (see Genin and Maybee 1970; Muszynska 1976; Tondl 1979).

The differential equation considered in this Chapter is

$$\ddot{z} + c_\alpha^2 z(\sqrt{z\bar{z}})^{\alpha-1} = \varepsilon Z(z, \dot{z}, cc), \tag{7.148}$$

where $z = x + iy$ is a complex function, $\bar{z} = x - iy$ is a complex conjugate function, α is a positive rational number (integer or non-integer), c_α^2 is a constant coefficient, ε is a small parameter, $Z = X + iY$ is a complex function $i = \sqrt{-1}$ is the imaginary unit and cc are the complex conjugate functions. The differential equation (7.148) is an ordinary second order differential equation with complex function where

$c_\alpha^2 z(\sqrt{z\bar{z}})^{\alpha-1}$ is the pure nonlinear term whose order of nonlinearity is determined by α and εZ is a small function which depends on complex function z, its time derivative $\dot{z}$ and complex conjugate functions cc. The general initial conditions are

$$z(0) = z_0, \qquad \dot{z}(0) = \dot{z}_0. \tag{7.149}$$

7.4.1 Adopted Krylov–Bogolubov Method

Cveticanin (2004a) suggested the adopted Krylov–Bogolubov method for obtaining of the approximate solution of (7.148). The method is based on the generating solution of the differential equation (7.148) for $\varepsilon = 1$ with arbitrary initial conditions (7.149). Perturbing that solution, the trial solution of (7.148) is formed. Using this assumed solution and some constraints the differential equation (7.148) is transformed into a new system of two coupled first order complex-valued differential equations. The approximate solution is obtained using the averaging procedure.

For $\varepsilon = 0$ the generating equation for (7.148) has the form

$$\ddot{z} + c_\alpha^2 z(\sqrt{z\bar{z}})^{\alpha-1} = 0. \tag{7.150}$$

Its solution is

$$z = f(K_1, K_2, K_3, K_4, t), \tag{7.151}$$

where K_1, K_2, K_3, K_4 are constants dependent on the initial conditions (7.149). Perturbing this generating solution (7.151) the trial solution of (7.148) is

$$z = f(K_1(t), K_2(t), K_3(t), K_4(t), t), \tag{7.152}$$

where $K_1(t), K_2(t), K_3(t), K_4(t)$ are real time dependent functions. The solution (7.152) has to satisfy the Eq. (7.148). It requires the following constraint to be introduced: the first time derivative of (7.152) must have the same form as the first time derivative of the generating solution (7.151) and it is

$$\dot{z} = \frac{\partial f}{\partial t}. \tag{7.153}$$

Comparing the time derivative of (7.152) and the relation (7.153) the following complex-valued equation is obtained

$$\sum_{j=1}^{j=4} \frac{\partial f}{\partial K_j} \dot{K}_j = 0. \tag{7.154}$$

It is a first order ordinary differential equation with real variables but with real and imaginary terms.

Introducing the solution (7.152), the relation (7.153) and its time derivative into (7.148) it is

$$\sum_{j=1}^{j=4} \frac{\partial}{\partial K_j}\left(\frac{\partial f}{\partial t}\right)\dot{K}_j = \varepsilon Z\left(f, \frac{\partial f}{\partial t}, cc\right). \tag{7.155}$$

The Eq. (7.155) is also a first order differential equation with real and imaginary terms which with (7.154) represent the transformed differential equation (7.148) with the new variables K_j. Separating the real and imaginary terms in the differential Eqs. (7.154) and (7.155) a system of four coupled first order differential equations is obtained

$$\dot{K}_j = \frac{\Delta_{Kj}}{\Delta}, \qquad j = 1, \ldots, 4, \tag{7.156}$$

where

$$\begin{aligned}
&\Delta = \left|a_{ij}\right|, \qquad i = 1, \ldots, 4, \qquad j = 1, \ldots, 4,\\
&a_{1j} = \mathrm{Re}\left(\frac{\partial f}{\partial K_j}\right), \quad a_{2j} = \mathrm{Im}\left(\frac{\partial f}{\partial K_j}\right),\\
&a_{3j} = \mathrm{Re}\left(\frac{\partial}{\partial K_j}\left(\frac{\partial f}{\partial t}\right)\right), \quad a_{4j} = \mathrm{Im}\left(\frac{\partial}{\partial K_j}\left(\frac{\partial f}{\partial t}\right)\right),
\end{aligned} \tag{7.157}$$

$$\Delta_{Kj} = \begin{vmatrix} b_{1p} \\ b_{2p} \\ b_{3p} \\ b_{4p} \end{vmatrix}, \qquad p = 1, \ldots, 4, \tag{7.158}$$

$$\begin{aligned}
&b_{1p} = 0 \text{ for } p = j, \qquad b_{1p} = a_{1j} \text{ for } p \neq j,\\
&b_{2p} = 0 \text{ for } p = j, \qquad b_{2p} = a_{2j} \text{ for } p \neq j,\\
&b_{3p} = \varepsilon\,\mathrm{Re}(Z) \text{ for } p = j, \qquad b_{3p} = a_{3j} \text{ for } p \neq j,\\
&b_{4p} = \varepsilon\,\mathrm{Im}(Z) \text{ for } p = j, \qquad b_{4p} = a_{4j} \text{ for } p \neq j.
\end{aligned} \tag{7.159}$$

Solving the differential equations (7.156) for $K_j(t)$ the exact solution (7.152) of the differential equation (7.148) is obtained.

Unfortunately, usually it is impossible to find the closed form solution for (7.156) in analytical form. As it is suggested by Krylov and Bogolubov (Nayfeh and Mook 1979) the Eq. (7.156) have to be averaged. Namely, as the motion is periodical with period T the averaged equations are

$$\dot{K}_j = \left\langle \frac{\Delta_{Kj}}{\Delta} \right\rangle, \qquad j = 1, \ldots, 4, \tag{7.160}$$

where $\langle \cdot \rangle \equiv \int_0^T (\cdot)\, dT$. Solving the averaged system of differential equations (7.160) and substituting the so obtained values for $K_j(t)$ into (7.152) the approximate analytical solution of (7.148) is obtained.

The advantage of the method is its simplicity. The solving method is based on the generating solution which is perturbed. Unfortunately, there is a strong difficulty to obtain the closed form generating solution of (7.150) for arbitrary initial conditions (7.149). The averaging which is introduced in solving process has a significant influence on the accuracy of the solution. Usually, the approximate averaged solution is on the top of the exact solution only for a short time interval and small values of nonlinearity.

7.4.2 Method Based on the First Integrals

The complex function z is defined in the polar form by introducing the polar coordinates ρ and θ

$$z = \rho \exp(i\theta) \tag{7.161}$$

where ρ is the position coordinate and θ the angle coordinate. Substituting the polar form of the complex function (7.161) into (7.148) and (7.149) and separating the real and imaginary terms a system of two second order differential equations is obtained

$$\begin{aligned} \ddot{\rho} - \rho\dot{\theta}^2 + c_1\rho + \rho F(\rho) &= \varepsilon\Phi_\rho(\rho, \dot{\rho}, \theta, \dot{\theta}), \\ \rho\ddot{\theta} + 2\dot{\rho}\dot{\theta} &= \varepsilon\Phi_\theta(\rho, \dot{\rho}, \theta, \dot{\theta}), \end{aligned} \tag{7.162}$$

with initial conditions

$$\rho(0) = A, \qquad \dot{\rho}(0) = E, \qquad \theta(0) = C, \qquad \dot{\theta}(0) = D, \tag{7.163}$$

where $\varepsilon\Phi_\rho$ and $\varepsilon\Phi_\theta$ are small real functions. The solving method in this Chapter considers the differential Eq. (7.162) and initial conditions (7.163).

The generating equation, when $\varepsilon = 0$, is

$$\ddot{\rho}_0 - \rho_0\dot{\theta}_0^2 + c_\alpha^2\rho_0 \left|\rho_0\right|^{\alpha-1} = 0, \qquad \rho_0\ddot{\theta}_0 + 2\dot{\rho}_0\dot{\theta}_0 = 0. \tag{7.164}$$

The Eq. $(7.164)_2$ represents a cyclic first integral

$$\rho_0^2\dot{\theta}_0 = A^2 D \equiv K_{10} = const., \tag{7.165}$$

for the cyclic coordinate θ_0. This integral represents the twice value of the so called sectorial velocity. Its value is constant for the system (7.164).

Introducing the relation (7.165) into $(7.164)_1$ it transforms to an another first integral of energy type

$$\frac{1}{2}\dot{\rho}_0^2 + \frac{1}{2}\frac{K_{10}^2}{\rho_0^2} + \frac{c_\alpha^2}{\alpha+1}|\rho_0|^{\alpha+1} = \frac{1}{2}E^2 + \frac{1}{2}\frac{K_{10}^2}{A^2} + \frac{c_\alpha^2}{\alpha+1}|A|^{\alpha+1} \equiv K_{20} = const. \tag{7.166}$$

It can be concluded that for the conservative system (7.164) the energy integral (7.166) is constant.

Solving the relation

$$\int_A^{\rho_0} \frac{d\rho_0}{\sqrt{2K_{20} - \frac{K_{10}^2}{\rho_0^2} - 2\frac{c_\alpha^2}{\alpha+1}|\rho_0|^{\alpha+1}}} = t, \tag{7.167}$$

the $\rho_0 = \rho_0(t)$ coordinate variation is obtained. Substituting the solution of (7.167) into (7.165) and integrating it is

$$\theta_0 = C + K_{10}^2 \int_0^t \frac{dt}{\rho_0^2(t)}. \tag{7.168}$$

Using the results of the generating equation (7.164) the differential equation (7.162) is transformed into a system of four coupled first order ordinary differential equations

$$\begin{aligned}
\frac{d\theta}{dt} &= \frac{K_1}{\rho^2},\\
\frac{d\rho}{dt} &= \sqrt{2K_2 - \frac{K_1^2}{\rho^2} - 2\frac{c_\alpha^2}{\alpha+1}|\rho|^{\alpha+1}},\\
\frac{dK_1}{dt} &= \varepsilon\rho\Phi_\theta(\rho,\theta,K_1,K_2),\\
\frac{dK_2}{dt} &= \varepsilon\left(2K_2 - \frac{K_1^2}{\rho^2} - 2\frac{c_\alpha^2}{\alpha+1}|\rho|^{\alpha+1}\right)^{1/2}\Phi_\rho(\rho,\theta,K_1,K_2),
\end{aligned} \tag{7.169}$$

with initial conditions

$$\rho(0) = A, \quad \theta(0) = C, \quad K_1(0) = K_{10}, \quad K_2(0) = K_{20}, \tag{7.170}$$

where

$$K_1 = \rho^2\dot{\theta}, \qquad K_2 = \frac{1}{2}\dot{\rho}^2 + \frac{1}{2}\frac{K_1^2}{\rho^2} + \frac{c_\alpha^2}{\alpha+1}\,|\rho|^{\alpha+1}, \tag{7.171}$$

are the perturbed time dependent first integrals. So the task of obtaining the solution $z(t)$ of Eq. (7.148) i.e., (7.162) has been transformed into the equivalent one of obtaining the four solutions $K_1(t)$, $K_2(t)$, $\rho(t)$, $\theta(t)$ of the system of Eq. (7.169). The procedure for approximate solving of the differential equations (7.169) is introduced. The straightforward series expansion for the small parameter ε is

$$\rho = \rho_0 + \varepsilon\rho_1 + \varepsilon^2\rho_2 + \cdots, \tag{7.172}$$

$$\theta = \theta_0 + \varepsilon\theta_1 + \varepsilon^2\theta_2 + \cdots, \tag{7.173}$$

$$K_1 = K_{10} + \varepsilon K_{11} + \varepsilon^2 K_{12} + \cdots, \tag{7.174}$$

$$K_2 = K_{20} + \varepsilon K_{21} + \varepsilon^2 K_{22} + \cdots \tag{7.175}$$

The relation (7.174) represents an adiabatic invariant and describes the slow variation of the sector velocity in time due to existence of small functions. The relation (7.175) is also an adiabatic invariant which considers the energy increase or decrease dependently on the properties of the small forces.

The Taylor series expansion for functions $\Phi_\theta(\rho, \theta, K_1, K_2)$, $\Phi_\rho(\rho, \theta, K_1, K_2)$ and $\mathcal{F}(\rho)$ about ρ_0, θ_0, K_{10} and K_{20}, which is signed as $()_0$, is

$$\begin{aligned}\Phi_\theta(\rho, \theta, K_1, K_2) &= (\Phi_\theta)_0 + \varepsilon\left[\rho_1\left(\frac{\partial\Phi_\theta}{\partial\rho}\right)_0 + \theta_1\left(\frac{\partial\Phi_\theta}{\partial\theta}\right)_0 + K_{11}\left(\frac{\partial\Phi_\theta}{\partial K_1}\right)_0 + K_{21}\left(\frac{\partial\Phi_\theta}{\partial K_2}\right)_0\right] \\ &\quad + \varepsilon^2\ldots,\end{aligned} \tag{7.176}$$

$$\begin{aligned}\Phi_\rho(\rho, \theta, K_1, K_2) &= (\Phi_\rho)_0 + \varepsilon\left[\rho_1\left(\frac{\partial\Phi_\rho}{\partial\rho}\right)_0 + \theta_1\left(\frac{\partial\Phi_\rho}{\partial\theta}\right)_0 + K_{11}\left(\frac{\partial\Phi_\rho}{\partial K_1}\right)_0 + K_{21}\left(\frac{\partial\Phi_\rho}{\partial K_2}\right)_0\right] \\ &\quad + \varepsilon^2\ldots,\end{aligned} \tag{7.177}$$

$$\begin{aligned}\mathcal{F}(\rho) = \frac{c_\alpha^2}{\alpha+1}\,|\rho|^{\alpha+1} &= (\mathcal{F})_0 + \varepsilon(\rho_1 + \varepsilon\rho_2 + \cdots)\left(\frac{d\mathcal{F}}{d\rho}\right)_0 \\ &\quad + \frac{\varepsilon^2}{2!}(\rho_1 + \varepsilon\rho_2 + \cdots)^2\left(\frac{d^2\mathcal{F}}{d\rho^2}\right)_0 + \cdots\end{aligned} \tag{7.178}$$

Substituting the functions (7.172)–(7.178) into (7.169) and separating the terms with the same order of the small parameter ε the following set of equations is obtained:

for ε^0, the Eqs. (7.165)–(7.168), and the initial conditions

$$K_{10}(0) = A^2 D, \quad \rho_0(0) = A, \quad \theta_0(0) = C,$$
$$K_{20}(0) = \frac{1}{2}(B^2 + A^2 D^2) + \left(\frac{c_\alpha^2}{\alpha+1}|A|^{\alpha+1}\right), \tag{7.179}$$

for ε^1, the first order equations are

$$\frac{dK_{11}}{dt} = (\Phi_\theta)_0\,\rho_0, \qquad \frac{dK_{21}}{dt} = (\Phi_\rho)_0\,\dot{\rho}_0,$$
$$\frac{d\rho_1}{dt} = \frac{1}{\dot{\rho}_0}\left[K_{21} + K_{10}^2\frac{\rho_1}{\rho_0^3} - \frac{K_{10}K_{11}}{\rho_0^2} - \rho_1 c_\alpha^2 \rho_0^\alpha\right],$$
$$\frac{d\theta_1}{dt} = \frac{K_{11}}{\rho_0^2} - 2K_{10}\frac{\rho_1}{\rho_0^3}, \tag{7.180}$$

with the initial conditions:

$$K_{11}(0) = 0, \quad K_{21}(0) = 0, \quad \rho_1(0) = 0, \quad \theta_1(0) = 0, \tag{7.181}$$

for ε^2, the equations are

$$\frac{dK_{12}}{dt} = (\Phi_\theta)_0\,\rho_1 + \rho_0\left[\rho_1\left(\frac{\partial\Phi_\theta}{\partial\rho}\right)_0 + \theta_1\left(\frac{\partial\Phi_\theta}{\partial\theta}\right)_0 + K_{11}\left(\frac{\partial\Phi_\theta}{\partial K_1}\right)_0 + K_{21}\left(\frac{\partial\Phi_\theta}{\partial K_2}\right)_0\right],$$

$$\frac{dK_{22}}{dt} = (\Phi_\rho)_0\,\dot{\rho}_1 + (\dot{\rho})_0\left[\rho_1\left(\frac{\partial\Phi_\theta}{\partial\rho}\right)_0 + \theta_1\left(\frac{\partial\Phi_\theta}{\partial\theta}\right)_0 + K_{11}\left(\frac{\partial\Phi_\theta}{\partial K_1}\right)_0 + K_{21}\left(\frac{\partial\Phi_\theta}{\partial K_2}\right)_0\right],$$

$$\frac{d\rho_2}{dt} = \frac{1}{2\dot{\rho}_0}\left[2K_{22} + \frac{K_{10}^2}{\rho_0^2}\left(\frac{\rho_1^2}{\rho_0^2} + 2\frac{\rho_2}{\rho_0}\right) + 4\frac{K_{10}K_{11}\rho_1}{\rho_0^3} - \frac{K_{11}^2 + 2K_{10}K_{12}}{\rho_0^2} - 2\rho_2\left(\frac{d\mathcal{F}}{d\rho}\right)_0 - \rho_1^2\left(\frac{d^2\mathcal{F}}{d\rho^2}\right)_0\right] - \frac{\dot{\rho}_1}{2\dot{\rho}_0},$$
$$\frac{d\theta_2}{dt} = \frac{K_{12}}{\rho_0^2} - 2K_{11}\frac{\rho_1}{\rho_0^3} - \frac{K_{10}}{\rho_0^2}\left(\frac{\rho_1^2}{\rho_0^2} + 2\frac{\rho_2}{\rho_0}\right), \tag{7.182}$$

with initial conditions

$$K_{12}(0) = 0, \quad K_{22}(0) = 0, \quad \rho_2(0) = 0, \quad \theta_2(0) = 0. \tag{7.183}$$

For ε^0 the solutions are ρ_0, θ_0, K_{10} and K_{20}. Substituting these solutions into the set of differential equations (7.180) with the small parameter ε the functions ρ_1, θ_1, K_{11} and K_{21} are obtained. Namely, integrating the first two relations of (7.180) for the corresponding initial conditions the functions K_{11} and K_{21} in the first approximation are obtained

$$\begin{aligned} K_{11}(t) &= \varepsilon \int_0^t [\rho \Phi_\theta(\rho, \theta, K_1, K_2)]_0 dt, \\ K_{21}(t) &= \varepsilon \int_0^t \left[\left(2K_2 - \frac{K_1^2}{\rho^2} + \frac{2c_\alpha^2}{\alpha+1} |\rho|^{\alpha+1} \right)^{1/2} \right. \\ &\qquad \left. \times \Phi_\rho(\rho, \theta, K_1, K_2) \right]_0 dt. \end{aligned} \tag{7.184}$$

Using the previously obtained functions ρ_0, ρ_1, θ_0, θ_1, K_{10}, K_{11}, K_{20} and K_{21} and solving the system of differential equations (7.182) the functions ρ_2, θ_2, K_{12} and K_{22} in the second approximation are obtained.

Usually for technical reasons the solution in the first approximation is satisfactory. The approximate analytic solution in the first approximation and in the complex form is

$$z = (\rho_0 + \varepsilon\rho_1) \exp(\theta_0 + \varepsilon\theta_1). \tag{7.185}$$

In comparison of the suggested procedure with first integrals with the previously mentioned method the first has an advantage as it gives the solutions of higher order of approximation. The main problem which appears for this method is that it requires the exact solution for the system of differential equations (7.179) which is very often connected with difficulties.

The procedure with first integrals is very convenient for solving of the systems with special initial conditions especially for the case of without impact initial conditions, when the first time derivatives of the polar coordinates are zero.

7.4.2.1 Initial Conditions Without Impact

For this special case of initial conditions

$$\rho(0) = A, \qquad \dot\rho(0) = 0, \qquad \theta(0) = C, \qquad \dot\theta(0) = 0, \tag{7.186}$$

i.e.,

$$\rho(0) = A, \qquad \theta(0) = C, \qquad K_1(0) = 0, \qquad K_2(0) = K_{20}, \tag{7.187}$$

the system of differential equations (7.179)–(7.182) simplifies. The corresponding solution of the set of (7.179) for $\varepsilon = 0$ is

$$\begin{aligned} K_{10} &= K_1(0) = 0, \qquad \theta_0 = \theta(0) = C, \\ K_{20} &= K_2(0) = \frac{1}{2}c_1 A^2 + \left(\int \rho F(\rho) d\rho \right)_A, \\ t &= \int_A^{\rho_0} \frac{d\rho_0}{\sqrt{2K_{20} - c_1\rho_0^2 - 2\mathcal{F}(\rho)_0}}. \end{aligned} \tag{7.188}$$

The functions in the first approximation are

$$\begin{aligned} K_1 &= \varepsilon K_{11}, \qquad K_2 = K_{20} + \varepsilon K_{21}, \\ \theta &= C + \varepsilon\theta_1, \qquad \rho = \rho_0 + \varepsilon\rho_1, \end{aligned} \tag{7.189}$$

where $K_{11}(t)$, $K_{21}(t)$, $\rho_1(t)$ and $\theta_1(t)$ are the solutions of the system of differential equations

$$\begin{aligned} \frac{dK_{11}}{dt} &= (\Phi_\theta)_0\,\rho_0, \qquad \frac{dK_{21}}{dt} = \left(\Phi_\rho\right)_0 \sqrt{2K_{20} - c_1\rho_0^2 - 2\mathcal{F}(\rho)_0}, \\ \frac{d\theta_1}{dt} &= \frac{K_{11}}{\rho_0^2}, \qquad \frac{d\rho_1}{dt} = \frac{1}{\dot{\rho}_0}\left[K_{21} - c_1\rho_0\rho_1 - \rho_1 \left(\frac{d\mathcal{F}}{d\rho} \right)_0 \right], \end{aligned} \tag{7.190}$$

with $K_{11}(0) = 0$, $K_{21}(0) = 0$, $\rho_1(0) = 0$, $\theta_1(0) = 0$. The expression $(7.190)_4$ is the Bernoulli first order ordinary differential equation

$$\frac{d\rho_1}{dt} + \rho_1 \left[\frac{c_1\rho_0}{\dot{\rho}_0} + \frac{1}{\dot{\rho}_0} \left(\frac{d\mathcal{F}}{d\rho} \right)_0 \right] = \frac{K_{21}}{\dot{\rho}_0}. \tag{7.191}$$

Using the polar form of the complex-valued solution, $z = \rho \exp(i\theta)$, and the previously obtained results it is

$$z = (\rho_0 + \varepsilon\rho_1) \exp(i(C + \varepsilon\theta_1)). \tag{7.192}$$

The small function with parameter ε in the differential equation (7.162) affects the both polar coordinates in spite of the fact that the initial conditions are without impact. Dependently on the type of the small nonlinear function the following two special cases may appear:

(a) For the case when the right side of the first equation (7.180) is zero it is $K_{11} = 0$ and $\theta_1 = 0$. The solution in the first approximation is

$$z = (\rho_0 + \varepsilon\rho_1)\exp(i\theta_0). \tag{7.193}$$

The small function with $\varepsilon = 0$ has an influence only on the deflection coordinate ρ, but not on the angle position θ which remains constant. The solution is periodical along a straight line whose angle is constant and has the value $\theta = \theta(0) = C$. The small function has an influence only on the energy variation of the system and the amplitude of vibration ρ.

(b) For the case when right-hand side terms of $(7.180)_2$ is zero, it is $K_{21} = 0$ and for the first approximation $K_2 = K_{20} = const$. It means that the small function with the parameter ε has no meaningful influence on energy variation in the first approximation. The deflection coordinate ρ does not depend on the small function in the first approximation and it is $\rho = \rho_0$. At the other side, this type of small nonlinearity introduces a perturbation in the angle position θ. The solution is in the first approximation is

$$z = A\exp\left[i\left(C + \varepsilon\theta_1\right)\right]. \tag{7.194}$$

To prove the correctness of the suggested procedure some examples are considered. For all of them it is common that the strong nonlinearity is of cubic type.

7.4.2.2 The Strong Nonlinearity Is Cubic

If the strong nonlinearity is of cubic type the differential equation (7.164) transforms to

$$\begin{aligned} \ddot{\rho}_0 - \rho_0\dot{\theta}_0^2 + c_3^2\rho_0^3 &= 0, \\ \rho_0\ddot{\theta}_0 + 2\dot{\rho}_0\dot{\theta}_0 &= 0, \end{aligned} \tag{7.195}$$

with initial conditions

$$\rho(0) = A, \qquad \dot{\rho}(0) = 0, \qquad \theta(0) = C, \qquad \dot{\theta}(0) = 0. \tag{7.196}$$

For the system (7.195) with initial conditions (7.196) according to (7.188) the first integrals are

$$K_{10} = K_1(0) = 0, \quad K_{20} = K_2(0) = c_3^2 A^4/4. \tag{7.197}$$

The polar coordinates are

$$\rho_0 = A cn(\omega t, k^2), \qquad \theta_0 = \theta(0) = C = const., \tag{7.198}$$

where

$$k^2 = 1/2 = const., \qquad \omega = A\sqrt{c_3^2}. \tag{7.199}$$

$cn(\omega t, k^2)$ is a Jacobi elliptic function, where k^2 is the modulus of the function and ωt its argument. Using the relations (7.198) the solution of (7.195) in the complex form is

$$z = Acn(\omega t, k^2)\exp(iC). \tag{7.200}$$

This solution is the exact closed form analytical solution for the initial conditions (7.196). The solution is periodical and it is along a straight line with constant inclination angle $\theta_0 = \theta(0) = C$.

For the numerical value $c_3^2 = 1$ and initial conditions

$$\rho(0) = A = 0.1, \qquad \dot{\rho}(0) = 0, \qquad \theta(0) = C = \pi/3, \qquad \dot{\theta}(0) = 0, \tag{7.201}$$

it is

$$z = 0.1cn(0.1t, 0.5)\exp(i\pi/3), \tag{7.202}$$

The suggested procedure enables to obtain the exact analytical solution of the differential equation with additional linear term

$$\begin{aligned} \ddot{\rho}_0 - \rho_0\dot{\theta}_0^2 + c_1\rho + c_3^2\rho_0^3 &= 0, \\ \rho_0\ddot{\theta}_0 + 2\dot{\rho}_0\dot{\theta}_0 &= 0. \end{aligned} \tag{7.203}$$

After some calculation, we have

$$\begin{aligned} K_{20} = K_2(0) &= \frac{c_1A^2}{2} + \frac{c_3^2A^4}{4}, \\ k^2 &= \frac{c_3^2A^2}{2(c_1 + c_3^2A^2)}, \quad \omega = \sqrt{c_1 + c_3^2A^2}. \end{aligned} \tag{7.204}$$

For the previously mentioned parameter values and $c_1 = 2$, the exact solution of (7.203) is

$$z = 0.1cn(1.41774t, 0.00284)\exp(i\pi/3). \tag{7.205}$$

where

$$\omega = 1.41774, \quad k^2 = 0.00284, \tag{7.206}$$

are the parameters of the cn Jacobi elliptic function.

7.4.2.3 Small Gyroscopic Function

For the oscillator, where the linear and cubic nonlinearity are considered, but also a small gyroscopic function, the differential equation is

$$\ddot{z} + c_1 z + c_3 z(z\bar{z}) = 2\varepsilon g i \dot{z}(1 + p(z\bar{z})), \tag{7.207}$$

where εg is the coefficient of the gyroscopic force and p is a small coefficient of nonlinearity. Transforming the small function on the right side into polar coordinates it is

$$\Phi_\rho = -2\varepsilon g \rho \dot{\theta}(1 + p\rho^2), \qquad \Phi_\theta = 2\varepsilon g \dot{\rho}(1 + p\rho^2). \tag{7.208}$$

Due to (7.190) it is evident that as $(\Phi_\rho)_0 = 0$ it is $K_{21} = 0$ and $\rho_1 = 0$. The functions K_{11} and θ_1 are

$$K_{11} = -\varepsilon g A^2(1 + \frac{pA^2}{2}) + \varepsilon g A^2 cn^2(\omega t, k^2)\left(1 + \frac{pA^2}{2} cn^2(\omega t, k^2)\right), \tag{7.209}$$

$$\theta_1 = \varepsilon g \left[\frac{E(\omega t, m)}{\omega(1 - k^2)}(1 + \frac{pA^2}{2k^2}) - \frac{1 + pA^2/2}{\omega(1 - k^2)} dn(\omega t, k^2) tn(\omega t, k^2) - \frac{pA^2}{2k^2} t \right], \tag{7.210}$$

where $k' = \sqrt{1 - k^2}$, $E(\omega t, k^2)$ is the Legendre's incomplete elliptic integral of the second kind, dn and tn are Jacobi elliptic functions (Byrd and Friedman 1954) with modulus k^2 and frequency ω (7.197). The solution in the first approximation is

$$\begin{aligned} z = Acn(\omega t, k^2) \exp \Bigg\{ iC + i\varepsilon g \Bigg[& \frac{E(\omega t, k^2)}{\omega(1 - k^2)} \left(1 + \frac{pA^2}{2k^2}\right) \\ & - \frac{1}{\omega(1 - k^2)} \left(1 + \frac{pA^2}{2}\right) dn(\omega t, k^2) tn(\omega t, k^2) - \frac{pA^2}{2k^2} t \Bigg] \Bigg\}. \end{aligned} \tag{7.211}$$

The solution is periodical and two different periods appear: ω and Ω which correspond to Jacobi elliptic function and the trigonometric function, respectively. For $\omega \approx \Omega$ the effect of fluttering is evident.

For the numerical values $c_1 = 2$, $c_3 = 1$, $\varepsilon g = 0.1$ and $p = 1$ and initial conditions (7.201) it is

$$\begin{aligned} z = 0.1cn(1, 41774t, 0.00284) \exp \Bigg\{ i \Bigg[& \frac{\pi}{3} + 0.19526E(1.41774t, 0.00284) \\ & - 0.17857t - 0.071089dn(1, 41774t, 0.00284) tn(1, 41774t, 0.00284) \Bigg] \Bigg\}. \end{aligned} \tag{7.212}$$

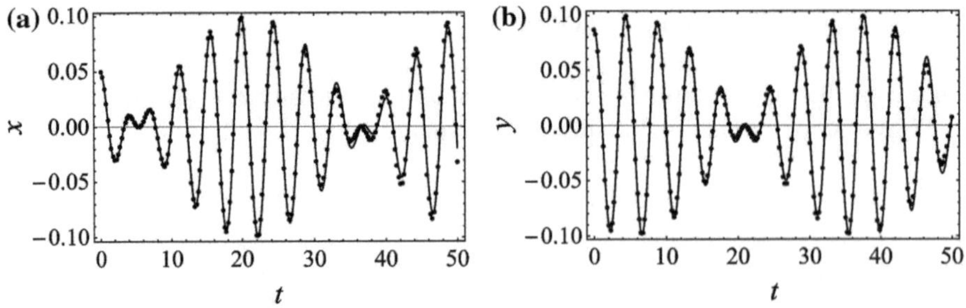

Fig. 7.10 **a** $x - t$ and **b** $y - t$ diagrams obtained analytically (*full line*) and numerically (*dotted line*)

Using the fact that the modulus k^2 (7.206) has the value approximately zero and after expansion in series the elliptic functions and elliptic integral E (see Abramowitz and Stegun 1979) it is

$$z \approx 0.1 \cos(1.41774t) \exp\left[i\left(\frac{\pi}{3} + 0.09826t\right)\right]. \tag{7.213}$$

To prove the correctness of the approximate analytical solution (7.213) it is compared with the exact numerical one. For the introduced numerical values the differential equation (7.207) is

$$\begin{aligned} \ddot{x} + 2x + x(x^2 + y^2) &= -2\varepsilon g\dot{y}(1 + p(x^2 + y^2)), \\ \ddot{y} + 2y + y(x^2 + y^2) &= 2\varepsilon g\dot{x}(1 + p(x^2 + y^2)), \end{aligned} \tag{7.214}$$

where the initial conditions (7.201) are

$$x(0) = 0.05, \quad \dot{x}(0) = 0, \quad y(0) = 0.0866, \quad \dot{y}(0) = 0. \tag{7.215}$$

Applying the Runge Kutta numerical procedure the exact solution of the system (7.214) is obtained (Fig. 7.10).

In Fig. 7.10. the analytical (7.213) and numerical $x - t$ and $y - t$ diagrams are plotted. It is evident that the analytical solution is on the top of the numerical one. The difference is negligible for a long time period and also for the small value to $\varepsilon = 0.1$.

7.4.2.4 Small Damping Function

The differential equation with small damping function is

$$\ddot{z} + c_1 z + c_3 z(z\bar{z}) = -\varepsilon b\dot{z}, \tag{7.216}$$

where εb is the coefficient of the damping function. Transforming the small function on the right side into polar coordinates it is

$$\Phi_\rho = -\varepsilon b\dot{\rho}, \qquad \Phi_\theta = -\varepsilon b\rho\dot{\theta}. \tag{7.217}$$

Due to (7.190) it is evident that as $(\Phi_\theta)_0 = 0$ it is $K_{11} = 0$ and $\theta_1 = 0$. The function K_{21} is

$$K_{21} = -\varepsilon bA^2\omega\left[\frac{1-k^2}{3k^2}\omega t + \frac{2k^2-1}{3k^2}E(\omega t, k^2) - \frac{1}{3}sn(\omega t, k^2)cn(\omega t, k^2)dn(\omega t, k^2)\right]. \tag{7.218}$$

According to (7.191) the polar coordinate ρ_1 is obtained from the Bernoulli equation

$$\frac{d\rho_1}{dt} + f(t)\rho_1 = g(t), \tag{7.219}$$

where

$$f(t) = -\frac{(c_1 + c_3A^2cn^2(\omega t, k^2))cn(\omega t, k^2)}{\omega sn(\omega t, k^2)dn(\omega t, k^2)},$$
$$g(t) = \varepsilon bA\left(\frac{(1-k^2)\omega t + (2k^2-1)E(\omega t, k^2)}{3k^2sn(\omega t, k^2)dn(\omega t, k^2)} - \frac{1}{3}cn(\omega t, k^2)\right). \tag{7.220}$$

For the parameter values $\varepsilon b = 0.1$, $c_1 = 2$ and $c_3 = 1$ and initial conditions (7.201) the modulus of the Jacobi elliptic function (7.206) is approximately zero and the function (7.218) simplifies to

$$K_{21} \approx -\frac{\varepsilon bA^2\omega}{2}\left(\omega t - \frac{\sin(2\omega t)}{2}\right). \tag{7.221}$$

The approximate solution of the Bernoulli differential equation (7.219) for (7.221) after some transformation and simplification is

$$\rho_1 \approx -\frac{\varepsilon b}{2\omega}At\cos(\omega t). \tag{7.222}$$

For the aforementioned numerical values the approximate analytic solution of (7.216) is

$$z \approx 0.1(1 - 0.035t)\cos(1.41774t)\exp(j\pi/3). \tag{7.223}$$

The differential equation (7.216) is solved numerically. Applying the Runge-Kutta numerical procedure the exact numerical solution is obtained. In Fig. 7.11 the approximate analytical $x(t)$ and $y(t)$ solutions (7.223) are compared with numericall obtained ones.

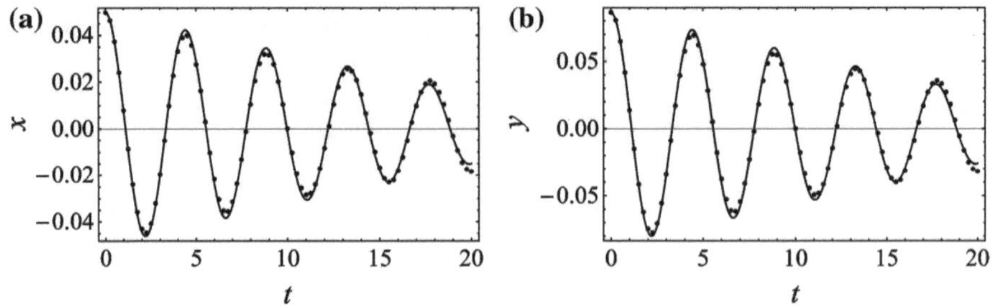

Fig. 7.11 **a** $x - t$ and **b** $y - t$ diagrams obtained analytically (*full line*) and numerically (*dotted line*)

It can be concluded that the analytical solution is on the top of the numerical solution. From the solution it can be seen that the small damping function which is a linear function of the time derivative of the complex function z decreases the amplitude ρ in time and tends to zero. The velocity of amplitude decrease depends not only on the value of coefficient of damping εb but also on the value of the initial amplitude A which affects the frequency of vibration ω.

7.4.3 Conclusion

From the previous considerations it can be concluded:

1. The generalized Krylov–Bogolubov solving procedure for the differential equation (7.148) with complex function is a very convenient one due to its simplicity. The obtained approximate analytical solution in the first approximation is quite satisfactory for technical reasons for the small value of the parameter ε and short time interval. The disadvantage of the method is that it is applicable if the generating closed form analytical solution is known.

2. The approximate analytic method based on the first integrals is especially appropriate for the special initial conditions where the initial conditions which are the first time derivatives of the polar coordinates are zero. Those initial conditions are named 'initial conditions without impact'. Using the straightforward series expansion the first and higher order approximate analytical solution is obtained. Comparing the analytical approximate solution with the exact numerical one it is concluded that the difference is negligible.

3. The mathematical model (7.148) considered in this paper represents the differential equation of vibration of the strong nonlinear Jeffcott rotor (symmetrically supported shaft-disc system) with small forces. The strong nonlinearity is usually caused by the nonlinear elastic force in the shaft. The analytical solutions obtained for the system with strong cubic nonlinearity, small gyroscopic force and small damping force considered in this Chapter give very useful information about the vibrations of the rotor and are of special interest for rotor dynamics.

References

Abe, Y., & Kato, N. (2013). Complex earthquake cycle simulations using a two-degree-of-freedom spring-block model with a rate- and state-friction law. *Pure and Applied Geophysics*, *170*, 745–765.

Abramowitz, M., & Stegun, I. A. (1979). *Handbook of mathematical functions with formulas, graphs and mathematical tables*. Moscow: Nauka. (in Russian).

Afsharfard, A., & Farshidianfar, A. (2012). Design of nonlinear impact dampers based on acoustic and damping behavior. *International Journal of Mechanical Sciences*, *65*, 125–133.

Alabudzev, P., Grichin, A., Kim, L., Migirenko, G., Chon, V., & Stepanov, P. (1989). *Vibration protecting and measuring systems with quasi-zero stiffness*. New York: Hemisphere Publishing.

Araky, Y., Asai, T., & Masui, T. (2010). Vertical vibration isolator having piecewise-constant restoring force. *Earthquake Engineering Structural Dynamics*, *38*, 1505–1523.

Bogolubov, N. N., & Mitropolski, Y. A. (1963). *Asimptoticheskie metodi v teorii nelinejnih kolebanij*. Moscow: Fiz.mat.giz.

Bondar, N. G. (1978). *Nonlinear vibrations excited by impulses*. Kiev-Doneck: Visca skola. (in Russian).

Burns, D. M., & Bright, V. M. (1997). Nonlinear flexures for stable deflection of an electrostatically actuated micromirror. In *Proceedings of SPIE: Microelectronics structures and MEMS for optical processing III* (vol. 3226, pp. 125–135).

Byrd, P. F., & Friedman, M. D. (1954). *Handbook of elliptic integrals for engineers and physicists*. Berlin: Springer.

Carela, A., Brennan, M. J., & Waters, T. P. (2007). Static analysis of a passive vibration isolator with quasi-zero stiffness characteristic. *Journal of Sound and Vibration*, *301*, 678–689.

Cortopassi, C., & Englander, O. (2010). *Nonlinear springs for increasing the maximum stable deflection of MEMS electrostatic gap closing actuators*. Berkeley: University of Berkeley.

Cveticanin, L. (1988). The stability of a textile machine rotor with increasing mass. *Mechanism and Machine Theory*, *23*, 275–278.

Cveticanin, L. (1992b). Approximate analytical solutions to a class of non-linear equations with complex functions. *Journal of Sound and Vibration*, *157*, 289–302.

Cveticanin, L. (1993). An asymptotic solution for weak nonlinear vibrations of the rotor. *Mechanism and Machine Theory*, *28*, 495–505.

Cveticanin, L. (1998a). Some particular solutions which describe the motion of the rotor. *Journal of Sound and Vibration*, *212*, 173–178.

Cveticanin, L. (1998b). Analytical methods for solving strongly non-linear differential equations. *Journal of Sound and Vibration*, *214*, 325–338.

Cveticanin, L. (2001a). Analytic approach for the solution of the complex valued strong non-linear differential equation of Duffing type. *Physica A*, *297*, 348–360.

Cveticanin, L. (2001b). Analytic solutions of the system of two coupled pure cubic non-linear oscillators equations. *Journal of Sound and Vibration*, *245*, 571–580.

Cveticanin, L. (2001c). Vibrations of a coupled two-degree-of-freedom system. *Journal of Sound and Vibration*, *247*, 279–292.

Cveticanin, L. (2002). The motion of a two mass system with non-linear connection. *Journal of Sound and Vibration*, *252*, 361–369.

Cveticanin, L. (2003). Forced non-linear vibrations of a symmetrical two-mass system. *Journal of Sound and Vibration*, *265*, 451–458.

Cveticanin, L. (2004a). A new approach for solving of a complex-valued differential equation. *Journal of Sound and Vibration*, *278*, 1181–1195.

Cveticanin, L. (2004b). Free vibration of a strong non-linear system described with complex function. *Journal of Sound and Vibration*, *277*, 815–824.

Cveticanin, L. (2004c). Vibrations of a free two-mass system with quadratic non-linearity and a constant excitation force. *Journal of Sound and Vibration*, *270*, 441–449.

Cveticanin, L. (2005). Approximate solution of a strongly non-linear complex differential equation. *Journal of Sound and Vibration, 284*, 503–512.

Cveticanin, L. (2012). Review on mathematical and mechanical models of the vocal cord. *Journal of Applied Mathematics,* Article ID 928591, 18 pages.

Cveticanin, L. (2015). A solution procedure based on the Ateb function for a two-degree-of-freedom oscillator. *Journal of Sound and Vibration, 346*(1), 298–313.

Cveticanin, L., & Pogany, T. (2012). Oscillator with a sum of non-integer order non-linearities. *Journal of Applied Mathematics,* Art. no. 649050, 20 pages.

Dimentberg, F. M. (1959). *Izgibnije kolabanija vrashchajushihsja valov*. Moscow: Izd. Akad. Nauk SSSR.

Gatti, G., Kovacic, I., & Brennan, M. J. (2010). On the response of a harmonically excited two degree-of-freedom system consisting of linear and nonlinear quasi-zero stiffness oscillators. *Journal of Sound and Vibration, 329*, 823–835.

Genin, J., & Maybee, J. S. (1970). External and material damped three dimensional rotor system. *International Journal of Non-linear Mechanics, 5*, 287–297.

Heris, H. K., Miri, A. K., Tripathy, U., Barthelat, F., & Mongeau, L. (2013). Nanoindentation of poroviscoelastic vocal fold tissue using an atomic force microscope. *Journal of the Mechanical Behavior of Biomedical Materials, 28*, 383–392.

Ishizaka, K., & Flangan, J. L. (1972). Synthesis of voiced sounds from a tw-mass model of the vocal cords. *The Bell System Technical Journal, 51*(6), 1233–1252.

Jones, T. B., & Nenadic, N. G. (2013). *Electromechanics and MEMS*. New York: Cambridge University Press.

Kovacic, I., Brennan, M. J., & Waters, T. P. (2008). A study of a non-linear vibration isolator with quasi-zero stiffness characteristic. *Journal of Sound and Vibration, 315*, 700–711.

Krylov, N., & Bogolubov, N. (1943). *Introduction to Nonlinear Mechanics*. Princeton: Princeton University Press.

Kucab, K., Gorski, G., & Mizia, J. (2013). Energy harvesting in the nonlinear two-masses piezoelasic system driven by harmonic excitations. *The European Physical Journal—Special Topics, 222*, 1607–1616.

Maccari, A. (2003). Periodic and quasiperioidc motion for complex nonlinear systems. *International Journal of Non-linear Mechanics, 38*, 575–584.

Machado, L. G., Savi, M. A., & Pacheco, P. M. C. L. (2013). Nonlinear dynamics and chaos in coupled shape memory oscillators. *International Journal of Solids and Structures, 40*, 5139–5156.

Mahmoud, G. M. (1998). Approximate solutions of a class of complex nonlinear dynamical systems. *Physica A, 253*, 211–222.

Mahmoud, G. M., & Shaban, A. H. A. (2000). On periodic solutions of parametrically excited complex non-linear dynamical systems. *Physica A, 278*, 390–404.

Manasevich, R., Mawhin, J., & Zanolin, F. (1999). Periodic solutions of some complex-valued Liénard and Rayleigh equations. *Nonlinear Analysis, 36*, 997–1014.

Mergel, P., Herzel, H., & Titze, I. R. (2000). Irregular vocal fold vibration—High-speed observation and modeling. *Journal of Acoustics Society of America, 108*, 2996–3002.

Muszynska, A. (1976). *On rotor dynamics*. Warszaw: PAN.

Nayfeh, A. H., & Mook, D. T. (1979). *Nonlinear Oscillations*. New York: Wiley.

Pham, T. T., Lamarque, C. H., & Pernot, S. (2011). Passive control of one degree of freedom nonlinear quadratic oscillator under combination resonance. *Communications in Nonlinear Science and Numerical Simulations, 16*, 2279–2288.

Pilipchuk, V. N. (2007). Strongly nonlinear vibration of damped oscillators with two nonsmooth limits. *Journal of Sound and Vibration, 302*, 398–402.

Pilipchuk, V. N. (2010). *Nonlinear dynamics: Between linear and impact limits*. New York: Springer.

Polo, M. F. P., Molina, M. P., & Chica, J. G. (2009). Chaotic dynamic and control for micro-electro-mechanical systems of massive storage with harmonic base excitation. *Chaos, Solitons and Fractals, 39*, 1356–1370.

Polo, M. F. P., Molina, M. P., & Chica, J. G. (2010). Self-oscillations and chaotic dynamic of a nonlinear controlled nano-oscillator. *Journal of Computational and Theoretical Nanoscience*, *7*, 2463–2477.

Rosenberg, R. M. (1963). The Ateb(h)-functions and their properties. *Quarterly of Applied Mathematics*, *21*, 37–47.

Sarkar, B. C., Dandapathak, M., Sarkar, S., & Banerjee, T. (2013). Studies on the dynamics of two bilaterally coupled periodic gunn oscillators using Melnikov technique. *Progress in Electromagnetics Research M*, *28*, 213–228.

Snamina, J., & Orkisz, P. (2014). Energy harvesting from vibrations of a two-degree-of-freedom mechanical system. *Acta Physica Polonica A*, *125*(4–A), A-174–A-178.

Sugiura, T., Hori, T., & Kawamura, Y. (2014). Synchronization of coupled stick-slip oscillators. *Nonlinear Processes in Geophysics*, *21*, 251–267.

Tondl, A. (1979). *Avtokolebanija mehanicheskih sistem*. Moscow: Mir. (in Russian).

Vakakis, A. F., & Rand, R. H. (1992a). Normal modes and global dynamics of a two-degree-of-freedom non-linear system. Part I: low energies. *International Journal of Non-linear Mechanics*, *27*, 861–874.

Vakakis, A. F., & Rand, R. H. (1992b). Normal modes and global dynamics of a two-degree-of-freedom non-linear system. Part II: high energies. *International Journal of Non-linear Mechanics*, *27*, 875–889.

Vance, J. M. (1988). *Rotor dynamics of turbomachinery*. New York: Wiley.

Wan, N., Peng, D. D., Sun, M., & Zhang, D. (2013). Nonlinear oscillation of pathological vocal folds during vocalization. *Science China: Physics, Mechanics and Astronomy*, *56*(7), 1324–1328.

Welte, J., Kniffka, T. J., & Ecker, H. (2013). Parameteric excitation in a two degree of freedom MEMS system. *Shock and Vibration*, *20*, 1113–1124.

Xing, J. T., Xiong, Y. P., & Price, W. G. (2005). Passive active vibration isolation systems to produce zero of infinite dynamic modulus: Theoretical and conceptual design strategies. *Journal of Sound and Vibration*, *286*, 615–636.

Xiong, Y. P., Xing, J. T., & Price, W. G. (2005). Interactive power flow characteristics of an integrated equipment—Nonlinear isolator—Travelling flexible ship excited by sea waves. *Journal of Sound and Vibration*, *287*, 245–276.

Zhu, Q., & Ishitoby, M. (2004). Chaos and bifurcations in a nonlinear vehicle model. *Journal of Sound and Vibration*, *275*, 1136–1146.

Chapter 8
Chaos in Oscillators

The problem of chaotic motion is widely investigated for the oscillators with negative linear and cubic nonlinear term

$$\ddot{x} - x + x^3 = \varepsilon(\gamma \cos \omega t - \delta \dot{x}), \tag{8.1}$$

where ω is the excitiation frequency and for $\varepsilon << 1$ the coefficients $\varepsilon\gamma$ and $\varepsilon\delta$ are small forcing and damping parameters. The region and properties of chaotic motion are investigated analytically by Tseng and Dugundji (1971), Holmes (1979), Holmes and Marsden (1981), but also numerically (see, for example, Guckenheimer and Holmes 1983; Wiggins 1988; Van Dooren and Janssen 1996). The model (8.1) corresponds to the ideal oscilllator with negative linear term excited with an ideal energy source (the motion of the oscillator does not affect the work of the energy source). Nevertheless, there exist also non-ideal oscillators excited with non-ideal energy sources where the forcing source has a limited energy supply. This source connected with an oscillator forms a non-ideal system. An example of a non-ideal system is the vibrating cart whose motion is induced by an in-board non-ideal motor driving an unbalanced rotor which is studied by Souza et al. (2005a, b). The forcing system has a limited energy source (an electric motor, for example) and the dynamics is influenced by that of the oscillating system being forced. In other words, the excitation is drawn on an energy source that is limited so that the excited system has an appreciable effect on it (see Dimentberg et al. 1997; Wiggins 1988). The special influence of the non-ideal energy source on the system is remarked during passage through resonance. It is caused by the special properties of the non-ideal source in this working regime. Namely, in the region before resonance as the power supplied to the source increases, the RPM of the energy source (motor) increases accordingly. But, the closed motor speed moves toward the resonance frequency the more power the source requires to increase the motor speed. Near resonance it appears that additional power supplied to the motor will only increase the amplitude of the response with little effect on the RPM of the motor, and the amplitude of vibration increases.

L. Cveticanin, *Strong Nonlinear Oscillators*, Mathematical Engineering,
DOI 10.1007/978-3-319-58826-1_8

In non-ideal vibrating systems the passage through resonance requires more input power than is available. As a consequence the vibrating system cannot pass through resonance or requires an intensive interaction between the vibrating system and the energy source to be able to do it. Strong interaction leads to fluctuating motor speed and fairly large vibration amplitudes appear. The motor may not have enough power to reach higher regimes with low energy consumption as most of its energy is applied to move the structure and not to accelerate the shaft. In fact, the vibrating response provides a certain energy sink. In the paper of Balthazar et al. (2003) an overview on non-ideal vibrations is shown.

In this section the dynamics and specially the chaotic motion of a generalized ideal oscillator and of a non-ideal oscillator is considered.

8.1 Chaos in Ideal Oscillator

The generalized model of an ideal oscillator is describec as

$$\ddot{x} - x + x\,|x|^{\alpha-1} = \varepsilon(\gamma \cos \omega t - \delta \dot{x}), \tag{8.2}$$

where $\alpha \in R$ and is is an integer or a non-integer. Equation (8.1) is only a special case of (8.2) for $\alpha = 3$. Dependence of parameters for chaos on the parameter α and the chaos control is done. The criteria of chaos is obtained analytically by generalization the Melnikov's procedure (Melnikov 1963) which was developed for the Duffing equation with cubic nonlinearity (8.1) (see Holmes 1979; Chacon 1999) and its modifications for the Duffing equation with complex function (Cveticanin 1993), Duffing equation with strong cubic and quintic term (Wang et al. 2006), a three-well Duffing system (Yang and Jing 2008), etc. Cveticanin and Zukovic (2009) applied the Melnikov's criteria for oscillator with rational order of nonlinearity. In this section the homoclinic orbit for (8.2) is given in the exact analytical form and according to the Melnikov's theorem the rate of change of the distance between stable and unstable manifolds for small perturbations is calculated. The parameter values producing chaotic behavior of the system are determined. The dependence of the critical parameters value on the value of α is explicitly shown. For certain values of parameter α, the numerical simulation of (8.2) is done. The phase plane diagrams and the Poincaré maps are plotted. In the paper the existence of the chaotic motion is proved by calculating the Lyapunov characteristic exponents. The bifurcation diagrams for these values of parameter α are also plotted and the critical values for chaotic motion are numerically determined. The analytically and numerically obtained critical values are compared. The delayed feedback control, the so called ' Pyragas method' (see Pyragas 1992, 1996, 2001, 2006; Ushio 1996; Chen and Yu 1999; Zhu and Tian 2005), is adopted for the chaos control of the differential equation (8.2). The suggested 'Pyragas method', due to its simplicity, is one of the most widely used chaos control procedures. The results obtained after control are compared with those before suppressing chaos.

8.1.1 Homoclinic Orbits in the Unperturbed System

For the unperturbed system with nonlinear order displacement, when $\varepsilon = 0$, the differential equation (8.2) simplifies to

$$\ddot{x} - x + x\,|x|^{\alpha-1} = 0. \tag{8.3}$$

The system has two centers at $(x, \dot{x}) = (\pm 1, 0)$ and a hyperbolic saddle at $(0, 0)$. The energy function for (8.3) is

$$\frac{\dot{x}^2}{2} - \frac{x^2}{2} + \frac{x^2\,|x|^{\alpha-1}}{\alpha+1} = K, \tag{8.4}$$

where K is the energy constant dependent on the initial amplitude $x(0) = x_0$, initial velocity $\dot{x}(0) = \dot{x}_0$ and the order α

$$K = \frac{\dot{x}_0^2}{2} - \frac{x_0^2}{2} + \frac{x_0^2\,|x_0|^{\alpha-1}}{\alpha+1}. \tag{8.5}$$

Dependently on K, the level sets are different. For all of them it is common that they form closed periodic orbits which surround the fixed points $(x, \dot{x}) = (1, 0)$ or $(x, \dot{x}) = (-1, 0)$ or all the three fixed points $(\pm 1, 0)$ and $(0, 0)$. The boundary between these two groups of orbits corresponds to $K = 0$, when

$$\dot{x}_0 = \pm\sqrt{x_0^2\left(1 - \frac{2}{\alpha+1}\,|x_0|^{\alpha-1}\right)}. \tag{8.6}$$

The level set

$$\frac{\dot{x}^2}{2} - \frac{x^2}{2} + \frac{x^2\,|x|^{\alpha-1}}{\alpha+1} = 0, \tag{8.7}$$

is composed of two homoclinic orbits

$$\Gamma_+^0(t) \equiv \left(x_+^0(t), \dot{x}_+^0(t)\right), \tag{8.8}$$

$$\Gamma_-^0(t) \equiv \left(x_-^0(t), \dot{x}_-^0(t)\right), \tag{8.9}$$

which connect the fixed hyperbolic saddle point $(0, 0)$ to itself and contain the stable and unstable manifolds. The functions $x_\pm^0(t)$ are the exact solution of (8.7) for

$$x_0 = \pm\left(\frac{\alpha+1}{2}\right)^{\frac{1}{\alpha-1}}, \quad \dot{x}_0 = 0, \tag{8.10}$$

i.e.,

Table 8.1 Extremal points of the homoclinic orbit for various α

α	4/3	3/2	5/3	2	3
x_{max}	1.5880	1.5625	1.5396	1.5000	1.4142
$\dot{x}_{max}$	0.37796	0.44721	0.50000	0.57735	0.70711

$$\begin{aligned}\Gamma^0_+(t) &= \sqrt[\alpha-1]{\frac{\alpha+1}{2}}\left[\sec h\left(\frac{\alpha-1}{2}t\right)\right]^{\frac{2}{\alpha-1}},\\ &\quad -\sqrt[\alpha-1]{\frac{\alpha+1}{2}}\left[\sec h\left(\frac{\alpha-1}{2}t\right)\right]^{\frac{2}{\alpha-1}}\tanh\left(\frac{\alpha-1}{2}t\right),\\ \Gamma^0_-(t) &= -\Gamma^0_+(t),\end{aligned} \tag{8.11}$$

where sec h and tanh are the hyperbolic functions. The relations (8.10) and (8.11) represent the closed form analytical solution for the homoclinic orbits of (8.2) for $\varepsilon = 0$. The relations are the general form of the homoclinic orbits for (8.2) and are valid for all values of parameter $\alpha > 1$, including not only all integer numbers, but also the whichever non-integers.

As the special case, let us determine the homoclinic orbit for the case of cubic nonlinearity. Substituting $\alpha = 3$ into (8.10) and (8.11) the homoclinic orbits for $x_0 = \pm\sqrt{2}$ and $\dot{x}_0 = 0$ have the form

$$\Gamma^0_+(t) = (\sqrt{2}\sec h(t), -\sqrt{2}\sec h(t)\tanh(t), \quad \Gamma^0_-(t) = -\Gamma^0_+(t). \tag{8.12}$$

Comparing (8.12) with the results given by Guckenheimer and Holmes (1983). it is concluded, that they are the same.

Parameters of the homoclinic orbits. The homoclinic orbits form a closed curve which connects the fixed saddle point (0, 0) to itself. The extreme characteristics x_{max} and $\dot{x}_{max}$ of the homoclinic orbit are according (8.10) and (8.7): $(\pm(\frac{\alpha+1}{2})^{\frac{1}{\alpha-1}}, 0)$ and $(\pm 1, \pm\sqrt{\frac{\alpha-1}{\alpha+1}})$. In Table 8.1 the extreme value x_{max} and $\dot{x}_{max}$ of the the homoclinic orbit $\Gamma^0_+(t)$ for various values of α is calculated.

Analyzing the extreme values in the homoclinic motion, it is concluded that the maximal values x_{max} and $\dot{x}_{max}$ depend on the order of nonlinearity α: the higher the α, the higher the value of $\dot{x}_{max}$ and smaller the value of x_{max}. For higher values of α both values tend to 1. For $\alpha \approx 1$ the maximal value $\dot{x}_{max}$ tends to zero.

The time period necessary for going from x_{max} to $\dot{x}_{max}$ in $x = 1$ also depends on the order of nonlinearity α. Namely, rewritten the relation (8.7) in integral form

$$\int_0^t dt = \int_{x_{max}}^x \frac{dx}{x\sqrt{1 - \frac{2}{\alpha+1}|x|^{\alpha-1}}}, \tag{8.13}$$

and evaluating (8.13) for $x(\dot{x}_{max}) = 1$ the following time period is obtained

Table 8.2 Time period for various α

α	4/3	3/2	5/3	2	3
T_1	0.39768	0.481 21	0.549 31	0.65848	0.88137

$$T_1 = \ln \frac{\sqrt{\frac{2}{\alpha+1}}}{1+\sqrt{1-\frac{2}{\alpha+1}}} = \ln\left(\sqrt{\frac{\alpha+1}{2}} + \sqrt{\frac{\alpha-1}{2}}\right). \tag{8.14}$$

In Table 8.2 the time period T_1 for various values of α is given.

Analyzing (8.14) and the values given in Table 8.2, it can be concluded that the higher the value of α the time period T_1 is longer. Besides, from (8.13) is evident that independently of α, the time necessary for homoclinic orbit to reach the saddle point i.e., $x \to 0$, is infinitely long, i.e., $t \to \infty$. The same result is given by Guckenheimer and Holmes (1983) for the cubic nonlinearity.

The homoclinic orbit separates the phase plane into two areas. Inside the separatrix curve the orbits are around one of the centers, and outside the separatrix curve the orbits surround the both centers and the saddle point. Physically it means that for certain initial conditions the oscillations are around one steady state position, and for others around all the steady state solutions (two stable and an unstable).

8.1.2 Melnikov's Criteria for Chaos

Let us form the Melnikov's function for (8.2) and $\Gamma_+^0(t)$ i.e. $\Gamma_-^0(t)$ given by (8.8) and (8.9)

$$M(t_0) = \int_{-\infty}^{+\infty} \dot{x}^0(t)\Big[\gamma\cos\omega(t+t_0) - \delta\dot{x}^0(t)\Big]dt, \tag{8.15}$$

where

$$\dot{x}^0(t) = -\sqrt[\alpha-1]{\frac{\alpha+1}{2}}\left[\sec h\left(\frac{\alpha-1}{2}t\right)\right]^{\frac{2}{\alpha-1}} \tanh\left(\frac{\alpha-1}{2}t\right),$$

$$x^0 = \sqrt[\alpha-1]{\frac{\alpha+1}{2}}\left[\sec h\left(\frac{\alpha-1}{2}t\right)\right]^{\frac{2}{\alpha-1}}. \tag{8.16}$$

Substituting (8.16) into (8.15) the transformed Melnikov's function is obtained

$$M(t_0) = -\gamma \left(\frac{\alpha+1}{2}\right)^{\frac{1}{\alpha-1}} \int_{-\infty}^{+\infty} \left[\sec h\left(\frac{\alpha-1}{2}t\right)\right]^{\frac{2}{\alpha-1}} \tanh\left(\frac{\alpha-1}{2}t\right) \cos\omega(t+t_0)dt$$
$$-\delta \left(\frac{\alpha+1}{2}\right)^{\frac{2}{\alpha-1}} \int_{-\infty}^{+\infty} \left[\sec h\left(\frac{\alpha-1}{2}t\right)\right]^{\frac{4}{\alpha-1}} \left[\tanh\left(\frac{\alpha-1}{2}t\right)\right]^2 dt, \tag{8.17}$$

i.e.,

$$M(t_0) = -\gamma \left(\frac{\alpha+1}{2}\right)^{\frac{1}{\alpha-1}} \cos\omega t_0 I_1 + \gamma \left(\frac{\alpha+1}{2}\right)^{\frac{1}{\alpha-1}} \sin\omega t_0 I_2 - \delta \left(\frac{\alpha+1}{2}\right)^{\frac{2}{\alpha-1}} I_3, \tag{8.18}$$

where

$$I_1 = \int_{-\infty}^{+\infty} \frac{\sinh(\frac{\alpha-1}{2}t)}{[\cosh(\frac{\alpha-1}{2}t)]^{\frac{\alpha+1}{\alpha-1}}} \cos\omega t dt, \tag{8.19}$$

$$I_2 = \int_{-\infty}^{+\infty} \frac{\sinh(\frac{\alpha-1}{2}t)}{[\cosh(\frac{\alpha-1}{2}t)]^{\frac{\alpha+1}{\alpha-1}}} \sin\omega t dt, \tag{8.20}$$

$$I_3 = \int_{-\infty}^{+\infty} \frac{[\sinh(\frac{\alpha-1}{2}t)]^2}{[\cosh(\frac{\alpha-1}{2}t)]^{\frac{2(\alpha+1)}{\alpha-1}}} dt. \tag{8.21}$$

The first integral I_1 in (8.19) is zero, since it is the product of an odd and an even function.

Integral I_2. Using the partial integration the integral I_2 is transformed into the following form

$$I_2 = 2\int_{0}^{+\infty} \frac{\sinh(\frac{\alpha-1}{2}t)}{[\cosh(\frac{\alpha-1}{2}t)]^{\frac{\alpha+1}{\alpha-1}}} \sin\omega t dt = \frac{4}{\alpha-1}\int_{0}^{+\infty} \frac{\sinh(x)}{[\cosh(x)]^{\frac{\alpha+1}{\alpha-1}}} \sin\left(\frac{2\omega x}{\alpha-1}\right) dx$$
$$= -2\left[\frac{\sin(\frac{2\omega x}{\alpha-1})}{[\cosh(x)]^{\frac{2}{\alpha-1}}}\right]_0^{\infty} + \frac{4\omega}{\alpha-1}\int_{0}^{+\infty} \frac{\cos(\frac{2\omega x}{\alpha-1})}{[\cosh(x)]^{\frac{2}{\alpha-1}}} dx. \tag{8.22}$$

The first term in the integral is zero and the second has the exact analytical solution for $\alpha > 1$ (see Gradstein and Rjizhik 1971)

$$I_2 = \frac{\omega}{\alpha-1} \frac{2^{\frac{2}{\alpha-1}}}{\Gamma(\frac{2}{\alpha-1})} \Gamma\left(\frac{1+i\omega}{\alpha-1}\right) \Gamma\left(\frac{1-i\omega}{\alpha-1}\right), \tag{8.23}$$

where Γ is the Euler Gamma function and $i = \sqrt{-1}$ is the imaginary unit.

Special cases:

1. For $\alpha = 3$, when $\Gamma(1) = 1$ and $\left|\Gamma(\frac{1}{2} + i\frac{\omega}{2})\right|^2 = \frac{\pi}{\cosh(\frac{\pi\omega}{2})}$, (see Gradstein and Rjizhik 1971), the value of the integral is

$$I_2(\alpha = 3) = \omega\pi \sec h\left(\frac{\pi\omega}{2}\right). \tag{8.24}$$

The result is equal to that obtained by the use of the method of residues (Guckenheimer and Holmes 1983).

2. For $\alpha = 2$ when $\Gamma(1 + i\omega)\Gamma(1 - i\omega) = \frac{\pi\omega}{\sinh(\pi\omega)}$ and $\Gamma(2) = 1$ (Gradstein and Rjizhik 1971)

$$I_2 = \frac{4\pi\omega^2}{\sinh(\pi\omega)}. \tag{8.25}$$

3. If $\frac{2}{\alpha-1}$ in (8.22) is an even integer number, i.e., $\frac{2}{\alpha-1} = 2n$, $n \geq 2$, the exact analytical solution for the integral I_2 is given by Gradstein and Rjizhik (1971)

$$I_2 = \frac{4^n \pi n^2 \omega^2}{(2n-1)! \sinh(n\pi\omega)} \prod_{k=1}^{n-1} (\omega^2 n^2 + k^2). \tag{8.26}$$

4. If $\frac{2}{\alpha-1}$ is an odd integer, i.e., $\frac{2}{\alpha-1} = 2n + 1$, $n \geq 1$, the analytical expression of the integral I_2 is

$$I_2 = \frac{2^{2n}(2n+1)\pi\omega}{(2n)! \cosh(\frac{\pi\omega(2n+1)}{2})} \prod_{k=1}^{n} \left[\frac{\omega^2(2n+1)^2}{4} + \left(\frac{2k-1}{2}\right)^2\right]. \tag{8.27}$$

Integral I_3. After partial integration and using the boundary conditions, the integral I_3 simplifies to

$$I_3 = \frac{2(\alpha-1)}{\alpha+3} \int_0^{+\infty} \frac{dt}{[\cosh(\frac{\alpha-1}{2}t)]^{\frac{4}{\alpha-1}}} = \frac{4}{\alpha+3} \int_0^{+\infty} \frac{dx}{\cosh^{\frac{4}{\alpha-1}}(x)}. \tag{8.28a}$$

Unfortunately, the integral (8.28a) has not a closed form analytical solution for all values of $\alpha > 1$. Some special cases are considered:

1. For $\frac{4}{\alpha-1} = 2n$, where $(2n)$ is a whole even number and $n \geq 2$, the solution of (8.28a) is

$$I_3 = \frac{(2n)}{(2n)^2 - 1} \frac{2^{n-1}(n-1)!}{(2n-3)!!}, \tag{8.29}$$

where

$$(n-1)! = 1 \cdot 2 \cdot \ldots \cdot (n-1), \qquad (2n-3)!! = 1 \cdot 3 \cdot \ldots \cdot (2n-3).$$

Table 8.3 Critical parameters for various values of α and $\omega = 1$

α	I_2	I_3	$\left(\frac{\gamma}{\delta}\right)_c$	$\left(\frac{\gamma}{\delta}\right)_c, \omega = 1$
4/3	$\frac{24\pi\omega^2(1+9\omega^2)(4+9\omega^2)}{5\sinh(3\pi\omega)}$	0.34099	$\frac{0.11281\sinh(3\pi\omega)}{\pi\omega^2(1+9\omega^2)(4+9\omega^2)}$	1.7114
7/5	$\frac{5\pi\omega(1+25\omega^2)(9+25\omega^2)}{24\cosh(\frac{5\pi\omega}{2})}$	0.36941	$\frac{2.7971\cosh(\frac{5\pi\omega}{2})}{\pi\omega(1+25\omega^2)(9+25\omega^2)}$	1.2972
3/2	$\frac{32\pi\omega^2(4\omega^2+1)}{3\sinh(2\pi\omega)}$	0.40635	$\frac{5.9524\times10^{-2}\sinh(2\pi\omega)}{\pi\omega^2(4\omega^2+1)}$	1.0146
5/3	$\frac{3\pi\omega(1+9\omega^2)}{2\cosh(\frac{3\pi\omega}{2})}$	0.45714	$\frac{0.46921\cosh(\frac{3\pi\omega}{2})}{\pi\omega(1+9\omega^2)}$	0.83135
2	$\frac{4\pi\omega^2}{\sinh(\pi\omega)}$	0.53333	$\frac{0.2\sinh(\pi\omega)}{\pi\omega^2}$	0.73521
3	$\frac{\pi\omega}{\cosh(\frac{\pi\omega}{2})}$	0.66667	$\frac{0.94281\cosh(\frac{\pi\omega}{2})}{\pi\omega}$	0.75302

According to (8.29), for the special case when $\alpha = 3$, i.e., $(2n) = 2$, we obtain $I_3 = 2/3$. This result corresponds to the value given in Guckenheimer and Holmes (1983).

2. If $\frac{4}{\alpha-1}$ is an odd number, i.e., $\frac{4}{\alpha-1} = (2n+1)$, where $n \geq 1$, the evaluation of the integral gives

$$I_3 = \frac{\pi(2n+1)(2n-1)!!}{4(n+1)(2n)!!}, \tag{8.30}$$

where $(2n)!! = 2 \cdot 4 \cdot 6...$, $(2n+1)!! = 1 \cdot 3 \cdot 5....$

Using the previous results and Melnikov's theorem (1963) the following is stated:

Proposition *If* $M(t_0)$ *has a simple zero and the corresponding critical parameter value is*

$$\left(\frac{\gamma}{\delta}\right)_{cr} = \left(\frac{\alpha+1}{2}\right)^{\frac{1}{\alpha-1}} \frac{I_3}{I_2}, \tag{8.31}$$

then in the system with nonlinear order of displacement (8.2) *the deterministic chaos may appear for certain parameter values which satisfy the relation*

$$\frac{\gamma}{\delta} > \left(\frac{\gamma}{\delta}\right)_{cr}. \tag{8.32}$$

Remark For (8.31) the distance between the stable and unstable manifolds of the homoclinic point (0, 0) is zero and the manifolds intersect transversely forming the transverse homoclinic orbits. The presence of such orbits implies that for certain parameters (8.32) the Poincaré map of the system with nonlinear order displacement (8.2) has the strange attractor and the countable infinity of unstable periodic orbits, an uncountable set of bounded nonperiodic orbits and a dense orbit which are the main characteristics of the chaotic motion.

The relation (8.32) with (8.31) represents the analytical criteria for chaos.

In Table 8.3 the critical parameters for various fraction values α, and specially when $\omega = 1$, are shown.

Analyzing the critical parameter $(\gamma/\delta)_c$ in Table 8.3 it can be concluded that it increases by decreasing the parameter α.

8.1.3 Numerical Simulation

Let us consider the set of three first order differential equations

$$\dot{x} = y, \qquad \dot{y} - x + x\,|x|^{\alpha-1} = \varepsilon(\gamma\cos\omega t - \delta y), \qquad \dot{t} = 1. \tag{8.33}$$

For $\omega = 1$ and

$$\frac{(\gamma/\delta)}{(\gamma/\delta)_c} = k, \tag{8.34}$$

the system of differential equations (8.33) is rewritten as

$$\dot{x} = y, \quad \dot{y} - x + x\,|x|^{\alpha-1} = \varepsilon(k\delta\left(\frac{\gamma}{\delta}\right)_c \cos t - \delta y), \quad \dot{t} = 1, \tag{8.35}$$

where the critical parameter value $(\gamma/\delta)_c$ depends on α.

To prove the accuracy of the analytical solving procedure, the numerical simulation is done. The phase plane diagrams and Poincaré maps are plotted for certain values of α. The parameter k is varied and δ is assumed as a fixed constant. Three examples are investigated.

1. For $\alpha = 2$ and $\varepsilon\delta = 0.3$ the differential equation of motion has the form

$$\ddot{x} + 0.3\dot{x} - x + x\,|x| = 0.220563k\cos t, \tag{8.36}$$

where the critical parameter value is $(\gamma/\delta)_c = 0.7351$ (see Table 8.3). In Fig. 8.1 the phase diagrams and the Poincaré maps for $k = 0.9$ and $k = 3.5$ and also $k = 3$ for initial conditions $x(0) = 1$ and $\dot{x}(0) = 0$ are plotted. The orbits in Fig. 8.1a are periodical with period $1T$.

In Fig. 8.1b the phase diagram corresponds to chaotic motion and the Poincaré map forms the strange attractor.

In Fig. 8.2. the Poincaré maps for (8.36), showing stable and unstable manifolds of the saddle point near (0,0) for $k = 0.9$, $k = 1$ and $k = 3$ are plotted. These are computed numerically. It can be seen that the first tangency appears to occur about $k = 1$, the value which according to (8.34) corresponds to the theoretically obtained value $(\gamma/\delta)_c = 0.7351$.

2. For the non-integer order $\alpha = 4/3$ and parameter values: $\varepsilon\delta = 0.25$ and $(\gamma/\delta)_c = 1.7114$ (see Table 8.3), the differential equation of motion is

$$\ddot{x} + 0.25\dot{x} - x + x\,|x|^{1/3} = 0.42785k\cos t. \tag{8.37}$$

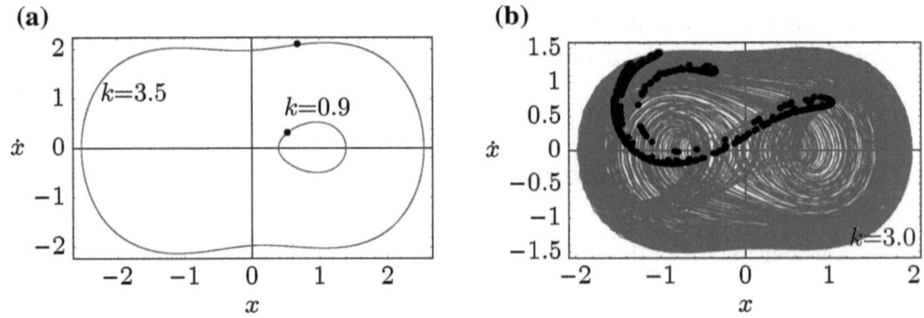

Fig. 8.1 Phase plane diagram and Poincaré map for $\alpha = 2$ and $\delta = 0.3$ and: **a** $k = 0.9$ and $k = 3.5$; **b** $k = 3$

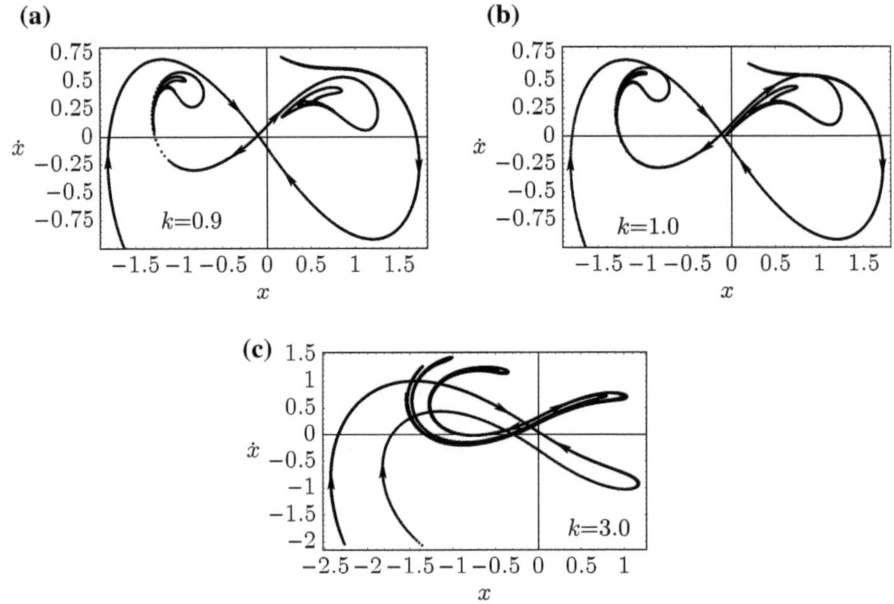

Fig. 8.2 Poincaré map for the Duffing equation, showing stable and unstable manifolds of the saddle point near (0,0) for $\alpha = 2$, $\varepsilon\delta = 0.3$ and: **a** $k = 0.9$, **b** $k = 1$ and **c** $k = 3$

In Fig. 8.3. the phase plane diagrams and the Poincaré maps are plotted. For $k = 0.8$ and $k = 2.2$ the motion is periodical (see Fig. 8.3a) and for $k = 1.3$ chaotic (see Fig. 8.3b).

3. If the non-integer order is $\alpha = 7/5$, the critical parameter $(\gamma/\delta)_c = 1.2972$ and $\varepsilon\delta = 0.3$, the following differential equation exists

$$\ddot{x} + 0.3\dot{x} - x + x\,|x|^{2/5}\,x = 0.38916k\cos t. \tag{8.38}$$

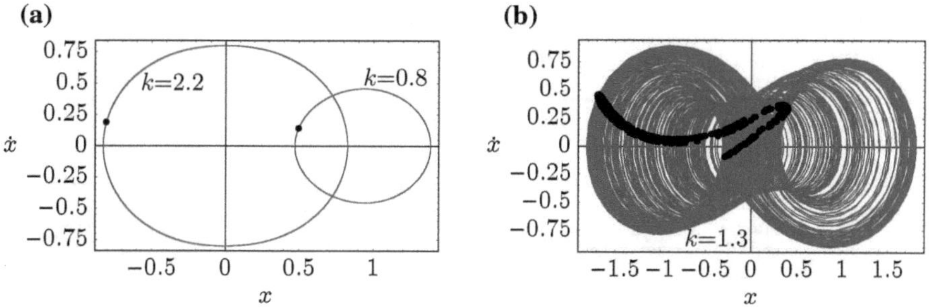

Fig. 8.3 Phase plane diagram and Poincare map for $\alpha = 4/3$ and $\varepsilon\delta = 0.25$ and: **a** $k = 0.8$ and $k = 2.2$; **b** $k = 1.3$

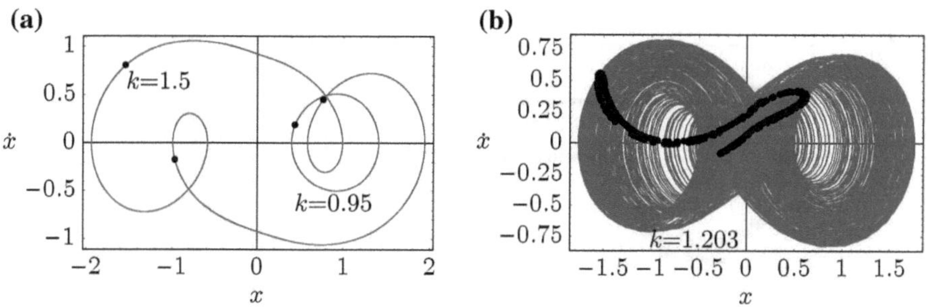

Fig. 8.4 Phase plane diagram and Poincare map for $\alpha = 7/5$ and $\varepsilon\delta = 0.3$ and: **a** $k = 0.95$ and $k = 1.5$; **b** $k = 1.203$

In Fig. 8.4 the phase plane diagrams and the Poincaré maps are plotted for $k = 0.95$, $k = 1.5$ and $k = 1.203$.

It is obvious that for $k = 0.95$ and $k = 1.5$ the motion is $1T$ and $3T$ periodical, respectively (Fig. 8.4a). For $k = 1.203$ the motion is chaotic (see Fig. 8.4b) and the strange attractor is evident.

From the previous consideration it can be concluded that chaotic motion appears for $k > 1$. Comparing this result of numerical simulation with the analytically obtained one, we see that they agree. For $k > 1$ the motion may be steady state chaotic.

8.1.4 Lyapunov Exponents and Bifurcation Diagrams

The chaotic motion of dynamic systems is usually studied numerically through the concept of Lyapunov characteristic exponents and bifurcation diagrams. Both approaches are applied for the dynamic system (8.35).

Lyapunov exponents quantify the chaotic behavior of the system. For the dynamic system (8.35) the Lyapunov spectrum is computed using Mathematica, as it is sug-

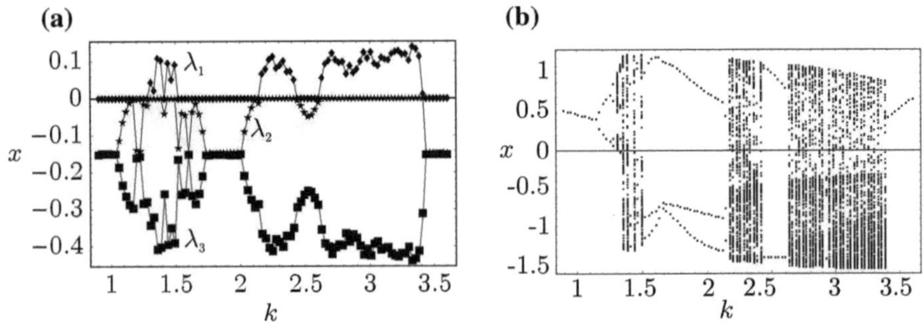

Fig. 8.5 **a** Lyapunov's spectra, **b** bifurcation diagram for $\alpha = 2$

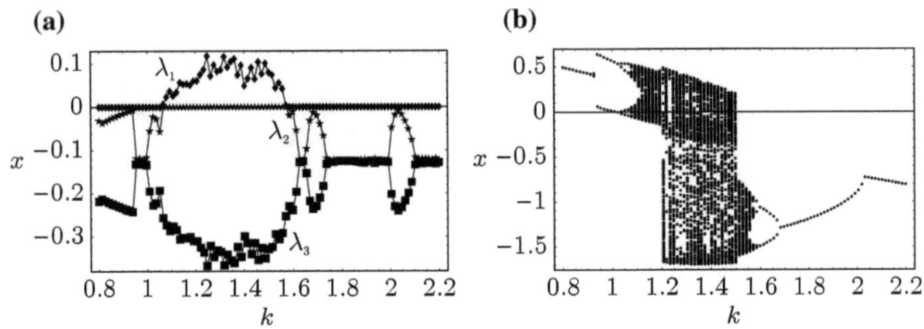

Fig. 8.6 **a** Lyapunov's spectra, **b** bifurcation diagram for $\alpha = 4/3$

gested by Sandri (1996). Varying the value of k the bifurcation $x - k$ diagram is obtained.

In Figs. 8.5, 8.6, and 8.7 the bifurcation diagrams and the Lyapunov spectrums for (8.36–8.38) are plotted. Analyzing the diagrams the following is concluded:

1. As the topological dimension of the system (8.35) is three, the chaotic motion appears for the case when one of the Lyapunov exponents is positive. For this nonautonomous model there is a spurious Lyapunov exponent which converges to zero. It corresponds to the additional trivial evolution equations $\dot{t} = 1$. The third Lyapunov exponent is negative.

2. The Lyapunov spectrum and the bifurcation diagram are in good agreement. Namely, the intervals of chaotic motion plotted in bifurcation diagrams correspond to those where one of the Lyapunov exponent is positive.

3. The chaos is the result of the period doubling bifurcation and is manifested for $k > 0$.

4. In Figs. 8.5, 8.6, and 8.7 is visible the widening of the chaotic attractor.

5. For $\alpha = 2$ and $k \in [1.3, 1.55] \wedge [2.1, 2.4] \wedge [2.65, 3.4]$ one of the Lyapunov exponents is positive and the bifurcation diagram shows the chaos. The same correlation is seen for $\alpha = 4/3$ and $k \in [1.1, 1.58]$ and also for $\alpha = 7/5$ and $k \in [1.05, 1.39]$.

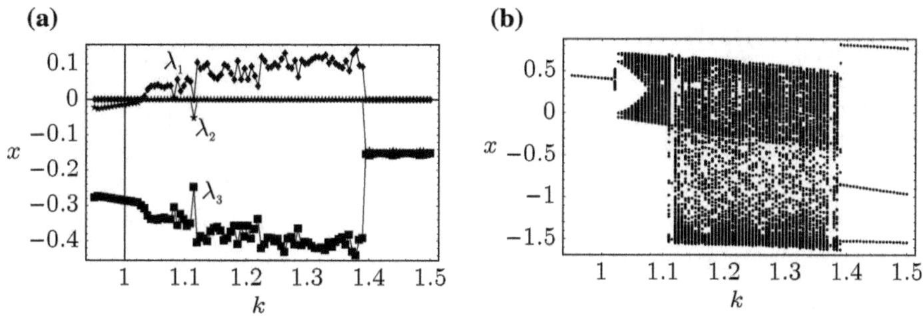

Fig. 8.7 **a** Lyapunov's spectra, **b** bifurcation diagram for $\alpha = 7/5$.

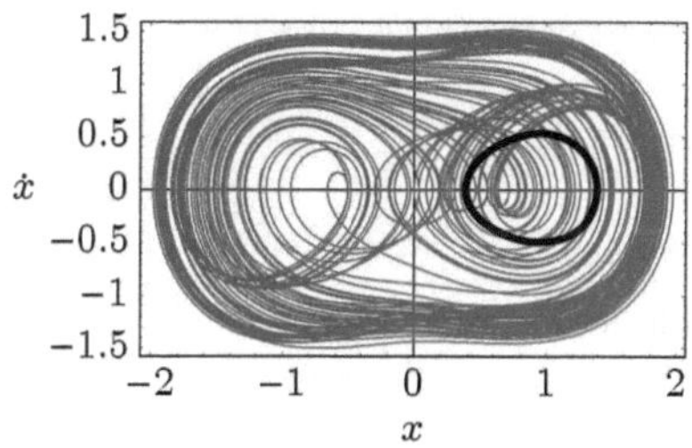

Fig. 8.8 The phase plane diagrams for $\alpha = 2$, $\varepsilon\delta = 0.3$ and $k = 3$ before (*grey curve*) and after chaos control (*black curve*)

6. For all of three examples it is evident that the motion is not chaotic for $k < 1$, i.e., for $(\gamma/\delta) < 0.7351$ when $\alpha = 2$, $(\gamma/\delta) < (\gamma/\delta)_c = 1.7114$ when $\alpha = 4/3$ and $(\gamma/\delta) < (\gamma/\delta)_c = 1.2972$ for $\alpha = 7/5$. The obtained values are in good agreement with analytical results obtained by using the Melnikov's criteria and with numerical simulations.

8.1.5 Control of Chaos

The delayed feedback control, i.e., the ' Pyragas method' (Pyragas 2006) is based on the idea of the stabilization of unstable periodic orbits embedded within a strange attractor. This is achieved by making a small time-dependent perturbation in the form of feedback to an accessible system parameter. The method turns the presence of chaos into an advantage. Due to the infinite number of different unstable periodic orbits embedded in a strange attractor, a chaotic system can be tuned to a large number of distinct periodic regimes by watching the temporal programming of small parameter perturbation to stabilize different periodic orbits.

The introduced simple control law in the paper of Pyragas (1992) is

$$F(t) = K[\dot{x}(t - \tau) - \dot{x}(t), \qquad (8.39)$$

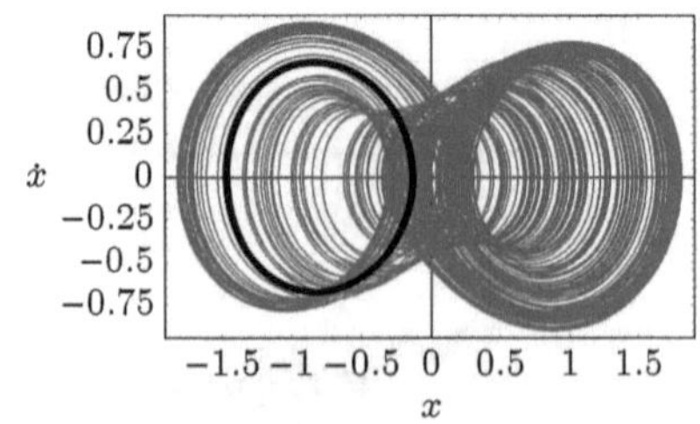

Fig. 8.9 The phase plane diagrams for $\alpha = 4/3$, $\varepsilon\delta = 0.25$ and $k = 1.3$ before (*grey curve*) and after chaos control (*black curve*)

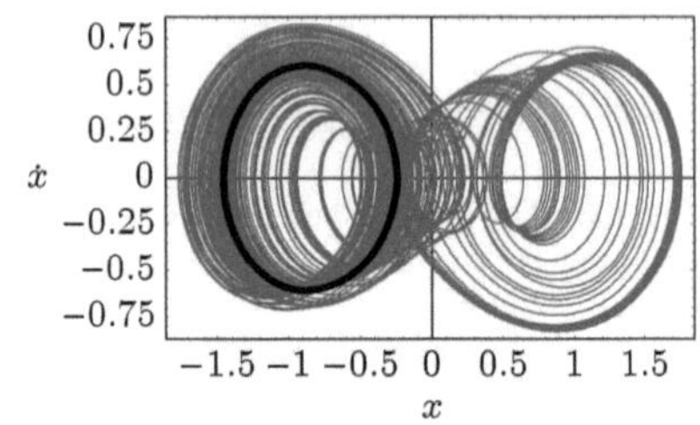

Fig. 8.10 The phase plane diagrams for $\alpha = 7/5$, $\varepsilon\delta = 0.3$ and $k = 1.203$ before (*grey curve*) and after chaos control (*black curve*)

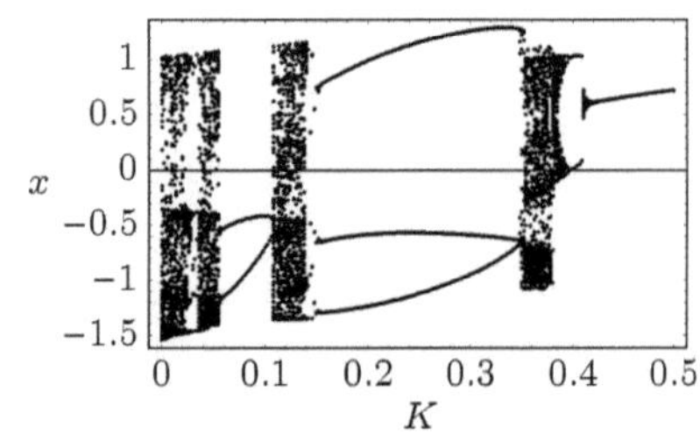

Fig. 8.11 $x - K$ diagram for $\alpha = 2$, $\varepsilon\delta = 0.3$, $k = 3$

where K is a constant called the weight of perturbation and τ is the time-delay. The system with the controlling function is

$$\ddot{x} + \delta\dot{x} - x + x\,|x|^{\alpha-1} = \gamma\cos(\omega t) + K[\dot{x}(t-\tau) - \dot{x}(t)]. \tag{8.40}$$

The control procedure depends on the parameters K and τ. As the system is T-periodic and the goal is to stabilize a forced T -periodic solution, we choose $\tau = T$. The weight K of the feedback is adjusted by numerical experiment.

The control procedure is applied to the three examples mentioned in the previous section. For $\alpha = 2$, $\delta = 0.3$, $k = 3$ the chaotic motion is transformed into a periodical one when the control parameter is $K = 0.5$ and the time delay $\tau = \pi$ (Fig. 8.8). The chaos control is achieved with the weight of perturbation $K = 0.5$ and the time delay $\tau = 2\pi$ in the system where $\alpha = 4/3$, $\delta = 0.25$, $k = 1.3$ (Fig. 8.9), and also $\alpha = 7/5$, $\delta = 0.3$, $k = 1.203$ (Fig. 8.10).

The numerical experiment shows that the control procedure is seriously sensitive to the value of parameter K. In Fig. 8.11, the $x - K$ diagram for $\alpha = 2$, $\delta = 0.3$, $k = 3$ is plotted.

It is evident that choosing an appropriate weight K of the feedback the stabilization is achieved, the unstable periodic orbit becomes stable and the chaotic motion is transformed into periodical one.

8.1.6 Conclusion

The following can be concluded:

1. In dynamical systems with nonlinear order of displacement, and small damping and excitation, and where the homoclinic orbit exists, it is shown that chaos may appear.

2. For the system with nonlinear order of displacement the exact analytical formulation of the homoclinic orbit is obtained. The homoclinic orbit strongly depends on the order of nonlinearity α. The homoclinic orbit of the cubic Duffing oscillator is one special case and can be obtained from the general analysis given.

3. Melnikov's procedure adopted for systems with any nonlineaer order of displacement gives analytical criteria for chaos depending on the relationship between the (δ/γ) and critical $(\delta/\gamma)_c$ parameter value. The exact analytical critical parameter values $(\delta/\gamma)_c$ can be obtained dependently on the order of nonlinearity α.

4. The numerical simulation shows the dependence of the motion type (periodic and chaotic) on the variation of parameter values. The phase plane diagrams and the Poincaré maps for certain parameter values give the results which were expected according to the analytical procedure.

5. The Lyapunov exponents and the bifurcation diagrams show the period-doubling bifurcation and the transformation from periodic to chaotic motion and vice versa. The results obtained show the correctness of the analysis.

6. In systems with any nonlinear order of displacement (integer or non-integer) chaos control by means of delayed self controlling feedback possible, and the motion is transformed from chaotic to periodic.

8.2 Chaos in Non-ideal Oscillator

The first kind of non-ideal problem to arise in the current literature is the so called Sommerfeld effect, discovered in 1904, commented in a book of Kononenko (1969) and described in the book of Nayfeh and Mook (1976). To explain the motion Tsuchida et al. (2003, 2005) modeled the non-ideal system with two coupled oscillators. The mathematical description is a system of two coupled nonlinear second order ordinary differential equations. Analyzing the solutions of the equations the motion properties of the system are investigated. It is concluded that for the certain values of parameters of the system in the resonant regime the jump phenomena occurs and chaos appears. In the papers of Cveticanin (1993), Zukovic and Cveticanin (2007), Mahmoud et al. (2001) and also Dantas and Balthazar (2006), the criteria for chaotic motion in the two-degree-of-freedom systems are determined. The obtained results refer only to the model where the main oscillator is of unstable Duffing type, often referred to as a buckled beam model.

In this paper an extension to the previous assumption is introduced and the stable Duffing oscillator with non-ideal excitation is considered. The model is a motor-structure system where the energy source is non-ideal. The structure represents a stable Duffing oscillator which is explored by Ueda (1985). Further investigations in dynamics of the stable Duffing oscillator under static plus large periodic excitation are presented in the paper of Fang and Dowell (1987) and Pezeshki and Dowell (1988). Using these obtained results and the investigation methodology for unstable Duffing oscillator with non-ideal excitation the dynamics of stable Duffing non-ideal system is considered. The motor with limited power supply and eccentric mass plays the part of the non-ideal perturbation source. The driving of the system comes from an unbalanced rotor linked to the oscillator fed by an electric motor. The driven system is taken as a consequence of the dynamics of the whole system (oscillator plus rotor). The system in the resonant region is specially considered. The purpose of the paper is to give the suggestion and recommendation to the designers and engineers how to drive the system through resonance.

8.2.1 *Modeling of the System*

Let us consider an electric motor operating on a structure. Figure 8.12 shows the model for this kind of problem, which is the goal of investigation of this paper. The structure of mass M is connected to a fixed basement by a nonlinear spring and a linear viscous damper (damping coefficient c). The nonlinear spring stiffness is given by $k_1x + k_2x^3$, where x denotes the structure displacement with respect to some equilibrium position in the absolute reference frame. The motion of the structure is due to an in-board non-ideal motor driving an unbalanced rotor. We denote by φ the

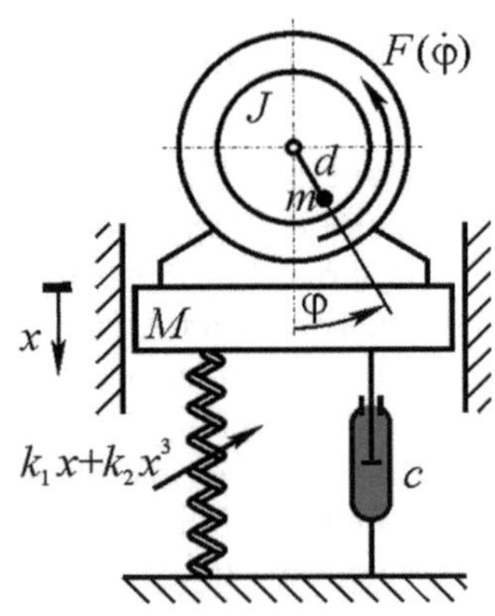

Fig. 8.12 Model of the motor-structure non-ideal system

angular displacement of the rotor unbalance, and model it as a particle of mass m and radial distance d from the rotating axis. The moment of inertia of the rotating part is J. For the resonant case the structure has an influence on the motor input or output. The forcing function is dependent of the system it acts on and the source is of non-ideal type.

The non-ideal problem has two-degrees of freedom, represented by the generalized coordinates x and φ. The kinetic energy T, potential energy V and the dissipative function Φ are expressed by

$$T = \frac{1}{2}M\dot{x}^2 + \frac{1}{2}m(\dot{x}^2 + d^2\dot{\varphi}^2 - 2d\dot{x}\dot{\varphi}\sin\varphi) + \frac{1}{2}J\dot{\varphi}^2, \tag{8.41}$$

$$V = \frac{1}{2}k_1x^2 + \frac{1}{4}k_2x^4 - (M+m)gx - mgd\cos\varphi, \quad \Phi = \frac{1}{2}c\dot{x}^2. \tag{8.42}$$

A dot denotes differentiation with respect to time t.

The differential equations of motion have the form

$$\begin{aligned}\ddot{x}(M+m) + c\dot{x} - md(\ddot{\varphi}\sin\varphi + \dot{\varphi}^2\cos\varphi) + k_1x + k_2x^3 \\ = (M+m)g, (J+md^2)\ddot{\varphi} - md\ddot{x}\sin\varphi + mgd\sin\varphi = F(\dot{\varphi}),\end{aligned} \tag{8.43}$$

where $F(\dot{\varphi}) = L(\dot{\varphi}) - R(\dot{\varphi})$ and $L(\dot{\varphi})$ and $R(\dot{\varphi})$ are the driving and resisting torques, respectively. The linear moment-speed relation (see Dimentberg et al. 1997) is often used

$$F(\dot{\varphi}) = \mathcal{M}\Big(1 - \frac{\dot{\varphi}}{\Omega}\Big), \tag{8.44}$$

where $\mathcal{M}$ and Ω are constant values.

It is convenient to normalize the coordinates and time according to

$$x \longrightarrow y = \frac{\Omega^2}{g}x, \qquad t \longrightarrow \tau = \Omega t, \tag{8.45}$$

where g is the gravity constant. By introducing (8.45) the differential equations (8.43) transform into

$$\begin{aligned}y'' + \alpha y' + p^2y + \gamma y^3 = 1 + q(\varphi''\sin\varphi + \varphi'^2\cos\varphi), \\ \varphi'' = \eta y''\sin\varphi - \eta\sin\varphi + F(1 - \varphi'),\end{aligned} \tag{8.46}$$

where

$$\begin{aligned}p = \frac{\omega^*}{\Omega}, \quad \omega^{*2} = \frac{k_1}{M+m}, \quad \gamma = \frac{k_3g^2}{(M+m)\Omega^6}, \quad \alpha = \frac{c}{\Omega(M+m)}, \\ q = \frac{m}{M+m}\frac{d\Omega^2}{g}, \quad \eta = \frac{gmd}{\Omega^2(J+md^2)}, \quad F = \frac{\mathcal{M}}{\Omega^2(J+md^2)},\end{aligned} \tag{8.47}$$

and prime denotes differentiation with respect to τ. The differential equations (8.46) are nonlinear and coupled.

8.2.2 Asymptotic Solving Method

In the regime near resonant the difference between the excitation frequency is close to the natural frequency. Introducing a small parameter $\varepsilon << 1$ one can write

$$\omega - p = \varepsilon\sigma, \tag{8.48}$$

where $\omega = \varphi'$. Expressing parameters of equations (8.47) by

$$\alpha = \varepsilon\tilde{\alpha}, \qquad \gamma = \varepsilon\tilde{\gamma}, \qquad q = \varepsilon\tilde{q}, \qquad \eta = \varepsilon\tilde{\eta}, \qquad F = \varepsilon\tilde{F}, \tag{8.49}$$

the differential equations of motions have the form

$$\begin{aligned} z'' + p^2 z &= \varepsilon\Big[\tilde{q}(\varphi'' \sin\varphi + \varphi'^2 \cos\varphi) - \tilde{\alpha} z' - \tilde{\gamma}\Big(z + \frac{1}{p}\Big)^3\Big], \\ \varphi'' &= \varepsilon[\tilde{\eta} z'' \sin\varphi - \tilde{\eta}\sin\varphi + \tilde{F}(1 - \varphi')], \end{aligned} \tag{8.50}$$

where the new variable is

$$z = y - \frac{1}{p^2}. \tag{8.51}$$

Expressing z'' and φ'' from the equations (8.50) and assuming only the terms to $O(\varepsilon^2)$ we obtain

$$\begin{aligned} z'' + p^2 z &= \varepsilon\Big[\tilde{q}\varphi'^2 \cos\varphi - \tilde{\alpha} z' - \tilde{\gamma}\Big(z + \frac{1}{p}\Big)^3\Big] + \varepsilon^2 \ldots, \\ \varphi'' &= \varepsilon\Big[\tilde{F}(1 - \varphi') - \tilde{\eta}(1 + p^2 z)\sin\varphi\Big] + \varepsilon^2 \ldots. \end{aligned} \tag{8.52}$$

Following the reference Warminski et al. (2001)

$$z = A\cos(\varphi + \psi), \tag{8.53}$$

where A and ψ are the new coordinates. The first derivative is

$$z' = -Ap\sin(\varphi + \psi), \tag{8.54}$$

when

$$A'\cos(\varphi + \psi) - A(\omega + \psi' - p)\sin(\varphi + \psi) = 0. \tag{8.55}$$

Determining the second derivative of z and substituting (8.53) and (8.54) into (8.52) we obtain

$$\begin{aligned} &- A'p\sin(\varphi+\psi) - Ap(\omega+\psi')\cos(\varphi+\psi) + p^2 A\cos(\varphi+\psi) \\ &= \varepsilon\Big\{\tilde{q}\omega^2\cos\varphi + Ap\tilde{\alpha}\sin(\varphi+\psi) - \tilde{\gamma}\Big[A\cos(\varphi+\psi) + \frac{1}{p}\Big]^3\Big\}, \\ \omega' &= \varepsilon\{\tilde{F}(1-\omega) - \tilde{\eta}[1 + Ap^2\cos(\varphi+\psi)]\sin\varphi\}. \end{aligned} \tag{8.56}$$

Equations (8.55) and (8.56) lead to the derivatives A', ψ', φ' and ω'

$$\begin{aligned} A' &= -\frac{\varepsilon}{p}\sin(\varphi+\psi)\Big\{\tilde{q}\omega^2\cos\varphi + Ap\tilde{\alpha}\sin(\varphi+\psi) - \tilde{\gamma}\Big[A\cos(\varphi+\psi) + \frac{1}{p}\Big]^3\Big\}, \\ A\psi' &= \varepsilon A\sigma - \frac{\varepsilon}{p}\cos(\varphi+\psi)\Big\{\tilde{q}\omega^2\cos\varphi + Ap\tilde{\alpha}\sin(\varphi+\psi) \\ &\quad -\tilde{\gamma}\Big[A\cos(\varphi+\psi) + \frac{1}{p}\Big]^3\Big\}, \end{aligned} \tag{8.57}$$

$$\omega' = \varepsilon\{\tilde{F}(1-\omega) - \tilde{\eta}[1 + Ap^2\cos(\varphi+\psi)]\sin\varphi\}, \quad \varphi' = \omega. \tag{8.58}$$

Unfortunately, the closed form analytical solution for (8.58) is complicate to be obtained and the numerical methods are convenient.

8.2.3 Stability and Sommerfeld Effect

Introducing the averaging procedure (Bogolubov and Mitropolski 1974)

$$A' = \frac{\varepsilon}{2\pi}\int_0^{2\pi} f_A d\varphi, \quad A\psi' = \frac{\varepsilon}{2\pi}\int_0^{2\pi} f_\psi d\varphi, \quad \omega' = \int_0^{2\pi} f_\omega d\varphi, \tag{8.59}$$

where

$$\begin{aligned} f_A &= -\frac{1}{p}\sin(\varphi+\psi)\Big\{\tilde{q}\omega^2\cos\varphi + Ap\tilde{\alpha}\sin(\varphi+\psi) - \tilde{\gamma}\Big[A\cos(\varphi+\psi) + \frac{1}{p}\Big]^3\Big\}, \\ f_\psi &= A\sigma - \frac{1}{p}\cos(\varphi+\psi)\Big\{\tilde{q}\omega^2\cos\varphi + Ap\tilde{\alpha}\sin(\varphi+\psi) - \tilde{\gamma}\Big[A\cos(\varphi+\psi) + \frac{1}{p}\Big]^3\Big\}, \\ f_\omega &= \tilde{F}(1-\omega) - \tilde{\eta}[1 + Ap^2\cos(\varphi+\psi)]\sin\varphi, \end{aligned} \tag{8.60}$$

the differential equations (8.58) are transformed into

$$\omega' = \varepsilon\Big[\tilde{F}(1-\omega) + \frac{1}{2}\tilde{\eta}Ap^2 \sin\psi\Big],$$
$$A' = -\frac{\varepsilon}{2}\Big(A\tilde{\alpha} + \frac{\tilde{q}\omega^2}{p}\sin\psi\Big),$$
$$\psi' = \varepsilon\sigma - \frac{\varepsilon}{2p}\Big(\frac{\tilde{q}\omega^2}{A}\cos\psi - \frac{3\tilde{\gamma}A^2}{4} - \frac{3\tilde{\gamma}}{p^2}\Big). \tag{8.61}$$

For the steady state response, equations (8.61) have the form

$$\tilde{F}(1-\omega_S) + \frac{1}{2}\tilde{\eta}A_S p^2 \sin\psi_S = 0,$$
$$A_S\tilde{\alpha} + \frac{\tilde{q}\omega_S^2}{p}\sin\psi_S = 0,$$
$$\sigma - \frac{1}{2p}\Big(\frac{\tilde{q}\omega_S^2}{A_S}\cos\psi_S - \frac{3\tilde{\gamma}A_S^2}{4} - \frac{3\tilde{\gamma}}{p^2}\Big) = 0, \tag{8.62}$$

where S denotes the steady state values. Combining the second and the third equation (8.62) we obtain

$$A_S^2\Big\{\frac{\tilde{\alpha}^2}{4} + \Big[\sigma + \frac{3\tilde{\gamma}}{2p}\Big(\frac{A_S^2}{4} + \frac{1}{p^2}\Big)\Big]^2\Big\} = \Big(\frac{\tilde{q}\omega_S^2}{2p}\Big)^2, \tag{8.63}$$

while combining the first and the second equation (8.62) yields

$$\tilde{F}\tilde{q}\omega_S^2(1-\omega_S) = \frac{1}{2}\tilde{\alpha}\tilde{\eta}A_S^2 p^3. \tag{8.64}$$

Expressing A_S in (8.64) and substituting into (8.63) the $\omega_S(\tilde{F})$ function is obtained

$$8\tilde{F}(1-\omega_S)\Big\{\frac{\tilde{\alpha}^2}{4} + \Big[\omega_S - p + \frac{3\tilde{\gamma}}{2p^5}\Big(\frac{\tilde{q}\omega_S^2\tilde{F}(1-\omega_S)}{2\tilde{\alpha}\tilde{\eta}p} + 1\Big)\Big]^2\Big\} - \tilde{q}\tilde{\alpha}\tilde{\eta}p\omega_S^2 = 0. \tag{8.65}$$

The number of real solution is one, two or three and it depends on the control parameter $\tilde{F}$. To determine which of the solutions actually correspond to a realizable motion, we need to consider the stability of the solutions. We determine the stability by determining the nature of the singular points of (8.62). To accomplish this, we let

$$A = A_S + A_1, \qquad \psi = \psi_S + \psi_1, \qquad \omega = \omega_S + \omega_1. \tag{8.66}$$

Substituting (8.66) into (8.62) and neglecting all but the linear terms in A_1, ψ_1, ω_1 we obtain

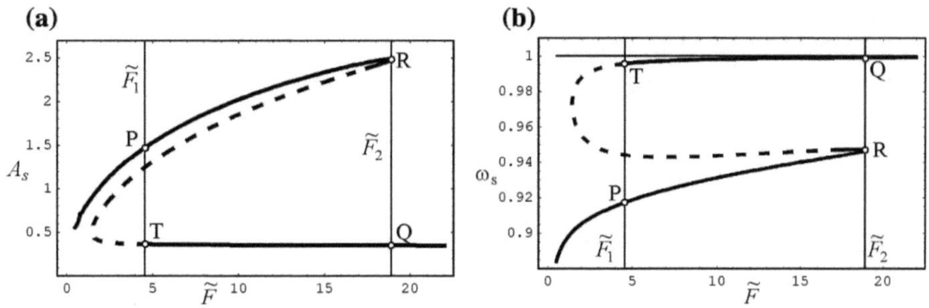

Fig. 8.13 Jump effect for: **a** amplitude-control parameter curve, **b** frequency-control parameter curve

$$
\begin{aligned}
\omega_1' &= \varepsilon\Big[-\tilde{F}\omega_1 + \frac{1}{2}\tilde{\eta}A_1 p^2 \sin\psi_S + \frac{\psi_1}{2}\tilde{\eta}A_S p^2 \cos\psi_S\Big],\\
A_1' &= -\frac{\varepsilon}{2}\Big(A_1\tilde{\alpha} + \psi_1\frac{\tilde{q}\omega_S^2}{p}\cos\psi_S + 2\frac{\tilde{q}\omega_S\omega_1}{p}\sin\psi_S\Big),\\
\psi_1' &= -\omega_1 - \frac{\varepsilon}{2p}\Big(2\frac{\tilde{q}\omega_S\omega_1}{A_S}\cos\psi_S - A_1\frac{\tilde{q}\omega_S^2}{A_S^2}\cos\psi_S\\
&\quad -\psi_1\frac{\tilde{q}\omega_S^2}{A_S}\sin\psi_S - \frac{3\tilde{\gamma}A_S A_1}{2}\Big).
\end{aligned}
\tag{8.67}
$$

Equations (8.67) are linear and have the solution in the form

$$
(A_1, \psi_1, \omega_1) = (A_{10}, \psi_{10}, \omega_{10})\exp(\lambda\tau),
$$

where λ is an eigenvalue of the coefficient of matrix. The solutions are stable and hence corresponding motions reliable, if the real part of each eigenvalue is negative or zero.

For the parameter values

$$
\varepsilon = 0.1, \quad \tilde{\alpha} = 0.2, \quad \tilde{\gamma} = 0.1, \quad \tilde{\eta} = 1.0, \quad \tilde{q} = 0.5, \tag{8.68}
$$

using the relations (8.63) and (8.64), i.e., (8.65) the response-control parameter $\tilde{F}$ and the frequency of vibration-control parameter $\tilde{F}$ diagrams are plotted (Fig. 8.13).

In Fig. 8.14 the frequency-response curves obtained numerically and analytically by solving the equations (8.63) and (8.64) are plotted. Comparing the solutions it can be concluded that the difference between "exact" numerical solution and approximate analytical solution is negligible. Analyzing the obtained curves and the relation (8.64) it is evident that the curve depends on the control parameter $\tilde{F}$. For $\omega_S = 2/3$ the maximal response

$$
A_{S\max} = \frac{2}{3p}\sqrt{\frac{2\tilde{q}}{3\tilde{\alpha}\tilde{\eta}p}\tilde{F}},
$$

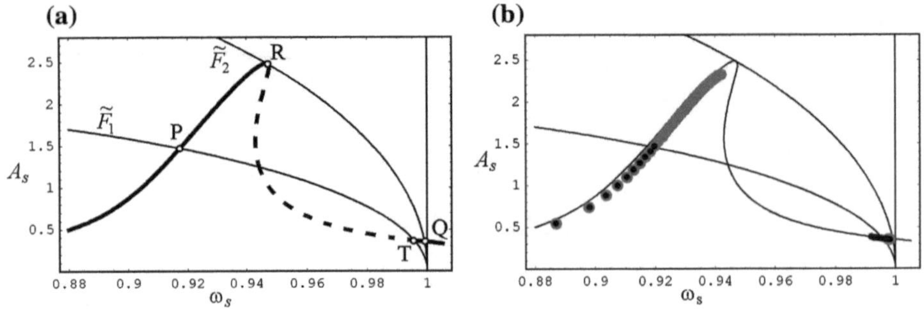

Fig. 8.14 Frequency-response curve obtained: **a** analytically, **b** numerically

is a function of $\tilde{F}$, but is not dependent on the parameter of nonlinearity $\tilde{\gamma}$.

Figures 8.13 and 8.14 show the characteristic points in the diagrams: point R and point T. Between these points the solution is unstable while all those outside this region are stable. Point R is the peak in $A_S - \omega_S$, $A_S - \tilde{F}$ and $\tilde{F} - \omega_S$ diagrams. Analyzing the relation (8.65) and equating the first derivative to zero ($d\tilde{F}/d\omega_S = 0$)

$$0 = -8\tilde{F}\Big\{\frac{\tilde{\alpha}^2}{4} + \Big[\omega_S - p + \frac{3\tilde{\gamma}}{2p^5}\Big(\frac{\tilde{q}\omega_S^2\tilde{F}(1-\omega_S)}{2\tilde{\alpha}\tilde{\eta}p} + 1\Big)\Big]^2\Big\} - 2\tilde{q}\tilde{\alpha}\tilde{\eta}p\omega_S$$
$$+2\tilde{F}(1-\omega_S)\Big[\omega_S - p + \frac{3\tilde{\gamma}}{2p^5}\Big(\frac{\tilde{q}\omega_S^2\tilde{F}(1-\omega_S)}{2\tilde{\alpha}\tilde{\eta}p} + 1\Big)\Big]\Big[1 + \frac{3\tilde{\gamma}}{2p^5}\frac{\tilde{q}\tilde{F}(2\omega_S - 3\omega_S^2)}{2\tilde{\alpha}\tilde{\eta}p}\Big].$$

the peak is obtained. The locus of the peak (point R) is according to (8.63)

$$\omega_{SR} = p - \frac{3\varepsilon\tilde{\gamma}}{2p}\Big(\frac{A_{SR}^2}{4} + \frac{1}{p^2}\Big), \tag{8.69}$$

and the amplitude is $A_{SR} = \tilde{q}\omega_{SR}^2/\tilde{\alpha}p$. Substituting (8.69) into the relation (8.64) we obtain the value of the control parameter for the peak amplitude

$$\tilde{F}_R = \frac{\tilde{\eta}p\tilde{q}\omega_{SR}^2}{2\tilde{\alpha}(1-\omega_{SR})}. \tag{8.70}$$

Substituting the parameter values (8.68) into (8.70) the numeric value of the control parameter is $\tilde{F}_R = 18.246$. For that calculated value of the control parameter and the known value of the parameter of nonlinearity ($\tilde{\gamma} = 0.1$) the amplitude and frequency of Q are obtained by solving the relations (8.63) and (8.64). For the control parameter $\tilde{F}_T = 4.57$ two stable solutions are obtained: T with A_{ST} and ω_{ST} and P with A_{SP} and ω_{SP}. We note that there are gaps in the diagrams (see Figs. 8.13 and 8.14) where no steady state response exists. The gaps are not the same for increasing and decreasing the control parameter $\tilde{F}$. Increasing the control parameter causes the increase of the amplitude and frequency of vibrations to R. Then the effect of jump to smaller

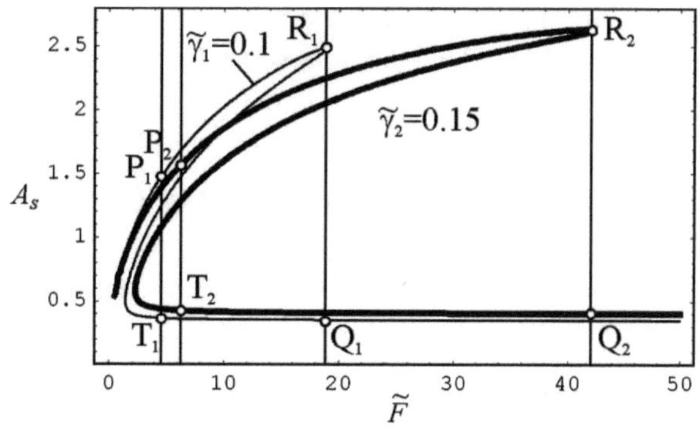

Fig. 8.15 Frequency-control parameter curve for various values of parameter of non-linearity

amplitude and higher frequency in Q appears. Decreasing the control parameter $\tilde{F}$ decreases the frequency and increases the amplitude to T and then jump into P with higher amplitude and smaller frequency occurs. The same hysteresis is seen in Fig. 8.14. This phenomena of jump is called the Sommerfeld effect in non-ideal systems.

To eliminate the Sommerfeld effect for the certain value of the control parameter $\tilde{F}_R$ the parameter of nonlinearity has to be calculated. The equation (8.63) indicates the peak amplitude

$$A_{SR} = \frac{\tilde{q}}{\tilde{\alpha} p} \omega_{SR}^2. \tag{8.71}$$

The function $A_{SR}(\omega_{SR})$ is independent on the parameter of nonlinearity $\tilde{\gamma}$. The relation (8.64) gives the dependence of amplitude A_S on ω_S for various values of control parameter $\tilde{F}$. Using (8.71) and (8.64) we obtain

$$\tilde{F}_R(1 - \omega_{SR}) - \frac{1}{2}\tilde{\eta}\frac{\tilde{q}}{\tilde{\alpha}}\omega_{SR}^2 p = 0. \tag{8.72a}$$

ω_{SR} is the solution of (8.72a) for the control parameter $\tilde{F}_R$. Substituting ω_{SR} into (8.71) the amplitude A_{SR} is calculated. Introducing the so obtained values ω_{SR} and A_{SR} into (8.63) the value of nonlinear parameter $\tilde{\gamma}_R$ is determined

$$\tilde{\gamma}_R = \frac{8}{3}\frac{p^3}{A_{SR}^2 p^2 + 4}(p - \omega_{SR}). \tag{8.73}$$

For $\tilde{\gamma} < \tilde{\gamma}_R$ only one stable solution exists for the control parameter $\tilde{F}_R$ and the motion is without jump.

In Fig. 8.15 the $A_S - \tilde{F}$ curve for two values of parameter of nonlinearity $\tilde{\gamma}$ is shown. It is seen that for $\tilde{\gamma} = 0.1$ the bending of the curve is smaller than for $\tilde{\gamma} = 0.15$. For $\tilde{F}_R = 42.1$ and $\tilde{\gamma}_R = 0.15$ two real values of amplitude exist and for $\tilde{\gamma} = 0.1$ only one. It can be concluded that for the real system with parameters (8.68) which works with control parameter $\tilde{F}_R = 42.1$ the parameter of nonlinearity of the structure has to be $\tilde{\gamma} < 0.15$. Then, the Sommerfeld effect is eliminated.

8.2.4 Numerical Simulation and Chaotic Behavior

To prove the analytically obtained results the numerical experiment is done. The system (8.46) is rewritten in the form

$$\begin{aligned}
y_1' &= y_2, \\
y_2' &= \frac{1}{1 - q\eta \sin^2 y_3}\Big\{ -\alpha y_2 - p y_1 - \gamma y_1^3 + 1 + q \sin y_3 \Big[q y_4^2 \cos y_3 \\
&\quad -\eta \sin y_3 + F(1 - y_4)\Big]\Big\}, \\
y_3' &= y_4, \\
y_4' &= \frac{\eta \sin y_3}{1 - q\eta \sin^2 y_3}\Big(-\alpha y_2 - p y_1 - \gamma y_1^3 + 1 + q y_4^2 \cos y_3 \\
&\quad +q \sin y_3(-\eta \sin y_3 + F(1 - y_4))\Big) \\
&\quad -\eta \sin y_3 + F(1 - y_4).
\end{aligned} \tag{8.74}$$

A number of numerical simulations are done for various control parameter F. Applying the Runge–Kutta numerical procedure with the fixed step lenght the system of four first order differential equations (8.74) is solved. The results are plotted in Fig. 8.16. The phase space of the system has out of four dimensions, but we were chiefly interested in position of the oscillator itself. We also plotted the Poincaré map which represents the surface of section $(y_1(\tau_n), y_2(\tau_n))$. The points $(y_1(\tau_n), y_2(\tau_n))$ are captured for $\tau_n = nT$, where $n = 1, 2, \ldots,$ with period $T = 2\pi/\overline{\Omega}$. The angular velocity $\overline{\Omega}$ is obtained numerically

$$\overline{\Omega} = \frac{\varphi(\tau_1) - \varphi(0)}{\tau_1} = \frac{y_3(\tau_1) - y_3(0)}{\tau_1}, \tag{8.75}$$

where τ_1 is a long time period for numerical calculation.

For the parameter values of the system, mentioned in the previous section, and the control parameter $F = 7.0$ the trajectory in the phase space and the Poincaré map are plotted (Fig. 8.16a). The motion is periodic with period 1 and frequency $\overline{\Omega} = 0.999$. For $F = 15$ the periodic motion is with period 2 (Fig. 8.16b) and for $F = 21$ with period 3 (Fig. 8.16c). Increasing the value of the control parameter the motion is periodic but the period number increases, too (see Fig.8.16d–8.16f). For high values of control parameter ($F = 80$ and $F = 10,000$) the strange attractor is obtained (Fig. 8.16g, h). The existence of strange attractor signifies chaos which is evident only if the certain criteria for the maximal (local) Lyapunov exponent are

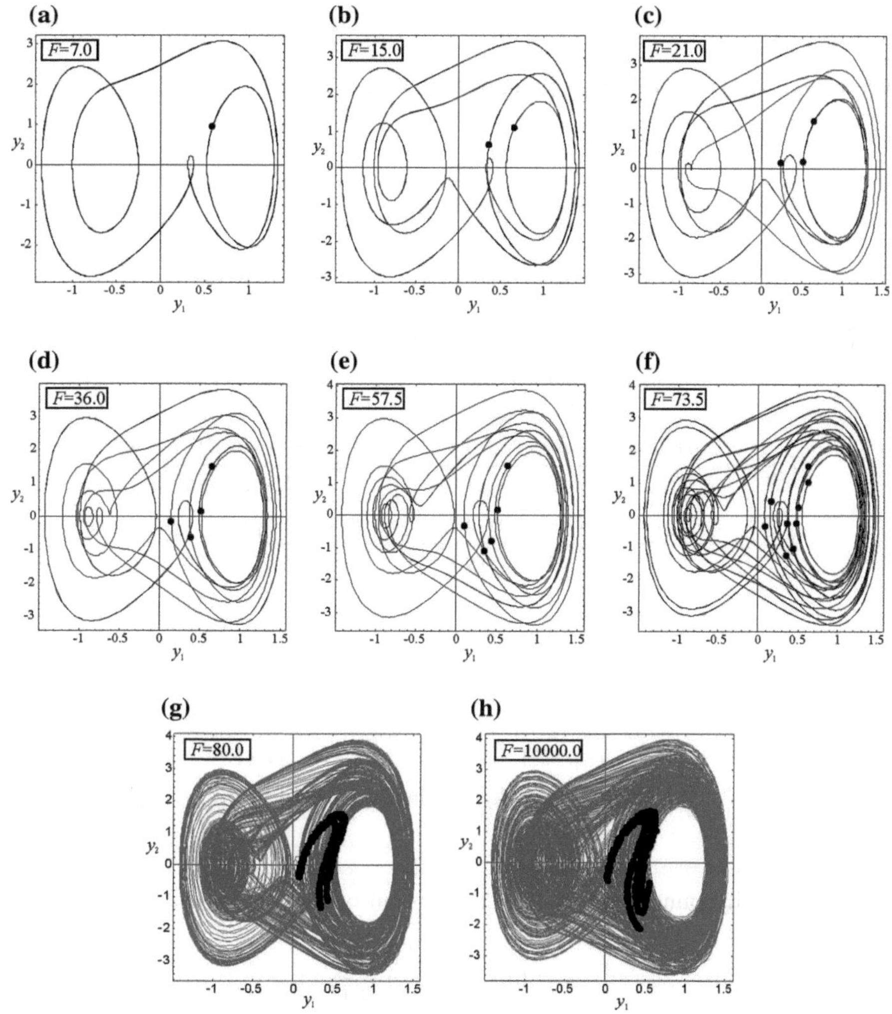

Fig. 8.16 Trajectories in the phase space for various values of the control parameter: **a** periodic solution with period 1, **b** periodic solution with period 2, **c** periodic solution with period 3, **d** periodic solution with period 4, **e** periodic solution with period 5, **f** periodic solution with period 9, **g, h** chaotic solution

satisfied. For computational reasons the vector notation for the system of equations (8.74) is introduced

$$Y' = G(Y, P), \tag{8.76}$$

where $Y = [y_1, y_2, y_3, y_4]^T$ is the state space vector, G=$[g_1, g_2, g_3, g_4]$, $P(p, \alpha, \eta, q, F)$ is set of parameters and $[...]^T$ is denoting transpose. The equations for small deviations δY from the trajectory $Y(t)$ are

$$\delta Y' = L_{ij}(Y(t))\delta Y, \quad i, j = 1, 2, \ldots 4, \tag{8.77}$$

where $L_{ij} = \partial g_i / \partial y_j$ is the Jacobian matrix of derivatives

$$\begin{aligned}
L_{11} &= L_{13} = L_{14} = 0, \quad L_{12} = 1, \\
L_{21} &= -\frac{p + 3\alpha y_1^2}{1 - q\eta \sin^2 y_3}, \quad L_{22} = -\frac{\alpha}{1 - q\eta \sin^2 y_3}, \\
L_{24} &= \frac{2qy_4 \cos x_3 - q\eta \sin y_3}{1 - q\eta \sin^2 y_3}, \quad L_{42} = -\frac{\alpha\eta \sin y_3}{1 - q\eta \sin^2 y_3}, \\
L_{23} &= \frac{-qy_4^2 \sin y_3 + q\eta(1 - y_4) \cos y_3}{1 - q\eta \sin^2 y_3} - \frac{Q_1}{(1 - q\eta \sin^2 y_3)^2}, \\
L_{31} &= L_{32} = L_{33} = 0, \quad L_{34} = 1, \\
L_{41} &= -\frac{\eta p \sin y_3 + 3\eta\gamma y_1^2 \sin y_3}{1 - q\eta \sin^2 y_3}, \quad L_{44} = \frac{q\eta y_4 \sin(2y_3) - F}{1 - q\eta \sin^2 y_3}, \\
L_{43} &= \frac{q\eta y_4^2 \cos(2y_3) - \eta \cos x_3(\alpha y_2 + py_1 + \gamma y_1^3)}{1 - q\eta \sin^2 y_3} \\
&\quad + \frac{Q_2}{(1 - q\eta \sin^2 y_3)^2},
\end{aligned} \tag{8.78}$$

where

$$\begin{aligned}
Q_1 &= -q\eta \sin(2y_3)[\alpha y_2 + py_1 + \gamma y_1^3 - qy_4^2 \cos y_3 - qF(1 - y_4) \sin y_3, \\
Q_2 &= q\eta \sin(2y_3)[F(1 - y_4) + 0.4q\eta y_4^2 \sin(2y_3) - \eta \sin y_3(\alpha y_2 + py_1 + \gamma y_1^3).
\end{aligned} \tag{8.79}$$

The maximal Lyapunov exponent of the system is then as defined by Wolf et al. (1985)

$$\lambda = \lim \frac{1}{t} \log \frac{\|\delta Y(t)\|}{\|\delta Y(0)\|}. \tag{8.80}$$

In Fig. 8.17 we present the local Lyapunov exponent for the control parameters $F = 80$ and $F = 10,000$. For the both parameter values the positive Lyapunov exponent is calculated. As it is discussed by Wolf et al. (1985), if the system contains at least one positive Lyapunov exponent the motion is chaotic. The two initially nearby orbits (or trajectories) diverge from each other and the separation of two nearby trajectories increases exponentially with time due to sensitive dependence on initial conditions.

The transitation from periodic motion to chaos is by periodic doubling. In Fig. 8.18 the $y_1 - F$ bifurcation diagram is plotted. The bifurcation diagram is obtained for the long time integration of the differential equations (8.74) of the motion. After decay of transient motion a steady state motion is established. The fixed values of the parameters are $\alpha = 0.1$, $\gamma = 9$, $p = 1$, $q = 8.373$, $\eta = 0.05$ and the control parameter

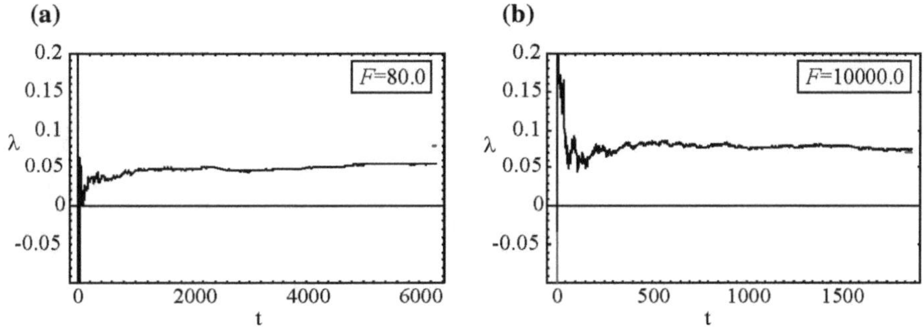

Fig. 8.17 Positive Lyapunov exponent for the control parameters: **a** $F = 80$ and **b** $F = 10,000$

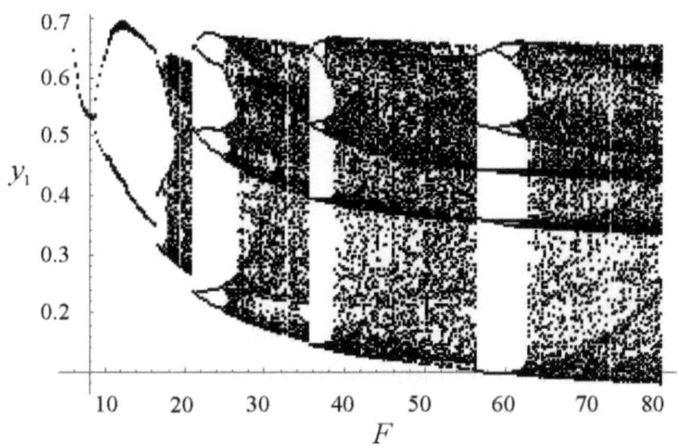

Fig. 8.18 Period doubling bifurcation diagram

we choose to work with is F. By increasing parameter F we found sequences of period doubling bifurcations (Fig. 8.18).

Diagram starts from $F = 6$ where periodic motion with period $n = 1$ exists. This periodic solution bifurcates on $F \approx 8$ onto period $2T$. Further period doubling bifurcation (period 4, 8, 16,...) leads to chaotic motion for $F \approx 20$. After this interval of chaos the periodic motion with period 3 exists. Further increase of parameter F leads to the new period doubling bifurcation with periods 5, 9,...and finally to chaos for $F \in [28, 33]$. The next regions of chaotic motion are for $F \in [40, 56]$ and $F > 62$. Comparing the bifurcation parameter values in Fig. 8.18 with the results of numerical experiment (see Fig. 8.16) it can be concluded that for the control parameter $F = 80$ chaotic motion exists.

8.2.5 Control of Chaos

It is of interest to control the chaos in the motion of the system (8.74) and specially of oscillator. There are many methods for chaos control. Dantas and Balthazar

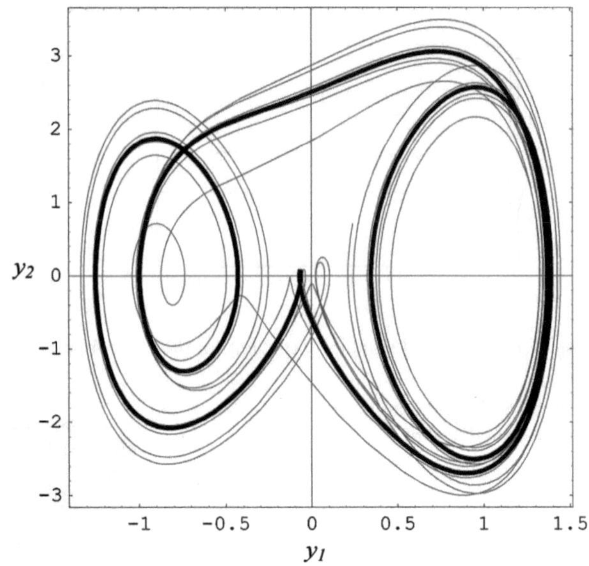

Fig. 8.19 The two-periodic solution: before stabilization (*gray line*) and after stabilization (*black line*)

(2003) show that if we use an appropriate damping coefficient the chaotic behavior is avoided. The method of Pyragas (1992, 1995) is based on the addition of a special kind of time-continuous perturbation (external force control), which does not change the form of the desired unstable periodic solution, but under certain conditions can stabilize it.

For the external force $\Phi(t)$ the model (8.74) becomes

$$\begin{aligned} y_1' &= y_2 + \Phi(t), \\ y_2' &= \frac{1}{1 - q\eta\sin^2 y_3}(-\alpha y_2 - p y_1 - \gamma y_1^3 + 1 + q\sin y_3[q y_4^2 \cos y_3 - \eta \sin y_3 \\ &\quad + F(1 - y_4)), \\ y_3' &= y_4, \\ y_4' &= \frac{\eta \sin y_3}{1 - q\eta\sin^2 y_3}(-\alpha y_2 - p y_1 - \gamma y_1^3 + 1 + q y_4^2 \cos y_3 + q\sin y_3(-\eta\sin y_3 \\ &\quad + F(1 - y_4))) - \eta\sin y_3 + F(1 - y_4). \end{aligned} \tag{8.81}$$

The external force $\Phi(t)$ is defined as

$$\Phi(t) = K[y_{up}(t) - y_1(t)], \tag{8.82}$$

where K is an adjustable weight of the perturbation $\Phi(t)$ and $y_{up}(t)$ is a component of the unstable periodic solution of (8.74) which we wish to stabilize. The function $y_{up}(t)$ is time periodical with period T. For $\Phi(t)$ zero the system has a strange attractor. However, by selecting the constant K one can achieve the desired stabilization.

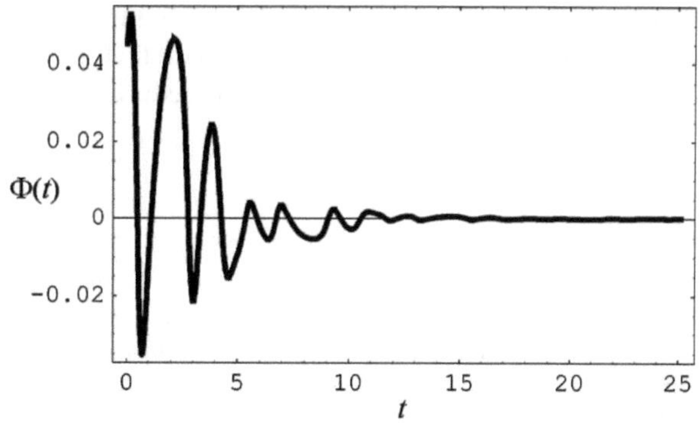

Fig. 8.20 External force-time history diagram

Using the shooting method suggested by Van Dooren and Janssen (1996) the unstable two periodic unstable solution is detected (Fig. 8.19). Varying the value of the constant K in the interval $[0.1, 2]$ it is concluded that for $K \in [0.3, 2]$ the stabilization is achieved (Fig. 8.19).

For $K = 2$ the function $\Phi(t)$ tends to a very small value (Fig. 8.20) and the component y_1, which is the solution after control, is very close to $y_{up}(t)$.

8.2.6 Conclusion

During passage through resonance of the motor-structure system which is modeled as a stable Duffing oscillator with non-ideal excitation severe vibrations appear. The energy of the system is not used for increasing of the rotation velocity, but is spent for vibrations which are harmful. Very often in the motion of the system near resonance the jump phenomena occurs: at the same value of the control parameter of the motor the amplitude of vibration skips to a higher value with lower frequency or to smaller amplitude with higher frequency. The manifestation depends on the direction of variation of the control setting. The jump phenomena and the increase in power required by a source operating near resonance are manifestations of a non-ideal energy source and are referred to as Sommerfeld effect. The Sommerfeld effect contributes to transform a regular vibration to an irregular chaotic one. In the paper it is concluded that in spite of the fact that the structure is modeled as the stable Duffing oscillator chaos appears. In the system the chaos is achieved by period doubling bifurcation. From engineering point of view it is necessary to eliminate the jump effect and the chaotic motion. The elimination of the jump phenomena for the certain control parameter is achieved by using the structure with coefficient of nonlinearity which is smaller than the critical value. Chaos is controlled using the external force control procedure where the added force does not change the form of the desired unstable periodic solution, but under certain conditions can stabilize it.

Comparing the results obtained applying the approximate analytic methods with those obtained numerically it is concluded that the difference is negligible. It proves the correctness of the used analytic procedure.

References

Balthazar, J. M., Mook, D. T., Weber, H. I., Brasil, R. M. I. R. F., Fenili, A., Beltano, D., et al. (2003). An overview on non-ideal vibrations. *Meccanica, 38*, 613–621.

Bogolubov, N. N., & Mitropolski, Ju A. (1974). *Asimptoticheskie metodi v teorii nelinejnih kolebanij*. Moscow: Nauka.

Chacon, R. (1999). General result on chaos suppression for biharmonically driven dissipative systems. *Physics Letters A, 257*, 293–300.

Chen, G. R., & Yu, X. H. (1999). On time-delayed feedback control of chaotic systems. *IEEE Transactions Circuit Systems, 46*, 767–772.

Cveticanin, L. (1993). Extension of Melnikov criterion for the differential equation with complex function. *Nonlinear Dynamics, 4*, 139–152.

Cveticanin, L., & Zukovic, M. (2009). Melnikov's criteria and chaos in systems with fraction order deflection. *Journal of Sound and Vibration, 326*, 768–779.

Dantas, M. H., & Balthazar, J. M. (2003). On the appearance of a Hopf bifurcation in a non-ideal mechanical system. *Mechanics Research Communications, 30*, 493–503.

Dantas, M. H., & Balthazar, J. M. (2006). A comment on a non-ideal centrifugal vibrator machine behavior with soft and hard springs. *International Journal of Bifurcation and Chaos, 16*, 1083–1088.

Dimentberg, M. F., McGovern, L., Norton, R. L., Chapdelaine, J., & Harrison, R. (1997). Dynamics of an unbalanced shaft interacting with a limited power supply. *Nonlinear Dynamics, 13*, 171–187.

Fang, T., & Dowell, E. H. (1987). Numerical simulations of periodic and chaotic responses in a stable Duffing system. *International Journal of Non-Linear Mechanics, 22*, 401–425.

Gradstein, I. S., & Rjizhik, I. M. (1971). *Tablici integralov, summ, rjadov i proizvedenij*. Moscow: Nauka.

Guckenheimer, J., & Holmes, P. (1983). *Nonlinear oscillations, dynamical systems, and bifurcations of vector fields*. New York: Springer.

Holmes, P. (1979). A nonlinear oscillator with a strange attractor. *Philosophical Transactions of the Royal Society of London, Series A, Mathematical and Physical Sciences, 292*, 419–448.

Holmes, P., & Marsden, J. (1981). A partial differential equation with infinitely many periodic orbits: Chaotic oscillations of a forced beam. *Archives for Rational Mechanics and Analysis, 76*, 135–166.

Kononenko, V. O. (1969). *Vibrating system with a limited power supply*. London: Illife.

Mahmoud, G. M., Mohamed, A. A., & Aly, S. A. (2001). Strange attractors and chaos control in periodically forced complex Duffing's oscillators. *Physica A, 292*, 193–206.

Melnikov, V. K. (1963). On the stability of the center for time periodic perturbations. *Transactions of the Moscow Mathematical Society, 12*, 1–57.

Nayfeh, A. H., & Mook, D. T. (1976). *Nonlinear oscillations*. New York: Wiley.

Pezeshki, C., & Dowell, E. H. (1988). On chaos and fractal behaviour in a general Duffing's system. *Physica D, 32*, 194–209.

Pyragas, K. (1992). Continuous control of chaos by self controlling feedback. *Physics Letters A, 170*, 421–428.

Pyragas, K. (1995). Control of chaos via extended delay feedback. *Physics Letters A, 206*, 323–330.

Pyragas, K. (1996). Continuous control of chaos by self-controlling feedback. *Controlling chaos* (pp. 118–123). San Diego: Academic Press.

Pyragas, K. (2001). Control of chaos via an unstable delayed feedback controller. *Physics Review Letters*, *86*, 2265–2268.

Pyragas, K. (2006). Delayed feedback control of chaos. *Philosophical Transactions of the Royal Society A: Mathematical, Physical and Engineering Sciences*, *364*, 2309–2334.

Sandri, M. (1996). Numerical calculation of Lyapunov exponents. *The Mathematical Journal*, *6*, 78–84.

Souza, S. L. T., Caldas, I. L., Viana, R. L., Balthazar, J. M., & Brasil, R. M. L. R. F. (2005a). Impact dampers for controlling chaos in systems with limited power supply. *Journal of Sound and Vibration*, *279*, 955–965.

Souza, S. L. T., Caldas, I. L., Viana, R. L., Balthazar, J. M., & Brasil, R. M. L. R. E. (2005b). Basins of attraction changes by amplitude constraining of oscillators with limited power supply. *Chaos, Solitons and Fractals*, *26*, 1211–1220.

Tseng, W. Y., & Dugundji, J. (1971). Nonlinear vibrations of a buckled beam under harmonic excitation. *Journal of Applied Mechanics*, *38*, 467–476.

Tsuchida, M., Guilherme, K. L., Balthazar, J. M., Silva, G. N., & Cheshankov, B. I. (2003). On regular and irregular vibrations of a non-ideal system with two degrees of freedom: 1:1 resonance. *Journal of Sound and Vibration*, *260*, 949–960.

Tsuchida, M., Guilherme, K. L., & Balthazar, J. M. (2005). On chaotic vibrations of a non-ideal system with two degree of freedom: 1:2 resonance and Sommerfeld effect. *Journal of Sound and Vibration*, *282*, 1201–1207.

Ueda, Y. (1985). Random phenomena resulting from nonlinearity in the system described by Duffing's equation. *International Journal of Non-Linear Mechanics*, *20*, 481–491.

Ushio, T. (1996). Limitation of delayed feedback control in nonlinear discrete-time systems. *EEE Transactions on Circuits and Systems*, *43*, 815–816.

Van Dooren, R., & Janssen, H. (1996). A continuation algorithm for discovering new chaotic motions in forced Duffing systems. *Journal of Computational and Applied Mathematics*, *66*, 527–541.

Wang, R., Deng, J., & Jing, Z. (2006). Chaos, control in Duffing system. *Chaos, Solitons and Fractals*, *23*, 249–257.

Warminski, J., Balthazar, J. M., & Brasil, R. M. L. R. F. (2001). Vibrations of a non-ideal parametrically and self-excited model. *Journal of Sound and Vibration*, *245*, 363–374.

Wiggins, S. (1988). *Global bifurcations and chaos: Analytical methods*. New York: Springer.

Wolf, A., Swift, J., Swinney, H., & Vastano, J. (1985). Determining Lyapunov exponents from a time series. *Physica D*, *16*, 285–317.

Yang, J., & Jing, Z. (2008). Control of chaos in a three-well duffing system. *Chaos, Solitons and Fractals*, *4*(13), 1311–1328. doi:10.1016/j.chaos.2008.05018.

Zhu, J., & Tian, Y.-P. (2005). Necessary and sufficient conditions for stabilizability of discrete-time systems via delayed feedback control. *Physics Letters A*, *343*, 95–107.

Zukovic, M., & Cveticanin, L. (2007). Chaotic responses in a stable Duffing system of non-ideal type. *Journal of Vibration and Control*, *13*, 751–767.

Chapter 9
Vibration of the Axially Purely Nonlinear Rod

In this Chapter the axial vibration of a rod with strong nonlinearity is considered. The model is a clamped-free rod with a strongly nonlinear elastic property. Axial vibration is described by a strong nonlinear partial differential equation with two linear boundary and two initial conditions. Axial vibration of the rod with small nonlinearity is investigated by Cveticanin and Uzelac (1999), but the developed procedure is not suitable for solving strong nonlinear systems. The pure nonlinear strong nonlinear axial vibration is considered in the paper of Cveticanin (2016). A solution of the equation is constructed for special initial conditions by using the method of separation of variables. The partial differential equation is separated into two uncoupled strongly nonlinear second order differential equations. Both equations, with displacement function and with time function are exactly determined. Exact solutions are given in the form of inverse incomplete and inverse complete Beta function. Using boundary and initial conditions, the frequency of vibration is obtained. It has to be mentioned that the determined frequency represents the exact analytic description for the axially vibrating purely nonlinear clamped-free rod. The procedure suggested in this Chapter is applied for calculation of the frequency of the longissimus dorsi muscle of a cow. In this muscle fibers are parallel to the direction of muscle compression. The influence of elasticity order and elasticity coefficient change, which occur due to alterations in the tissue, on the frequency property is tested.

9.1 Model of the Axially Vibrating Rod

Let us consider the vibrations of a clamped-free rod (Fig. 9.1).

The length of the rod is l and its cross-section is S. Cross-section properties of the rod are smaller than its length. Rod has axial vibrations. Deflection u of the rod depends on time t and position x. Elementary part of the rod has length dx and mass

L. Cveticanin, *Strong Nonlinear Oscillators*, Mathematical Engineering,
DOI 10.1007/978-3-319-58826-1_9

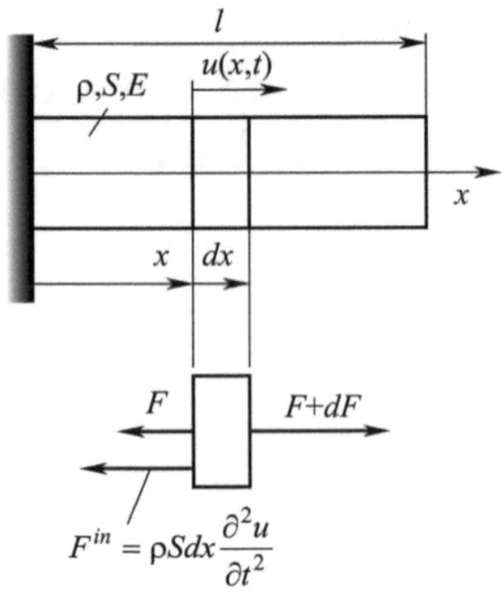

Fig. 9.1 Model of the clamped-free rod

$\rho A dx$, where ρ is density of material. Inertial force of the elementary part is the product of elementary mass and acceleration: $\rho A dx(\partial^2 u/\partial t^2)$. Material of rod has strongly nonlinear properties and its stress-strain rheological model is given as

$$\sigma = E\varepsilon\,|\varepsilon|^{\alpha-1} = E\frac{\partial u}{\partial x}|\frac{\partial u}{\partial x}|^{(\alpha-1)}, \tag{9.1}$$

where E is the elasticity coefficient, ε is the deformation, $|\partial u/\partial x|$ is positive strain and $\alpha \in [1, \infty)$ is the order of non-linearity. Usually, the order of nonlinearity is obtained experimentally and may be integer or non-integer.

Elastic force F is

$$F = \sigma S = ES\frac{\partial u}{\partial x}|\frac{\partial u}{\partial x}|^{(\alpha-1)}. \tag{9.2}$$

Corresponding elementary elastic force dF is

$$dF = ES\frac{\partial}{\partial x}(\frac{\partial u}{\partial x}|\frac{\partial u}{\partial x}|^{\alpha-1}).$$

Using the mathematical definition of the absolute value of the function $|\partial u/\partial x|^{(\alpha-1)} = \sqrt{(\partial u/\partial x)^{2(\alpha-1)}}$ and its derivation, the expression of elastic force is rewritten as

$$dF = ES\frac{\partial^2 u}{\partial x^2}|\frac{\partial u}{\partial x}|^{\alpha-1}.$$

Equating elementary elastic force dF and elementary inertial force, we obtain the equation for longitudinal vibration

$$\rho S\frac{\partial^2 u}{\partial t^2} = ES\alpha\frac{\partial^2 u}{\partial x^2}|\frac{\partial u}{\partial x}|^{\alpha-1}. \tag{9.3}$$

Boundary conditions for the clamped free rod are

$$u(0,t)=0, \qquad F(l,t)\equiv ES(\frac{\partial u}{\partial x}|\frac{\partial u}{\partial x}|^{(\alpha-1)})_{l,t}=0. \tag{9.4}$$

Since the problem (9.3) is nonlinear the superposition principle can not be applied. Only one solution of the partial differential equation (9.3) will be considered for which initial conditions for deflection and velocity are

$$u(x,0)=U(x), \qquad \frac{\partial u(x,0)}{\partial t}=0, \tag{9.5}$$

where $U(x)$ is the initial position function. Mathematical model (9.3) is a second order partial differential equation with strong nonlinearity. To give a valid analysis of the rod motion, it is necessary to solve Eq. (9.3) according to the boundary (9.4) and initial conditions (9.5).

9.2 Solving Procedure

One of the simplest procedures for solving of partial differential equations is the method of separation of variables. This method is applicable for solving of the Eq. (9.3) and the solution is introduced in the form

$$u(x,t)=X(x)T(t), \tag{9.6}$$

where $X(x)$ is a deflection function and $T(t)$ is a time function. Substituting (9.6) into (9.3), we have

$$\rho X\ddot{T}=E\alpha X''T\sqrt{(X'T)^{2(\alpha-1)}}, \tag{9.7}$$

where $\ddot{T}=d^2T/dt^2$, $X'=dX/dx$ and $X_{\text{J}}=d^2X/dx^2$. Introducing the absolute values of functions into (9.7), it is

$$\frac{\rho}{E}X\ddot{T}=\alpha T|T|^{\alpha-1}|X'|^{\alpha-1}X'', \tag{9.8}$$

It is obvious that we can separate variables in (9.8)

$$\frac{\rho}{\alpha E}\frac{\ddot{T}}{T|T|^{\alpha-1}}=\frac{|X'|^{\alpha-1}X''}{X}=-k^2=const. \tag{9.9}$$

and we obtain

$$\ddot{T}+c_1^2T|T|^{\alpha-1}=0, \qquad |X'|^{\alpha-1}X''+k^2X=0, \tag{9.10}$$

where k^2 is an unknown constant value and

$$c_1^2 = k^2 \frac{\alpha E}{\rho}. \tag{9.11}$$

Due to (9.6), the boundary conditions (9.4) transform into

$$X(0) = 0, \qquad X'(l) = 0. \tag{9.12}$$

For $X(x) \neq 0$ the initial conditions are

$$X(x)T(0) = U(x), \qquad \dot{T}(0) = 0. \tag{9.13}$$

There is a significant number of publications dealing with the problem of solving a strong nonlinear differential equation $(9.10)_1$ [see for example Waluya and Horssen (2003), Mickens (2006), Andrianov and Horssen (2006), Pilipchuk (2007), Kovacic and Zukovic (2012) and Blocher et al. (2013)]. In this problem two ordinary strongly nonlinear uncoupled second order differential equations $(9.10)_1$ and $(9.10)_2$ have to be solved. Our major task is to determine frequency of axial vibration of the rod, based on the constants k and c_1.

9.2.1 Solving of the Equation with Displacement Function

Equation $(9.10)_2$ is strongly nonlinear, but has a first integral

$$\frac{|X'|}{\alpha+1}^{\alpha+1} + \frac{k^2 X^2}{2} = const. = K_1, \tag{9.14}$$

where K_1 is an arbitrary constant. Using the relation (9.14) and the boundary condition $(9.12)_2$ we have

$$A = X(l) = \sqrt{\frac{2K_1}{k^2}}, \tag{9.15}$$

where A is the maximal displacement at the position $x = l$. At this point, it is the aim to determine the constant k as a function of the displacement A.

Let us rewrite (9.13) into the form

$$X' = \frac{dX}{dx} = (\alpha+1)^{1/(\alpha+1)} (K_1 - \frac{k^2}{2} X^2)^{1/(\alpha+1)},$$

i.e.,

$$dx = \frac{dX}{(\alpha+1)^{1/(\alpha+1)} (K_1 - \frac{k^2}{2} X^2)^{1/(\alpha+1)}}. \tag{9.16}$$

Integrating (9.16) along the length of the rod and using (9.15), we have

$$\int_0^l dx = \frac{1}{(\alpha+1)^{1/(\alpha+1)} K_1^{1/(\alpha+1)}} \int_0^{X(l)=A} \frac{dX}{(1-\frac{k^2}{2K_1}X^2)^{1/(\alpha+1)}}. \tag{9.17}$$

Introducing the new variable

$$Z = \frac{k^2X^2}{2K_1}, \tag{9.18}$$

and corresponding boundaries of integration

$$Z(0) = \frac{k^2X^2(0)}{2K_1} = 0, \qquad Z(l) = \frac{k^2X^2(l)}{2K_1} = 1, \tag{9.19}$$

relation (9.17) transforms into

$$l = \frac{1}{(\alpha+1)^{1/(\alpha+1)} K_1^{1/(\alpha+1)}} \sqrt{\frac{K_1}{2k^2}} \int_0^1 \frac{dZ}{\sqrt{Z}(1-Z)^{1/(\alpha+1)}}. \tag{9.20}$$

Having in mind the definition of the complete Beta function

$$B(m,n) = \int_0^1 (1-z)^{n-1} z^{m-1} dz, \tag{9.21}$$

we have

$$\int_0^1 \frac{dZ}{\sqrt{Z}(1-Z)^{1/(\alpha+1)}} = B(\frac{\alpha}{\alpha+1}, \frac{1}{2}). \tag{9.22}$$

Substituting (9.22) into (9.20) it is

$$l = B(\frac{\alpha}{\alpha+1}, \frac{1}{2}) \frac{1}{(\alpha+1)^{1/(\alpha+1)} K_1^{1/(\alpha+1)}} \sqrt{\frac{K_1}{2k^2}}. \tag{9.23}$$

Using (9.15), relation (9.23) is transformed into the more familiar form

$$l = \frac{A^{(\alpha-1)/(\alpha+1)}}{(2^\alpha(\alpha+1)k^2)^{1/(\alpha+1)}} B(\frac{\alpha}{\alpha+1}, \frac{1}{2}), \tag{9.24}$$

which gives the value of the constant k as

$$k = \sqrt{\frac{2}{\alpha+1}} \left(\frac{1}{2l} B \left(\frac{\alpha}{\alpha+1}, \frac{1}{2} \right) \right)^{(\alpha+1)/2} A^{(\alpha-1)/2}. \tag{9.25}$$

Constant k depends on the order of nonlinearity α and on constant A. For the linear case, when $\alpha = 1$, the constant k is independent on A and has the value $k = \pi/2l$, as it is already known.

Remark Very often, for computational reasons the complete Beta function is substituted with Gamma function Γ, and it is

$$B(\frac{\alpha}{\alpha+1}, \frac{1}{2}) = \frac{\Gamma(\frac{\alpha}{\alpha+1})\Gamma(\frac{1}{2})}{\Gamma(\frac{\alpha}{\alpha+1} + \frac{1}{2})}, \tag{9.26}$$

where $\Gamma(1/2) = \sqrt{\pi}$.

Substituting (9.25) into (9.13), the displacement function is rewritten as

$$X' = \left(\frac{k^2}{2} (A^2 - X^2 \right) (\alpha+1))^{1/(\alpha+1)}, \tag{9.27}$$

i.e.,

$$X' = \frac{1}{2l} B \left(\frac{\alpha}{\alpha+1}, \frac{1}{2} \right) A \left(1 - \frac{X^2}{A^2} \right)^{1/(\alpha+1)}. \tag{9.28}$$

Let us solve the Eq. (9.28). Introducing the new variable

$$Y = 1 - \frac{X^2}{A^2}, \tag{9.29}$$

Eq. (9.28) transforms into

$$(1-Y)^{-1/2} Y^{-1/(\alpha+1)} dY = - \left(\frac{1}{l} B \left(\frac{\alpha}{\alpha+1}, \frac{1}{2} \right) \right) dx. \tag{9.30}$$

Integrating the both sides

$$\int_0^Y (1-Y)^{-1/2)} Y^{-1/(\alpha+1)} dY = - \int_0^x \frac{1}{l} B \left(\frac{\alpha}{\alpha+1}, \frac{1}{2} \right) dx, \tag{9.31}$$

and using the definition of the incomplete Beta function

$$B_Y(m,n) = \int_0^Y (1-Y)^{n-1} Y dY,$$

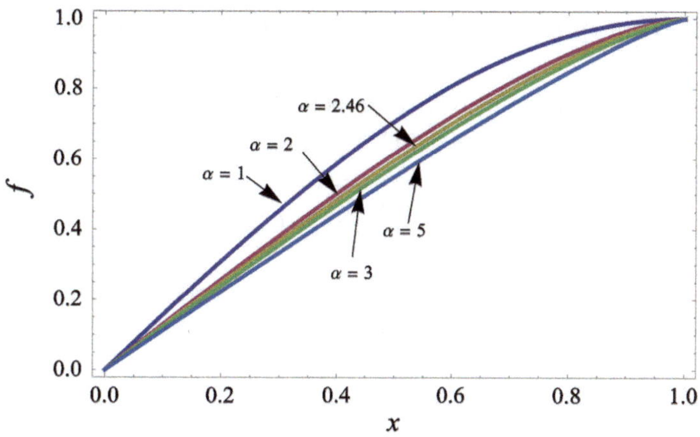

Fig. 9.2 $f - x$ curves for various values of the nonlinear order α

we obtain

$$B_{(1-X^2/A^2)} = -\frac{1}{l} B\left(\frac{\alpha}{\alpha+1}, \frac{1}{2}\right) x. \tag{9.32}$$

Applying the procedure suggested by Didonato (1991) for obtaining the inverse incomplete Beta function B^{-1}, we have

$$X = A\sqrt{(1 - B^{-1}(-(\frac{1}{l} B(\frac{\alpha}{\alpha+1}, \frac{1}{2})x))}. \tag{9.33}$$

Thus, the solution (9.33) has the form

$$X = A f(\alpha, x), \tag{9.34}$$

where

$$f(\alpha, x) = \sqrt{1 - B^{-1}(-(\frac{1}{l} B(\frac{\alpha}{\alpha+1}, \frac{1}{2})x))}. \tag{9.35}$$

It means, that the displacement function is a product of the amplitude A and of the function $f(\alpha, x)$ which is independent on A, but depends on the order of nonlinearity α. In Appendix E, the function $f(\alpha, x)$ for various values of α is presented. Analyzing the data in the Table, we conclude that the function $f(\alpha, x)$ increases from $f(\alpha, 0) = 0$ to $f(\alpha, 1) = 1$. In Fig. 9.2, $f(\alpha, x)$ for various values of α is plotted.

Remark The inverse incomplete Beta function is the solution of the differential equation (9.2) which satisfies the boundary conditions $X(0) = 0$ and $X(l) = 1$.

For the linear case, when $\alpha = 1$, mathematical description of the curve is $f(1, x) = sin(\pi x/2)$. Besides, the higher is the value of α in comparison to 1, the bending of the curve is smaller.

9.2.2 Solving of the Equation with Time Function

The time variable equation $(9.10)_1$ is a strongly nonlinear second order differential equation whose exact analytic solution is (see Cveticanin and Pogany 2012)

$$T = Cca(\alpha, 1, \sqrt{\frac{\alpha+1}{2}} c_1 t C^{(\alpha-1)/2} + \gamma_2), \tag{9.36}$$

where ca is the cosine Ateb function, C and γ_2 are integration constants. Due to the properties of Ateb function (Rosenberg 1963)

$$ca(\alpha, 1, z) = -\frac{2}{\alpha+1} sa(1, \alpha, z), \tag{9.37}$$

where sa is sine Ateb function, the first time derivative of (9.36) is

$$\dot{T} = -\sqrt{\frac{2}{\alpha+1}} c_1 C^{(\alpha-1)/2} sa(1, \alpha, \sqrt{\frac{\alpha+1}{2}} c_1 C^{(\alpha-1)/2} t + \gamma_2). \tag{9.38}$$

Solution (9.36) is periodical. Frequency of the Ateb function, which describes the motion, is $\sqrt{\frac{\alpha+1}{2}} c_1 C^{(\alpha-1)/2}$. It depends on the constant c_1. According to (9.11) and (9.25), the $c_1 - A$ relation follows as

$$c_1 = k\sqrt{\frac{E\alpha}{\rho}} = \sqrt{\frac{2}{\alpha+1}} \sqrt{\frac{E\alpha}{\rho}} \left(\frac{1}{2l} B\left(\frac{\alpha}{\alpha+1}, \frac{1}{2}\right)\right)^{(\alpha+1)/2} A^{(\alpha-1)/2}. \tag{9.39}$$

Substituting (9.39) into (9.36), solution of the time equation $(9.10)_1$ is

$$T = Cca(\alpha, 1, \sqrt{\frac{E\alpha}{\rho}} \left(\frac{1}{2l} B\left(\frac{\alpha}{\alpha+1}, \frac{1}{2}\right)\right)^{(\alpha+1)/2} (AC)^{(\alpha-1)/2} t + \gamma_2). \tag{9.40}$$

For the initial condition $(9.13)_2$ the arbitrary constant is $\gamma_2 = 0$ and the relation (9.40) transforms into

$$T = Cca(\alpha, 1, \sqrt{\frac{E\alpha}{\rho}} \left(\frac{1}{2l} B\left(\frac{\alpha}{\alpha+1}, \frac{1}{2}\right)\right)^{(\alpha+1)/2} (AC)^{(\alpha-1)/2} t). \tag{9.41}$$

Introducing solutions for the displacement and time functions, (9.34) and (9.41), into the initial condition $(9.13)_1$, we have

$$U(x) = ACf(\alpha, x), \tag{9.42}$$

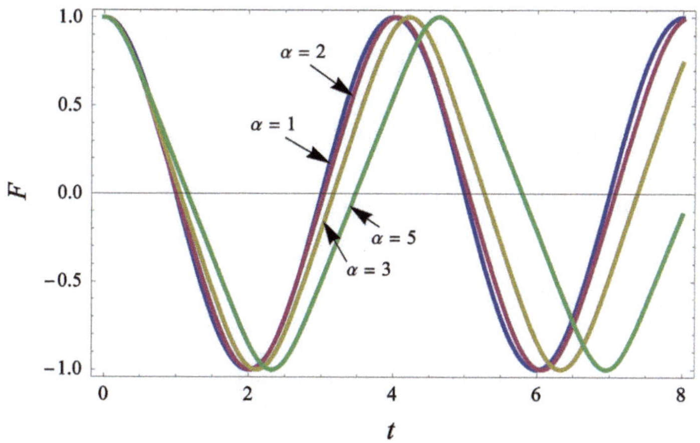

Fig. 9.3 $F - t$ curves for various values of nonlinearity order α and $D = 1$

which for boundary value $x = l$ gives $U(l) = AC$. The value of the constant $D = AC = U(l)$ represents the amplitude of vibration of the rod. Thus,

$$T = Cca(\alpha, 1, \sqrt{\frac{E\alpha}{\rho}} \left(\frac{1}{2l} B\left(\frac{\alpha}{\alpha+1}, \frac{1}{2}\right)\right)^{(\alpha+1)/2} D^{(\alpha-1)/2} t), \tag{9.43}$$

i.e.,

$$T = CF(\alpha, D, t),$$

where

$$F(\alpha, D, t) = ca(\alpha, 1, \sqrt{\frac{E\alpha}{\rho}} \left(\frac{1}{2l} B\left(\frac{\alpha}{\alpha+1}, \frac{1}{2}\right)\right)^{(\alpha+1)/2} D^{(\alpha-1)/2} t). \tag{9.44}$$

In Fig. 9.3, $F(\alpha, D, t)$ curves for a certain initial amplitude ($D = 1$) and various values of order of nonlinearity ($\alpha = 1, 2, 3$ and 5), are plotted.

Period of vibration of the ca Ateb function is longer for $\alpha = 2, 3$ and 5 than for the linear case when $\alpha = 1$.

9.3 Frequency of Axial Vibration

Analyzing the relation (9.44), we conclude that the frequency of the Ateb function is

$$\Omega = \sqrt{\frac{E\alpha}{\rho}} D^{(\alpha-1)/2} \left(\frac{1}{2l} B \left(\frac{\alpha}{\alpha+1}, 1/2 \right) \right)^{(\alpha+1)/2}. \tag{9.45}$$

Using the periodic property of the Ateb function, the period of vibration is obtained as

$$P = \frac{2\Pi}{\Omega} = \sqrt{\frac{\rho}{E\alpha}} D^{(1-\alpha)/2} \frac{2B(\frac{1}{\alpha+1}, \frac{1}{2})}{\left(\frac{1}{2l} B(\frac{\alpha}{\alpha+1}, 1/2)\right)^{(\alpha+1)/2}}, \tag{9.46}$$

where the period of Ateb function is

$$2\Pi = 2B \left(\frac{1}{\alpha+1}, \frac{1}{2} \right). \tag{9.47}$$

Based on the period of vibration (9.46), the frequency of vibration is

$$\omega = \frac{2\pi}{P} = \frac{\pi}{B(\frac{1}{\alpha+1}, \frac{1}{2})} \sqrt{\frac{E\alpha}{\rho}} D^{(\alpha-1)/2} \left(\frac{1}{2l} B \left(\frac{\alpha}{\alpha+1}, 1/2 \right) \right)^{(\alpha+1)/2}. \tag{9.48}$$

Frequency depends on the order of nonlinearity α and parameters of the rod E and ρ. Besides, frequency depends on D, i.e., on the amplitude of vibration, what was not the case for the linear oscillator. For the linear case, when $\alpha = 1$, the frequency of vibration corresponds to the already known value

$$\omega = \frac{\pi}{2l} \sqrt{\frac{E}{\rho}}.$$

We rewrite (9.48) into the form

$$\omega = K_\alpha (\frac{D}{l})^{(\alpha-1)/2} \sqrt{\frac{E}{\rho l^2}}, \tag{9.49}$$

where the so called 'frequency constant' is

$$K_\alpha = \frac{\pi\sqrt{\alpha}}{B(\frac{1}{\alpha+1}, \frac{1}{2})} \left(\frac{1}{2} B \left(\frac{\alpha}{\alpha+1}, 1/2 \right) \right)^{(\alpha+1)/2}, \tag{9.50}$$

i.e., according to (9.26),

$$K_\alpha = (\sqrt{\alpha\pi} \frac{\Gamma(\frac{1}{\alpha+1} + \frac{1}{2})}{\Gamma(\frac{1}{\alpha+1})} \left(\frac{\sqrt{\pi}}{2} \frac{\Gamma(\frac{\alpha}{\alpha+1})}{\Gamma(\frac{\alpha}{\alpha+1} + \frac{1}{2})} \right)^{(\alpha+1)/2}. \tag{9.51}$$

Table 9.1 Coefficient K_α for various values of α

A	1.0	1.1	1.2	1.3	1.4	1.5	1.6	1.7
$K\alpha$	1.5708	1.5748	1.5769	1.5773	1.5765	1.5745	1.5717	1.5681
A	2	3	5	10	100	10000		
$K\alpha$	1.5539	1.4895	1.3555	1.1144	0.4178	4.2689×10^{-2}		

Frequency of vibration is, in general, a product of three terms: a constant K_α which depends only on the order of nonlinearity in rod, a constant $\sqrt{E/(\rho l^2)}$ which takes into consideration the physical and geometric characteristic of the rod and a mixed term $(D/l)^{(\alpha-1)/2}$ which depends not only on the order of nonlinearity α but also on the initial amplitude of vibration.

Frequency of oscillation, independently on the order of nonlinearity, directly depends on

$$\sqrt{\frac{E}{\rho l^2}} = const., \tag{9.52}$$

where E and ρ are physical properties of the rod (elasticity and density of material, respectively) and l is the length of the rod.

Relation (9.51) represents the dimensionless frequency constant which depends on the order of nonlinearity α. In the Table 9.1, values of coefficient K_α for various values of α is calculated.

For $\alpha \in (1, 1.6260)$, the value of constant K_α is higher than $\pi/2$, while for $\alpha = 1$ and $\alpha = 1.6260$ it is $\pi/2$. For $\alpha \in [1, 1.3)$ the value of the constant increases up to the maximal value $K_\alpha = 1.5773$. For orders of nonlinearity $\alpha > 1.3$ the constant is smaller than the maximal value: the higher the order of α, the smaller the value of coefficient. For high orders of α, we have $K\alpha \approx 1/2(\alpha + 1)/2$ and for $\alpha \to \infty$ the constant tends to zero.

The mixed term of the product $(D/l)^{(\alpha-1)/2}$ is the most complex one. It is dimensionless and depends on the order of nonlinearity α and initial amplitude of vibration D. Due to this term, the frequency of vibration for any nonlinear rod depends on the initial condition. Only for the linear rod the frequency is independent on the initial amplitude.

Let us rewrite (9.49) into the form

$$\omega = K_\alpha D^{(\alpha-1)/2}\sqrt{\frac{E}{\rho l^{\alpha+1}}}. \tag{9.53}$$

and compare it with the frequency obtained for the one-degree-of-freedom truly nonlinear oscillator with order α

$$\ddot{u} + c_1^2 u|u|^{\alpha-1} = 0, \tag{9.54}$$

where c_1 is the rigidity constant. Frequency of vibration is (see Cveticanin 2009)

$$\omega = QD^{(\alpha-1)/2}c_1, \tag{9.55}$$

where $Q = \frac{\pi}{B(\frac{1}{\alpha+1},\frac{1}{2})}\sqrt{\frac{\alpha+1}{2}}$ and D is the initial amplitude of vibration. It can be seen that the expressions (9.53) and (9.55) have the same form and for both cases frequencies depend on initial amplitude with the same order $(\alpha-1)/2$. Values of constants K_α and Q are different.

9.4 Solution Illustration and Simulation

For numerical simulation of vibration we solve the system of two ordinary differential equations $(9.10)_1$ and (9.28) which corresponds to $(9.10)_2$.

Equation $(9.10)_1$ is a second order differential equation where the parameter c1 depends on the order of nonlinearity α and has the form (9.39). For calculation, the following numerical data are applied: $A = 1$, $E/\rho l = 1$ and $l = 1$. Differential equation $(9.10)_1$ with (9.39) and initial conditions $T(0) = 1$ and $\dot{T}(0) = 0$ is solved using the software *Mathematica*.

Equation (9.28) is a first order differential equation which corresponds to $(9.10)_2$. For the aforementioned numerical data the equation is also solved numerically applying *Mathematica* when $X(0) = 0$. Both mentioned equations are solved for various values of parameter α: $\alpha = 1$, $\alpha = 2$, $\alpha = 3$ and $\alpha = 5$.

According to the expression (9.7) numerical solutions for equations $(9.10)_1$ and (9.28) are multiplied and the vibration-displacement-time relation is obtained. In Fig. 9.4 a-d, vibration-displacement-time diagrams for $A = 1$, $E/\rho l = 1$, $l = 1$ and $\alpha = 1$, $\alpha = 2$, $\alpha = 3$ and $\alpha = 5$ are plotted.

The numerical solving procedure for $(9.10)_1$ and (9.28) is repeated for the following values: $\alpha = 3$, $A = 2$, $E/\rho l = 1$ and $l = 1$ when the initial condition for $(9.10)_1$ is $T(0) = 1$ and $\dot{T}(0) = 0$. The obtained result is plotted as the displacement-time diagram in Fig. 9.4f. The result of numerical simulation is compared with the analytical result for vibration

$$u(x,t) = D\sqrt{(1 - B^{-1}\left(-(\frac{1}{l}B(\frac{\alpha}{\alpha+1}, 1/2)x))\right)}$$
$$ca(\alpha, 1, \sqrt{\frac{E\alpha}{\rho}}t\left(\frac{1}{2l}B(\frac{\alpha}{\alpha+1}, \frac{1}{2})\right)^{(\alpha+1)/2} D^{(\alpha-1)/2)}), \tag{9.56}$$

when $E/\rho l = 1$ and $l = 1$. It is concluded that for $D = 1$ and $\alpha = 1$ (linear rod), $\alpha = 2$ (rod with quadratic nonlinearity), $\alpha = 3$ (rod with cubic nonlinearity) and $\alpha = 5$ (rod with quantic nonlinearity) the analytical results are equal to the numerical ones. The same conclusion yields for $D = 2$ and $\alpha = 3$.

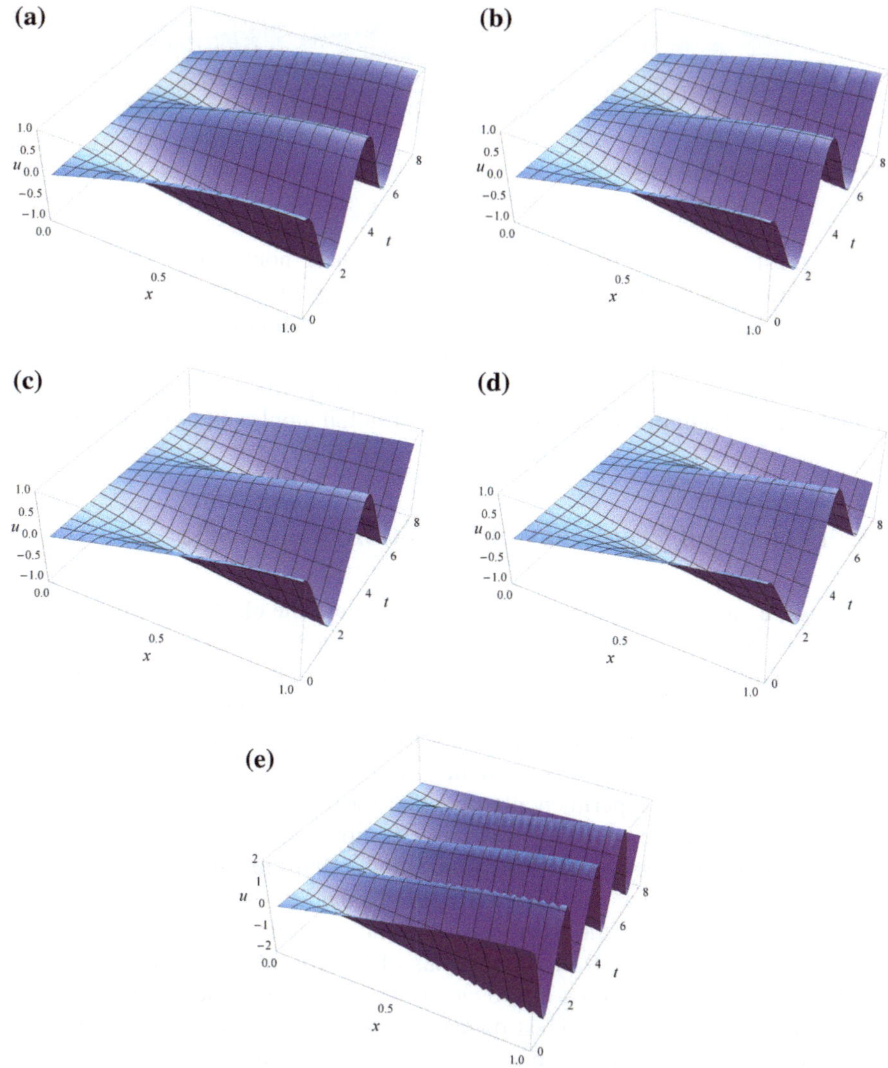

Fig. 9.4 $u - x - t$ diagrams for: **a** $\alpha = 1$ and $D = 1$; **b** $\alpha = 2$ and $D = 1$; **c** $\alpha = 3$ and $D = 1$; **d** $\alpha = 5$ and $D = 1$; **e** $\alpha = 3$ and $D = 2$

Comparing the results shown in Fig. 9.4a–d, it is seen that the order of nonlinearity has a significant influence on the period and frequency of oscillation: for $\alpha = 2$ the period of vibration is longer than for $\alpha = 1$; for $\alpha = 3$ the period is longer than for $\alpha = 2$ and for $\alpha = 5$ it is longer than for $\alpha = 3$. Namely, the higher the order of nonlinearity, the period of vibration is longer and the frequency of vibration is shorter (Fig. 9.4).

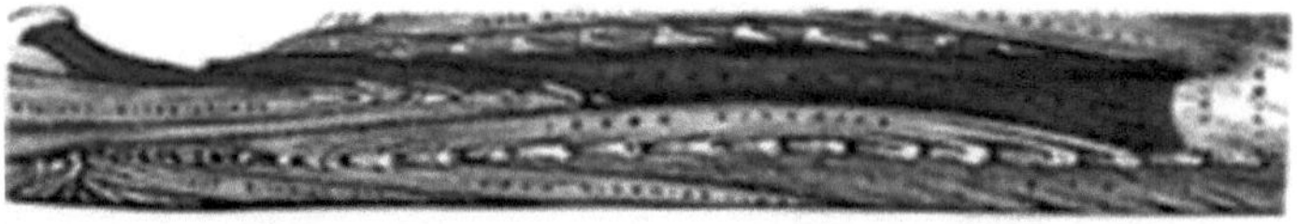

Fig. 9.5 Longissimus muscle (*black*)

In Fig. 9.4c, e axial vibrations of a rod with cubic nonlinearity ($\alpha = 3$) and different initial amplitudes are illustrated. Comparing Fig. 9.4c with Fig. 9.4e, we see that the initial amplitude has an influence on vibration properties of the rod. Namely, for higher value of initial amplitude ($D = 2$) the amplitude of vibration is higher than for initial amplitude $D = 1$. Besides, the period of vibration is shorter for higher initial amplitude than for the smaller initial amplitude (see Fig. 9.4c, e).

9.5 Period and Frequency of Vibration of a Muscle

Nowadays, a significant number of investigations into the elasticity of muscles are carried out, as it is believed that tissue elasticity has direct relevance in the early detection of diseases (Han et al. 2014; Andre et al. 2014). Namely, there are numerous diseases which significantly alter tissue elasticity, for example cancer. This is the reason that various methods for measuring for tendon, heart, skin, cartilage, liver, fat and muscle are developed (see Stanley et al. 1972; Ophir et al. 1991, 1994). Specially, elasticity is experimentally obtained for those muscles whose fibers are parallel to the direction of muscle stretch or compression (Chen et al. 1996), i.e., deformation orientation is axial. Thus, Chen et al. (1996) investigated the elasticity of the beef longissumus dorsi muscle smaple in axial direction. The longissimus is the longest subdivision of the erector spinae which in animals and humans extends the vertebral column (Fig. 9.5). Dimensions of the considered muscle are: length $l = 40\ cm$ and averaged radius $R = 6\ cm$. The dimensions of the sample of muscle which was settled on a standard Instron load cell tensile testing machine were $4\ cm\ x$ $4\ cm$ in width and lenght and with thicknesses ranging from $1 - 2\ cm$. The obtained stress-strain diagram is plotted in Fig. 9.6. From the Fig. 9.6, it can be seen that the elasticity function is strongly nonlinear and the muscle has to be treated as a system with strongly nonlinear property. It has to be addressed that the stress-strain diagram given by Chen et al. (1996) is for compression and for this type of biological tissue Cowin and Doty (2007) state that the strains are only with one sign.

A muscle with parallel fibers has to be treated as an axially vibrating rod with nonlinear elastic properties. Namely, the rod is assumed to be a clamped-free one. Free vibrations of the rod are analyzed. Then, the effect of nonlinearity and the elastic properties on the frequency of vibration is obtained.

Based on previous consideration, in this section the frequency of vibration of a longissumus dorsi muscle of a cow is calculated. Due to Fig. 9.6, the coefficient of

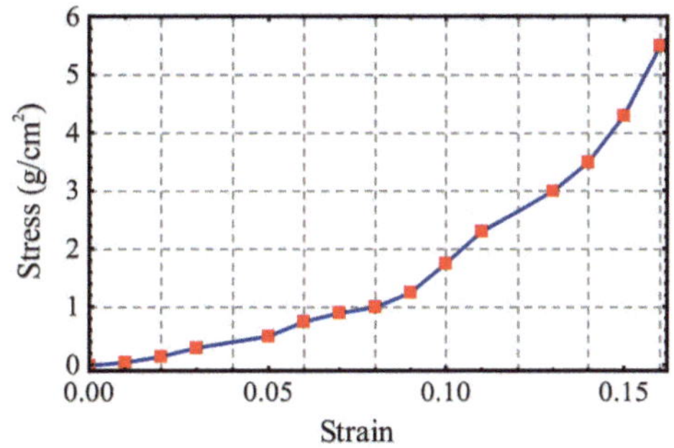

Fig. 9.6 Stress-strain diagram of the longissimus muscle (Chen et al. 1996)

elasticity is $E = 4.986\,kPa$, order of nonlinearity is $\alpha = 2.4594$ and the stress-strain function is according to Chen et al. (1996)

$$\sigma = 4986\varepsilon^{2.4594} \quad (\text{N/m}^2). \tag{9.57}$$

Density of tissue is $\rho = 1060$ kg/m^3. The initial position distribution in the muscle is assumed to be a linear function

$$U(x) = 0.125x. \tag{9.58}$$

The amplitude of vibration at the free end of the muscle is $D = 0.05\,m$. For numerical values the frequency constant is $K_\alpha = 1.52620$, the value of the additional constant is $\sqrt{E/(\rho l^{3.4594})} = 10.581$, and the frequency of vibration (9.52) is

$$\omega = 16.149 D^{0.7297}. \tag{9.59}$$

Frequency and period of vibration are, respectively,

$$\omega_{2.4594} = 1.8146\,\text{rad/s}, \qquad P_{2.4594} = \frac{2\pi}{\omega_{2.4594}} = 3.4626\,\text{s}. \tag{9.60}$$

If the elasticity of the muscle decreases and $\alpha = 2$, the frequency coefficient is $K_\alpha = 1.5539$ and the additional constant is $\sqrt{E/\rho l^3} = 8.573$, frequency increases to $\omega_2 = 2.9788$ rad/s and the period of vibration is $P_2 = 2.1028$ s.

It can be concluded that decrease of α causes a significant decrease of the period of vibration. Change of 23% of order of nonlinearity causes change of period of even 64.5%.

If the density of the muscle is increased or the modulus of elasticity is decreased, the frequency of vibration decreases, too. For example, if the order of nonlinearity is not varying, but ratio E/ρ decreases for 7.8%, the frequency of vibration decreases while the period increases for 4.14% comparing to the previous value. The same relation is evident for linear model when $\alpha = 1$.

Usually, the change in coefficient of elasticity and density is connected with the order of elasticity. If coefficient and order of elasticity decrease and density of muscle increases, frequency of muscle decreases and period increases. Influence of order of nonlinearity is much more significant than the change of coefficient of elasticity and density. For example, if $\alpha = 2$, $E = 4.8$ kPa and $\rho = 1100$ kg/m^3, the frequency function is $\omega = 2.8691$ rad/s, and the approximate period of vibration is $P = 2.1899$ s.

Based on this result, it can be concluded that decreasing of α for 23% and of tissue elasticity coefficient for 7.8% causes period variation of approximately 58%.

Due to this consideration, it can be concluded that change of the elastic property of muscle tissue has a significant influence on the period of vibration. Monitoring frequency property of the muscle gives us an opportunity to predict the change of elastic property and to signify the perturbed health.

9.6 Conclusion

In this Chapter a procedure for calculating of the frequency and period of axial vibration for a strongly nonlinear clamped-free rod is developed. Nonlinearity in the rod is due to nonlinear elastic property of rod material. Motion is described with a partial differential equation with two linear boundary and two initial conditions. The method of separation of variables is introduced. Two strongly nonlinear second order differential equations are obtained: one, which is with displacement function and other, with time function. Analytical solutions of equations in the form of inverse incomplete Beta function and Ateb function are determined. Using solutions of both aforementioned ordinary differential equations and also initial and boundary conditions, the constants of integration are obtained. Using these results, frequency and period of vibration are calculated. It is obvious that they depend on initial and boundary conditions. Numerical simulation of the axial vibration of strongly nonlinear rod is shown. The influence of order of nonlinearity and initial amplitude on the vibration is presented. Finally, an example of a longissimus muscle of a cow is considered. The solving procedure suggested in the paper is applied for obtaining of the frequency of vibration.

Based on the investigation, the following is concluded:

1. If the stress-strain relation is pure nonlinear and the order of nonlinearity is any positive number not smaller than one, the partial differential equation of axial vibration of the rod has two terms: one, which corresponds to the inertial force and a nonlinear term due to displacement.
2. For solving of the Eq. (9.3), method of separation of variables (time and displacement) is applicable. The constant of separation depends not only on the boundary conditions, but also on the order of nonlinearity.
3. Frequency of vibration depends not only on the constant of elasticity and density of the rod, but also on the order of nonlinearity and initial and boundary conditions. It

is quite a new result. Using to the case that the considered oscillator is conservative, the amplitude of vibration is determined according to initial conditions.

4. Dependence of the frequency of axial vibration on the initial amplitude is of the same order as for the one-degree-of-freedom truly nonlinear oscillator.
5. Influence of order of nonlinearity of elastic force is much stronger than influence of change of coefficient of elasticity or density of material. Increasing the order of nonlinearity causes the period of vibration to increase. Period of vibration increases also if density of material is increased and the coefficient of elasticity is decreased.
7. Based on (9.57), variation of the frequency and period for the longissimus dorsi muscle can be predicted. The obtained result can be applied for early detection of diseases where the elasticity of the muscle is changing.

References

Andre, M. P., Han, A., Heba, E., Erdman, J. W., & O'Brien, W. D. (2014). Accurate diagnosis of non-alcoholic fatty liver disease in human participants via quantitative ultrasound. *IEEE International Ultrasonics Symposium IUS, 6931973*, 2375–2377.

Andrianov, I. V., & van Horssen, W. T. (2006). Analytical approximations of the period of a generalized nonlinear van der Pol oscillator. *Journal of Sound and Vibration, 295*, 1099–1104.

Blocher, D., Rand, R. H., & Zehnder, A. T. (2013). Multiple limit cycles in laser interference transduced resonators. *International Journal of Non-linear Mechanics, 52*, 119–126.

Chen, E. J., Novakofski, J., Jenkins, W. K., & O'Brien, W. D, Jr. (1996). Young's modulus measurements of soft tissues with application to elasticity imaging. *IEEE Transactions on Ultrasonics, Ferroelectronics and Frequency Control, 43*(1), 191–194.

Cowin, S. C., & Doty, S. B. (2007). *Tissue mechanics*. Berlin: Springer.

Cveticanin, L., & Uzelac, Z. (1999). Nonlinear longitudinal vibrations of a rod. *Journal of Vibration and Control, 5*(6), 827–849.

Cveticanin, L. (2009). Oscillator with fraction order restoring force. *Journal of Sound and Vibration, 320*, 1064–1077.

Cveticanin, L. (2016). Period of vibration of axially vibrating truly nonlinear rod. *Journal of Sound and Vibration, 374*, 199–210.

Cveticanin, L., & Pogany, T. (2012). Oscillator with sum of non-integer order non-linearities. *Journal of Applied Mathematics*. doi:10.1155/2012/649050.

Didonato, A. (1991). An inverse of the incomplete Beta function (F-(Variance ratio) distribution function), *NSWCDD/TR-05/91*. www.dtic.mil/dtic/tr/fulltext/u2/a467901.pdf.

Han, A., Erdman, J. W., Simpson, D. G., Andre, M. P., & O'Brien, W. D. (2014). Early detection of fatty liver disease in mice via quantitative ultrasound. *IEEE International Ultrasonics Symposium IUS, 6931971*, 2362–2366.

Kovacic, I., & Zukovic, M. (2012). Oscillators with a power-form restoring force and fractional derivative damping. *Research Communications, 41*, 37–43.

Mickens, R. E. (2006). Iteration method solutions for conservative and limit-cycle x1/3 force oscillator. *Journal of Sound and Vibration, 292*, 964–968.

Ophir, J., Cespedes, I., Ponnekanti, H., Yazdi, Y., & Li, X. (1991). Elastography: A quantitative method for imaging the elasticity of biological tissues. *Ultrasound Imaging, 13*, 111–134.

Ophir, J., Miller, R. K., Ponnekanti, H., Cespedes, I., & Whittaker, A. D. (1994). Elastography and beef muscle. *Meat Science, 36*, 239–250.

Pilipchuk, V. N. (2007). Strongly nonlinear vibration of damped oscillators with two nonsmooth limits. *Journal of Sound and Vibration*, *302*, 398–402.

Rosenberg, R. M. (1963). The ateb(h)-functions and their properties. *Quarterly of Applied Matehmatics*, *21*, 37–47.

Stanley, D. W., McKnight, L. M., Hines, W. G., Usborne, W. R., & Deman, J. M. (1972). Predicting meat tenderness frm muscle tensile properties. *Journal of Texture Studies*, *3*, 51–68.

Waluya, S. B., & van Horssen, W. T. (2003). On the periodic solutions of a generalized non-linear Van der Pol oscillator. *Journal of Sound and Vibration*, *268*, 209–215.

Appendix A
Periodical Ateb Functions

In his classical paper Rosenberg 1963, introduced the so-called periodic Ateb functions concerning the problem of inversion of the half of the incomplete Beta function

$$z \mapsto \frac{1}{2} B_z(a, b) = \frac{1}{2} \int_0^{0 \le z \le 1} t^{a-1}(1-t)^{b-1} dt \,. \tag{A.1}$$

Obviously, we are interested in the case where $a = \frac{1}{\alpha+1}$ and $b = \frac{1}{2}$, i.e.,

$$z \mapsto \frac{1}{2} B_z \left(\frac{1}{\alpha+1}, \frac{1}{2} \right) = \frac{1}{2} \int_0^{0 \le z \le 1} \frac{dt}{(1-t)^{1/2} t^{\alpha/(\alpha+1)}} \,. \tag{A.2}$$

Senik in his article in 1969 shows that the Ateb functions are the solutions of the ordinary differential equations

$$\begin{aligned} \dot{v} - u^{\alpha} &= 0, \\ \dot{u} + \frac{2}{\alpha+1} v &= 0 \,, \end{aligned} \tag{A.3}$$

Namely,

$$v(z) = sa(1, \alpha, z), \quad u(z) = ca(\alpha, 1, z) \,. \tag{A.4}$$

It can be easily verified that the inverse of $\frac{1}{2} B_z(\frac{1}{2}, \frac{1}{\alpha+1})$ and $v(z)$ coincide on $[-\frac{1}{2}\Pi_{\alpha}, \frac{1}{2}\Pi_{\alpha}]$, where

$$\Pi_{\alpha} := B \left(\frac{1}{\alpha+1}, \frac{1}{2} \right) \,. \tag{A.5}$$

Having in mind the following set of properties:

L. Cveticanin, *Strong Nonlinear Oscillators*, Mathematical Engineering,
DOI 10.1007/978-3-319-58826-1

$$sa(1, \alpha, z) = \begin{cases} -sa(1, \alpha, -z) \\ \mp ca\left(\alpha, 1, \frac{1}{2}\Pi_\alpha \pm x\right) \\ \pm sa(1, \alpha, \Pi_\alpha \pm z) \\ \mp sa(1, \alpha, 2\Pi_\alpha \mp z) \end{cases}, \tag{A.6}$$

we see that $sa(\alpha, 1, z)$ is an odd function of $z \in \mathbb{R}$; it is the so-called $2\Pi_\alpha$-periodic sine Ateb, i.e., sa function. Also there holds

$$sa^2(1, \alpha, z) + ca^{\alpha+1}(\alpha, 1, z) = 1, \tag{A.7}$$

and cosine Ateb, that is $ca(1, \alpha, z)$ function, is even and $2\Pi_\alpha$ periodic having properties:

$$ca(\alpha, 1, z) = \begin{cases} ca(\alpha, 1, -z) \\ sa\left(1, \alpha, \frac{1}{2}\Pi_\alpha \pm z\right) \\ -ca(\alpha, 1, \Pi_\alpha \pm z) \\ ca(\alpha, 1, 2\Pi_\alpha \pm z) \end{cases}. \tag{A.8}$$

By these two sets of relations we see that functions sa,ca are defined on the whole range of $\mathbb{R}$.

The first derivatives of the ca and sa Ateb functions are

$$\begin{aligned} \frac{d}{dz}\, ca(\alpha, 1, z) &= -\frac{2}{\alpha+1}\, sa(1, \alpha, z) \\ \frac{d}{dz}\, sa(1, \alpha, z) &= ca^\alpha(\alpha, 1, z). \end{aligned} \tag{A.9}$$

Being cosine Ateb $ca(n, m, z)$ even $2\Pi_\alpha$ periodic function, it is a perfect candidate for a cosine Fourier series expansion. Let us mention that finite Fourier series approximation has been discussed in Droniuk and Nazarkevich $(2010)_1$, while in Droniuk and Nazarkevich $(2010)_2$ the sine Ateb $sa(n, m, z)$ has been approximated by its sine Fourier series. Applying the there exposed method to $ca(1, \alpha, z)$, we conclude that

$$ca(\alpha, 1, z) = \sum_{n=1}^{\infty} a_n \cos \frac{\pi n\, z}{\Pi_\alpha}, \tag{A.10}$$

since obviously $a_0 = 0$ and

$$a_n = \frac{2}{\Pi_\alpha} \int_0^{\Pi_\alpha} ca(\alpha, 1, z) \cdot \cos \frac{\pi n\, z}{\Pi_\alpha}\, dz = \frac{2}{\Pi_\alpha} \int_0^{\Pi_\alpha} \cos \frac{\pi n\, z}{\Pi_\alpha} \left\{ \int_z^1 \frac{d\overline{u}}{(1-\overline{u}^2)^{\frac{\alpha}{\alpha+1}}} \right\} dz. \tag{A.11}$$

The a_n values we compute numerically according to the prescribed accuracy. Of course, it is enough to compute $ca(\alpha, 1, z)$ for $z \in [0, \Pi_\alpha/2]$, another values we calculate by means of formula (A.8) (see also Gricik et al. 2009). Another model in approximating Ateb functions is the Taylor series expansion, such that corresponds to the investigations by Gricik and Nazarkevich in 2007.

Now, inverting the half of the incomplete Beta function in (3.26):

$$\frac{1}{2} B_{\left(\frac{|x|}{A}\right)^{\alpha+1}}\left(\frac{1}{\alpha+1}, \frac{1}{2}\right) = \frac{\Pi_\alpha}{2} + \frac{\sqrt{\alpha+1}\,|c_\alpha|}{\sqrt{2}} A^{(\alpha-1)/2}\, t\,,$$

we clearly deduce

$$x(t) = A \cdot sa\left(1, \alpha, \frac{\Pi_\alpha}{2} + \frac{\sqrt{\alpha+1}\,|c_\alpha|}{\sqrt{2}} A^{(\alpha-1)/2}\, t\right). \tag{A.12}$$

Having in mind the quarter period expansion formula, we arrive at

$$x(t) = A \cdot ca\left(\alpha, 1, \frac{\sqrt{\alpha+1}\,|c_\alpha|}{\sqrt{2}} A^{(\alpha-1)/2}\, t\right), \qquad t \in \mathbb{R}\,. \tag{A.13}$$

By $ca(\alpha, 1, 0) = 1$, we see that the initial condition $x(0) = A$ is satisfied as well.

Moreover, we have to point out a restricting characteristics of Rosenberg's and Senik's inversion (1969). Rosenberg (1963) pointed our the rule:

> "Exponents $n = \frac{\alpha+1}{2}$ and $1/n$ behave like odd integers", while Senik's restriction to some rational values of α constitutes the set of permitted α–values:

Approximate methods by numerically obtained evaluations of Ateb functions have been performed by Droniuk and Nazarkevich $(2010)_1$ and $(2010)_2$, and the references therein. Further study on Ateb function integral was realized by Senik in 1969 (see Table A.1) and Drogomirecka in 1997.

Table A.1 The values of $\alpha = \frac{2\nu+1}{2\mu+1}$, $\mu, \nu \in \mathbb{N}_0$, by Senik's traces

$\mu\backslash\nu$	0	1	2	3	4	5	6	7	…
0	1	3	5	7	9	11	13	15	
1	1/3	1	5/3	7/3	3	11/3	13/3	5	
2	1/5	3/5	1	7/5	9/5	11/5	13/5	3	
3	1/7	3/7	5/7	1	9/7	11/7	13/7	15/7	
4	1/9	1/3	5/9	7/9	1	11/9	13/9	5/3	
5	1/11	3/11	5/11	7/11	9/11	1	13/11	15/11	
6	1/13	3/13	5/13	7/13	9/13	11/13	1	15/13	
7	1/15	1/5	1/3	7/15	3/5	11/15	13/15	1	
⋮									⋱

References

Drogomirecka, H.T. (1997). Integrating a special Ateb–function. *Visnik Lvivskogo Universitetu. Serija mehaniko–matematichna, 46*, 108–110. (in Ukrainian)

Droniuk, I., & Nazarkevich, M. $(2010)_1$. Modeling nonlinear oscillatory system under disturbance by means of Ateb–functions for the Internet in: T. In: *Proceeding of the Sixth International Working Conference on Performance Modeling and Evaluation of Heterogeneous Networks HET-NETs 2010, Zakopane, Poland* (pp. 325–334).

Dronyuk, I.M., Nazarkevich, M.A., & Thir, V. $(2010)_2$. Evaluation of results of modelling Ateb–functions for information protection. *Visnik nacionaljnogo universitetu Lvivska politehnika, 663*, 112–126. (in Ukrainian)

Gricik, V.V., Dronyuk, I.M., & Nazarkevich, M.A. (2009). Document protection information technologies by means of Ateb–functions I. Ateb–function base consistency for document protection. *Problemy upravleniya i avtomatiki, 2*, 139–152. (in Ukrainian)

Gricik, V.V., & Nazarkevich, M.A. (2007). Mathematical models algorythms and computation of Ateb–functions. *Dopovidi NAN Ukraini Seriji A, 12*, 37–43. (in Ukrainian)

Rosenberg, R.M. (1963). The Ateb(h)-functions and their properties. *Quarterly of Applied Mathematics, 21*, 37–47.

Senik, P.M. (1969). Inversion of the incomplete Beta-function. *Ukr. Mat. Zh., 21*, 325–333; *Ukrainian Mathematical Journal, 21*, 271–278.

Appendix B
Fourier Series of the ca Ateb Function

Since the ca function is odd, its Fourier series comprises odd harmonics only, and it can be expressed as

$$\mathrm{ca}\,(\alpha, 1, t) = \sum_{N=1}^{\infty} C_{2N-1}\,(\alpha) \cos\left[(2N-1)\frac{2\pi}{T}t\right], \tag{B.1}$$

where the Fourier coefficients C_{2N-1} depend on the parameter α, and are defined by

$$C_{2N-1}\,(\alpha) = \frac{4}{T}\int_0^{T/2} \mathrm{ca}\,(\alpha, 1, t) \cos\left[(2N-1)\frac{2\pi}{T}t\right] \mathrm{d}t, \tag{B.2}$$

where T is the period.

To write this expression in a suitable form for further calculation, the procedure recently proposed in Belendez et al. (2015) is utilised. As a first step, the displacement is rescaled by the initial amplitude, $X = x/A$, yielding

$$C_{2N-1}\,(\alpha) = \frac{8}{T}\int_0^{T/4} X\,(\alpha, t) \cos\left[(2N-1)\frac{2\pi}{T}t\right] \mathrm{d}t. \tag{B.3}$$

Now, to find the expression for $\mathrm{d}t$, the first integral is composed, and the following is derived

$$\mathrm{d}t = \sqrt{\frac{\alpha+1}{2c_\alpha^2}}\,|A|^{(1-\alpha)/2} \frac{\mathrm{d}X}{\sqrt{1-|X|^{\alpha+1}}}. \tag{B.4}$$

This expression gives the possibility to determine how t depends on X (noting that this holds for $X \geq 0$):

$$t\,(X) = \sqrt{\frac{\alpha+1}{2c_\alpha^2}}\,|A|^{(1-\alpha)/2} \int_X^1 \frac{\mathrm{d}y}{\sqrt{1-y^{\alpha+1}}}. \tag{B.5}$$

L. Cveticanin, *Strong Nonlinear Oscillators*, Mathematical Engineering,
DOI 10.1007/978-3-319-58826-1

Performing some transformations, one can derive (see Belendez et al. 2015)

$$t(X) = \sqrt{\frac{\pi}{2c_\alpha^2(\alpha+1)}} \frac{\Gamma\left(\frac{1}{\alpha+1}\right)}{\Gamma\left(\frac{\alpha+3}{2(\alpha+1)}\right)} |A|^{(1-\alpha)/2} I\left(1 - X^{\alpha+1}, \frac{1}{2}, \frac{1}{\alpha+1}\right), \quad \text{(B.6)}$$

where I stands for the regularized incomplete beta function.

Finally, substituting (B.4) into (B.3) as well as (B.6) into the argument of the cosine function in (B.3), one derives

$$C_{2N-1}(\alpha) = \frac{2(\alpha+1)\Gamma\left(\frac{\alpha+3}{2(\alpha+1)}\right)}{\sqrt{\pi}\Gamma\left(\frac{1}{\alpha+1}\right)} \int_0^1 \frac{X}{\sqrt{1-X^{\alpha+1}}} \cos\left[\frac{(2n-1)\pi}{2} I\left(1 - X^{\alpha+1}, \frac{1}{2}, \frac{1}{\alpha+1}\right)\right] \mathrm{d}X. \quad \text{(B.7)}$$

By using the substitution $z = 1 - X^{\alpha+1}$, the following expression for the Fourier coefficients is obtained:

$$C_{2N-1}(\alpha) = \frac{2\Gamma\left(\frac{\alpha+3}{2(\alpha+1)}\right)}{\sqrt{\pi}\Gamma\left(\frac{1}{\alpha+1}\right)} \int_0^1 \frac{(1-z)^{(1-\alpha)/(1+\alpha)}}{\sqrt{z}} \cos\left[\frac{(2N-1)\pi}{2} I\left(z, \frac{1}{2}, \frac{1}{\alpha+1}\right)\right] \mathrm{d}z. \quad \text{(B.8)}$$

These values can be calculated by carrying out numerical integration. First four Fourier coefficients are calculated in this way by using (B.8) and plotted in Fig. B.1 as a function of the power α. It is seen that: C_1 decreases from unity as α increases; C_3 and C_7 are positive; C_5 is negative for $1 < \alpha < 2.34$, and positive otherwise.

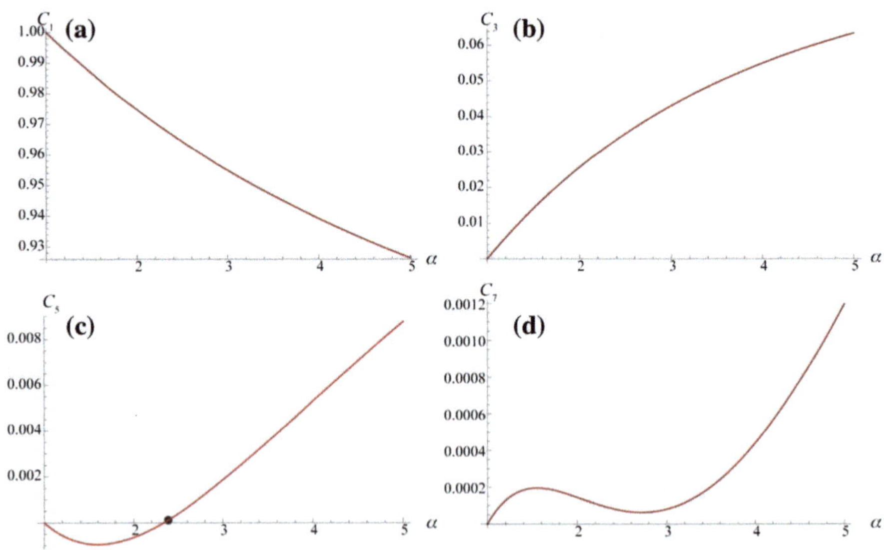

Fig. B.1 Fourier coefficients for ca(α, 1, t) versus order of nonlinearity α: (a) first C_1, (b) second C_2, (c) third C_3, (d) fourth C_4

Reference

Beléndez, A., Francés, J., Beléndez, T., Bleda, S., Pascual, C., & Arribas, E. (2015). Nonlinear oscillator with power-form elastic-term: Fourier series expansion of the exact solution. *Communications in Nonlinear Science and Numerical Simulation*, *22*, 134–148.

Appendix C
Averaging of Ateb Functions

1. For $c_\alpha^2 = 1$ and $A = 1$, the first integral (3.12) transforms into

$$\frac{\dot{x}^2}{2} = \frac{1}{\alpha+1}(1 - x^{\alpha+1}). \tag{C.1}$$

Assuming

$$x = ca(\alpha, 1, \psi) \equiv ca, \tag{C.2}$$

and substituting it into the first integral (C.1), we have

$$\frac{\dot{x}^2}{2} = \frac{1}{\alpha+1}(1 - ca^{\alpha+1}). \tag{C.3}$$

Using the relation for the sine and cosine Ateb functions

$$sa^2 + ca^{\alpha+1} = 1, \tag{C.4}$$

where $sa \equiv sa(1, \alpha, \psi)$, the expression (C.3) transforms into

$$sa^2 = \frac{\alpha+1}{2}\dot{x}^2. \tag{C.5}$$

Averaging of the function sa^2 is done in the time interval $[0, \bar{T}/4]$, i.e., for the positive displacement x in the interval $[0, 1]$

$$\langle sa^2 \rangle = \frac{1}{(\bar{T}/4)} \int_0^{\bar{T}/4} \frac{\alpha+1}{2}\dot{x}^2 dt = \frac{1}{(\bar{T}/4)} \int_0^1 \frac{\alpha+1}{2}\dot{x} dx. \tag{C.6}$$

where integrating the relation (C.3)

L. Cveticanin, *Strong Nonlinear Oscillators*, Mathematical Engineering,
DOI 10.1007/978-3-319-58826-1

$$\int_1^0 \frac{dx}{-\sqrt{\frac{2}{\alpha+1}(1-x^{\alpha+1})}} = \int_0^{\bar{T}/4} dt, \tag{C.7}$$

we obtain the quarter period of vibration (see Cveticanin and Pogany 2012)

$$\frac{\bar{T}}{4} = \sqrt{\frac{1}{2(1+\alpha)}} B\left(\frac{1}{\alpha+1}, \frac{1}{2}\right). \tag{C.8}$$

Substituting (C.8) and (C.2) into (C.7) and after some transformation we obtain, finally,

$$a_1 = \langle sa^2 \rangle = \frac{1+\alpha}{3+\alpha}. \tag{C.9}$$

For $\alpha \in (0, \infty)$, the parameter a_1 increases in the interval $a_1 \in (1/3, 1)$.

2. According to (C.3) and (C.5) we have

$$sa^2 ca^2 = \frac{\alpha+1}{2}\dot{x}^2 x^2 = (1 - x^{\alpha+1})x^2. \tag{C.10}$$

The averaged product sa^2ca^2 over the time period of 0 to $\bar{T}/4$, i.e., in the displacement interval [0, 1] is

$$\langle sa^2 ca^2 \rangle = \frac{1}{(\bar{T}/4)} \int_0^{\bar{T}/4} \frac{\alpha+1}{2}\dot{x}^2 x^2 dt = \frac{1}{(\bar{T}/4)} \int_0^1 \frac{\alpha+1}{2}\dot{x} x^2 dx. \tag{C.11}$$

Using (C.8), integrating the relation (C.11) and after some calculation we obtain

$$a_2 = \langle sa^2 ca^2 \rangle = \frac{B\left(\frac{3}{2}, \frac{3}{1+\alpha}\right)}{B\left(\frac{1}{2}, \frac{1}{1+\alpha}\right)}. \tag{C.12}$$

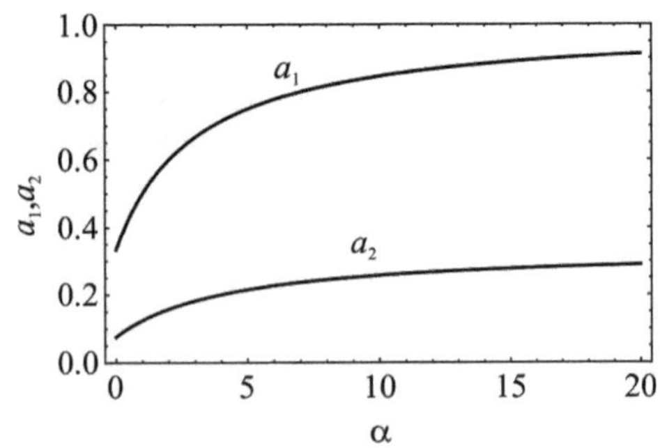

Fig. C.1 $a_1 - \alpha$ and $a_2 - \alpha$ curves

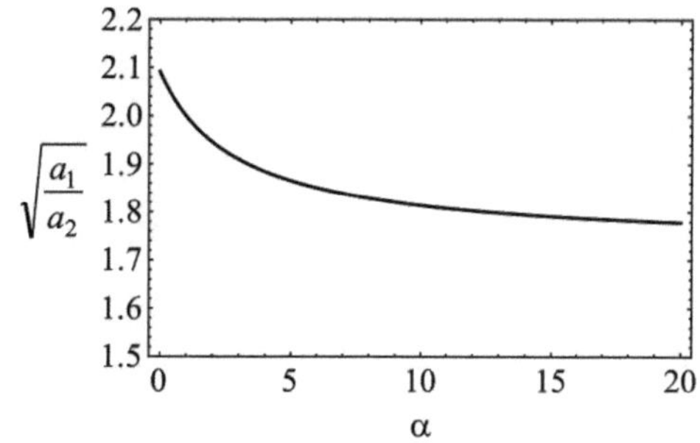

Fig. C.2 $\sqrt{a_1/a_2} - \alpha$ curve

In Fig. C.1, the $a_1 - \alpha$ and $a_2 - \alpha$ curves are plotted. It is shown that for $\alpha \in (0, \infty)$, the both parameters, a_1 and a_2 increase, but a_2 slower than a_1.

In Fig. C.2, the $\sqrt{a_1/a_2} - \alpha$ relation is plotted.

The curve decreases with increase of the nonlinearity order α from zero to infinity in a bounded region.

Reference

Cveticanin, L., & Pogany, T. (2012). Oscillator with a sum of non-integer order non-linearities. *Journal of Applied Mathematics,* Article D 649050, 20 p, doi:10.1155/2012/649050.2012.

Appendix D
Jacobi Elliptic Functions

Jacobian elliptic functions are doubtless periodic functions defined over the complex plane. They represent the special case of periodical Ateb function, as is shown in Sect. 3.2. The fundamental three elliptic functions are the Jacobi elliptic sine ($sn(\psi, k^2) \equiv sn$), cosine ($cn(\psi, k^2) \equiv cn$) and delta ($dn(\psi, k^2) \equiv dn$) functions with argument ψ and modulus k^2. The elliptic functions sn and cn may be thought of as generalizations of sine and cosine trigonometric functions where their period depends on the modulus k^2. For $k^2 = 0$, the Jacobi elliptic functions transform into trigonometric ones

$$sn(\psi, 0) = \sin\psi \qquad cn(\psi, 0) = \cos\psi \qquad dn(\psi, 0) = 1. \tag{D.1}$$

The period of the sn and cn Jacobi elliptic functions is $4K(k)$, while of the function dn it is $2K(k)$, where $K(k)$ is the complete elliptic integral of the first kind. The cn and dn Jacobi elliptic functions are even functions, while sn is an odd function. In Fig. D.1 the $sn(t, 1/2)$, $cn(t, 1/2)$ and $dn(t, 1/2)$ Jacobi elliptic functions are plotted.

The elliptic functions satisfy the following identities

$$ca^2 + sa^2 = 1, \qquad dn^2 + k^2sn^2 = 1, \qquad 1 - k^2 + k^2cn^2 = dn^2. \tag{D.2}$$

Only two of these three relations are independent.

The first time derivatives of the functions for the argument ψ are

$$\begin{aligned} &\frac{\partial}{\partial\psi}(cn) \equiv cn_\psi = -sndn, \qquad \frac{\partial}{\partial\psi}(sn) \equiv sn_\psi = cndn, \\ &\frac{\partial}{\partial\psi}(dn) \equiv dn_\psi = -k^2sncn. \end{aligned} \tag{D.3}$$

More about the Jacobi elliptic functions, the reader can find in the literature of the special functions (see Byrd and Friedman 1954, or Abramowitz and Stegun 1971).

L. Cveticanin, *Strong Nonlinear Oscillators*, Mathematical Engineering,
DOI 10.1007/978-3-319-58826-1

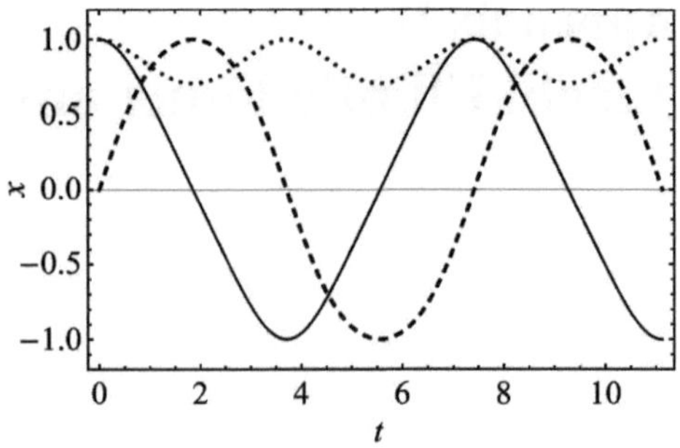

Fig. D.1 Jacobi elliptic functions: $sn(t, 1/2)$ (*dotted line*), $cn(t, 1/2)$ (*full line*) and $dn(t, 1/2)$ (*dashed line*)

References

Abramowitz, M., & Stegun, I.A. (1979). *Handbook of mathematical functions with formulas, graphs and mathematical tables*. Moscow: Nauka. (in Russian)

Byrd, P.F., & Friedman, M.D. (1954). *Handbook of elliptic integrals for engineers and physicists*. Berlin: Springer.

Appendix E
Euler's Integrals of the First and Second Kind

Euler's integral of the first kind also named Beta function, $B(p,q)$, is defined as (see Gradstein and Rjizhik 1971)

$$B(p,q) = \int_0^1 u^{p-1}(1-u)^{q-1} du, \tag{E.1}$$

which exists for

$$\mathrm{Re}(p) > 0, \qquad \mathrm{Re}(q) > 0. \tag{E.2}$$

Introducing the new variable $x = 1 - u$ into (E.1), the Beta function is expressed as

$$B(p,q) = -\int_1^0 x^{q-1}(1-x)^{p-1} dx = \int_0^1 x^{q-1}(1-x)^{p-1} dx = B(q,p). \tag{E.3}$$

The Beta function is symmetric in (p,q).

Euler's integral of the second kind also called Gamma function is (see Mickens, 2004)

$$\Gamma(p) = \int_0^\infty u^{p-1} e^{-u} du, \tag{E.4}$$

where p satisfies the relation (E.2). The connection between the Euler's integrals of the first and second kind is

$$B(p,q) = \frac{\Gamma(p)\Gamma(q)}{\Gamma(p+q)}. \tag{E.5}$$

For $(p-1) = n$, where n is a whole positive number, the relation (E.4) modifies into

L. Cveticanin, *Strong Nonlinear Oscillators*, Mathematical Engineering,
DOI 10.1007/978-3-319-58826-1

$$\Gamma(n+1) = \int_0^\infty u^n e^{-u} du = n! \tag{E.6}$$

Thus,

$$\Gamma(n) = (n-1)! \tag{E.7}$$

and the relation between (E.6) and (E.7) is

$$\Gamma(n+1) = n(n-1)! = n\Gamma(n). \tag{E.8}$$

Generalizing (E.6) for any value of p we have

$$\Gamma(p+1) = p! \tag{E.9}$$

and the corresponding relations

$$\Gamma(p) = (p-1)! \tag{E.10}$$

and

$$\Gamma(p+1) = p\Gamma(p). \tag{E.11}$$

Substituting (E.10) into (E.3) the transformed version of the Beta function is

$$B(p,q) = \frac{\Gamma(p)\Gamma(q)}{\Gamma(p+q)} = \frac{(p-1)!(q-1)!}{(p+q-1)!}, \tag{E.12}$$

which is suitable for calculation.

References

Gradstein, I.S., & Rjizhik, I.M. (1971). *Tablici integralov, summ, rjadov i proizvedenij*. Moscow: Nauka.

Mickens, R.E. (2004). *Mathematical Methods for the Natural and Engineering Sciences*. New Jersey: World Scientific.

Appendix F
Inverse Incomplete Beta Function

In this Appendix the Table of the inverse incomplete Beta function $f(\alpha, x)$ for various values of parameter α is given.

$x\backslash\alpha$	1	2	3	5	10
0.00	0.0000	0.0000	0.0000	0.0000	0.0000
0.05	0.0785	0.0646	0.0599	0.0560	0.0530
0.10	0.1564	0.1291	0.1197	0.1119	0.1060
0.15	0.2334	0.1932	0.1792	0.1678	0.1590
0.20	0.3090	0.2568	0.2385	0.2234	0.2118
0.25	0.3827	0.3196	0.2973	0.2788	0.2646
0.30	0.4540	0.3815	0.3555	0.3339	0.3172
0.35	0.5225	0.4423	0.4131	0.3886	0.3696
0.40	0.5878	0.5017	0.4698	0.4429	0.4219
0.45	0.6494	0.5596	0.5256	0.4966	0.4738
0.50	0.7071	0.6158	0.5803	0.5497	0.5255
0.55	0.7604	0.6699	0.6337	0.6021	0.5768
0.60	0.8090	0.7218	0.6856	0.6536	0.6277
0.65	0.8526	0.7710	0.7358	0.7041	0.6781
0.70	0.8910	0.8174	0.7841	0.7535	0.7280
0.75	0.9239	0.8604	0.8301	0.8015	0.7771
0.80	0.9511	0.8997	0.8734	0.8478	0.8254
0.85	0.9724	0.9346	0.9135	0.8920	0.8726
0.90	0.9877	0.9643	0.9495	0.9335	0.9184
0.95	0.9969	0.9874	0.9799	0.9710	0.9619
1.00	1.0000	1.0000	1.0000	1.0000	1.0000

L. Cveticanin, *Strong Nonlinear Oscillators*, Mathematical Engineering,
DOI 10.1007/978-3-319-58826-1

Index

L. Cveticanin, *Strong Nonlinear Oscillators*, Mathematical Engineering,
DOI 10.1007/978-3-319-58826-1

MIX
Papier aus verantwortungsvollen Quellen
Paper from responsible sources
FSC® C105338

If you have any concerns about our products,
you can contact us on
ProductSafety@springernature.com

In case Publisher is established outside the EU,
the EU authorized representative is:
Springer Nature Customer Service Center GmbH
Europaplatz 3, 69115 Heidelberg, Germany

Printed by Libri Plureos GmbH
in Hamburg, Germany